Adsorbents for Gas Adsorption and Environmental Applications

Adsorbents for Gas Adsorption and Environmental Applications

Dr. Vikas Mittal
Editor and Lead Author

CWP
Central West Publishing

This edition has been published by Central West Publishing, Australia

For more information about the books published by Central West Publishing, please visit https://centralwestpublishing.com

Disclaimer
Every effort has been made by the publisher, editor and authors while preparing this book, however, no warranties are made regarding the accuracy and completeness of the content. The publisher, editor and authors disclaim without any limitation all warranties as well as any implied warranties about sales, along with fitness of the content for a particular purpose. Citation of any website and other information sources does not mean any endorsement from the publisher and authors. For ascertaining the suitability of the contents contained herein for a particular lab or commercial use, consultation with the subject expert is needed. In addition, while using the information and methods contained herein, the practitioners and researchers need to be mindful for their own safety, along with the safety of others, including the professional parties and premises for whom they have professional responsibility. To the fullest extent of law, the publisher, editor and authors are not liable in all circumstances (special, incidental, and consequential) for any injury and/or damage to persons and property, along with any potential loss of profit and other commercial damages due to the use of any methods, products, guidelines, procedures contained in the material herein.

NATIONAL LIBRARY OF AUSTRALIA

A catalogue record for this book is available from the National Library of Australia

ISBN (print): 978-1-922617-15-6

Contents

Preface

Nanomaterials have impacted all engineering, science and technology sectors owing to their superior properties and performance profiles as compared to the conventional materials. As a result, a large number of advanced materials have been developed for specific applications. In this respect, adsorbents represent a functional class of materials which are of tremendous benefit for gas adsorption, gas separation, water treatment, pollutant removal, oil spill clean up, etc. The book aims to present a variety of such advanced adsorbent materials by providing both basic information as well as recent developments reported for these materials. All aspects related to synthesis, properties and applications of these materials have been reviewed.

In Chapter 1, various polymer incorporated adsorption systems have been reviewed for effective separation of acid gases from natural gas. In Chapter 2, the main focus is to elucidate the role of metal organic frameworks (MOFs) and metal oxides for the adsorptive removal of hazardous gases like H_2S, NH_3, CO_2 and other hydrocarbons present in the atmosphere as well as the dissolved heavy metal salts or the organic pollutants present in the polluted water. In Chapter 3, novel aerogels were prepared using nanofibrillated cellulose (NFC). Specific properties of these aerogels were subsequently studied, which include density, porosity and pore size characteristics, along with mechanical properties. In Chapter 4, recent developments and prospects of polymer nanocomposites for wastewater treatment are presented. Four types of nanofillers, namely graphene, carbon nanotubes (CNTs), clay and metal oxides are analyzed for their effect on the properties and applications of diverse polymers. In Chapter 5, various polymer-based sponges/foams, aerogels, fibers and membranes have been reviewed for the removal of oil from water. Chapter 6 reports graphene-phosphazene nanotubes (PZTs) based carbonaceous monolithic cryogels containing cobalt ferrite nanoparticles as an efficient sorbent. Crude oil and vegetable oil were selected as model pollutants for evaluating the oil sorption performance under laboratory conditions. Chapter 7 focuses on the recent advances in the synthesis and properties of the functional magnetic polymer nanocomposite materials, along with summarizing their cutting-edge environmental applications. In Chapter 8, the 3-D carbonaceous monolithic cryogel nanocomposites have been employed for the removal of phenol-based pollutants from wastewater. In Chapter 9, the adsorption of N_2O on B-, N-, Si-, P-, Ga-, Cr- and Mn-doped graphene sheets has been investigated by performing first-principles calculations based on density functional theory (DFT). In Chapter 10, the state-of-the-art research on first-principles theoretical studies on gas molecular adsorption on graphene and chemically functionalized graphene has been reviewed. In Chapter 11, hybridization of both Cu and ZnO on the graphene sheets has been demonstrated. The photocatalytic efficiency of the hybrids was studied using the photodegradation of methylene blue (MB) solution under visible light medium. In Chapter 12, the adsorption performance of various adsorbents based on silica, clay, activated carbon, graphene, graphite oxide, nanotubes, zeolites and polymers has been thoroughly compared. In Chapter 13, graphene-based carbonaceous monolithic material containing cobalt ferrite nanoparticles has been synthesized for adsorption applications.

The support of all the chapter contributors is greatly acknowledged towards the successful accomplishment of the book. The book is dedicated to my family for their constant support.

Vikas MITTAL

1

Polymer Modified/Enhanced Adsorbents for Gas Adsorption and Sweetening

1.1 Introduction

In order to satisfy the increase in global energy demand along with the need to reduce the environmental impacts of global greenhouse gas emissions, it is continuously strived to develop promising energy substitutes. Thus, some of the concerns of the present day energy technologies are production of adequate energy with good quality, environmental feasibility and economical viability. Natural gas, which is predominantly hydrocarbon in nature, is one of the prominent energy sources for many decades, with its usage as a fuel progressively increasing worldwide [1]. Commercial natural gas has several advantages such as high heating value and low carbon content per unit mass. However, along with methane, raw natural gas contains contaminants such as higher hydrocarbons, acidic gases (hydrogen sulfide (H_2S) and carbon dioxide (CO_2)), water and several other minute impurities. The existence of acid gases in the natural gas leads to shrinkage in natural gas heating value and also causes corrosion in facilities [2]. Hence, the removal of acid gases from natural gas is necessary prior to its transportation through pipelines [3,4]. The ultimate level of CO_2 and H_2S are generally limited to 2 mol% and 4 ppm respectively by the regulations [5]. Thus, in order to meet the sales gas specifications and to keep the values of acid gases within the limits, the natural gas streams are treated using various processes. The acid gas removal process, which is generally termed as gas sweetening process, is employed for natural gas purification from acid gases like H_2S and CO_2. In this stage, the acid gases are separated using various technologies, depending on the operating conditions and levels of H_2S and CO_2 present in the natural gas [2].

Separation of acid gases from natural gas streams is very challenging and demands the use of special materials which can resist corrosive environments. Different technologies commonly used for natural gas purification include adsorption and absorption processes,

Haleema Saleem and Vikas Mittal, Khalifa University of Science and Technology, Abu Dhabi, UAE

membrane separation and cryogenic condensation. Various factors to be considered for the selection of appropriate technology are nature and quantity of contaminants present in feed gas, quantity of hydrocarbons in the gas, targeted acid gas removal capacity, pipeline specification, desired selectivity, operating cost and feed gas processing conditions [6]. Conventional liquid amine scrubbing process (chemical absorption) is one of the mostly used acid gas removal processes, where the separation of CO_2 and H_2S from gas mixtures like natural gas takes place through chemical reactions of amines with acid gases [7]. In this process, the sour gas is introduced into an absorber unit where it comes in contact with an aqueous solution of primary or secondary alkylamine e. g. monoethanolamine (MEA). As a result, CO_2 reacts with amine to form a solution containing corresponding alkylammonium carbamate salt [8]. MEA has the ability to separate both H_2S as well as CO_2 from gas streams. In spite of the fact that amine scrubbing offers high selectivity of CO_2 and H_2S, the major limitation of this process is very high energy demand because of the need of regeneration of amine solution. This requirement is compounded further due to the strong interactions between acid gas molecules and amine. The energy penalty caused by MEA absorption process is approximately 15-37% of energy production of an ordinary power plant [9]. Due to the high capital and operational cost of this technology, the development of alternate low cost and environmentally effective technologies has become very important for the effective removal of acid gases.

Adsorption is regarded as another promising technology for the effective removal of CO_2 and H_2S from natural gas, especially to remove ppm level contents. Here, the acid gases interact with the surface of highly porous solid material having high surface area. The gases can adsorb physically (physisorbents) as well as chemically (chemisorbents) on the surface of adsorbents. Physisorbents can remarkably reduce capital and operating cost as these can reversibly adsorb and desorb CO_2 at near ambient conditions. Activated carbons are commonly used adsorbents for gases and vapors. The major advantages of using adsorbents are operation simplicity and the opportunity of concurrent acid removal as well as gas dehydration. Surface modifications of these adsorbents with polymers and organic molecules is generally carried out in order to enhance their performance.

In this chapter, various polymer incorporated adsorption systems have been reviewed for effective separation of acid gases from natural gas. Special emphasis has also been given to the effect of polymers with basic nitrogen functionalities on the efficiency and selectivity of the adsorbents for acid gases. Various observations in terms of advantages and disadvantages of using different polymers have also been made. Thus, an up-to-date analysis of developing concepts, experimental results as well as theoretical predictions of the polymer enhanced adsorbents for acid gas removal, with particular emphasis on CO_2, H_2S, SO_x and NO_x, are highlighted.

1.2 Polymer Enhanced Adsorbents for CO_2 Capture

For ideal CO_2 capturing technology, Khatri *et al.* [10] suggested three key aspects as (1) high CO_2 adsorption capacity which should be greater than 1 mmol.g^{-1} (2) energy necessity for regeneration should be less compared to aqueous amine process, (3) long term regeneration capacity in a power plant flue gas atmosphere. Several research studies have been performed on different types of adsorbents for the separation of CO_2, including amine-supported adsorbents [11], metal organic frameworks (MOF) [12], carbon based adsorbents [13] and zeolites [14]. Carbon based adsorbents offer plentiful advantages such as low regeneration energy, fast adsorption kinetics, high adsorption capacity and low cost initial sources [15]. Nano-casting is considered as one of the dominant methods for the preparation of mesoporous carbon materials, where the mesopores contribute remarkably to the adsorption process.

For CO_2 adsorption, the presence of basic nitrogen functionalities is very important, thus, requiring to maximize the number of nitrogen functional groups in the surface modification of the adsorbents. As a result, amine functionalized adsorbents have gained considerable research interest because of their high efficiency as well as selectivity for CO_2 removal from gas mixtures. Generally, amine based CO_2 removal functionalities are added into porous supports by employing two approaches. First method is based on changing the carbon matrix surface chemistry by consolidating hetero-atoms like nitrogen to increase the characteristic adsorbate-adsorbent interactions. On the other hand, the second method employs the modification of the porous substrate surface chemistry by impregnation with liquid organic polymers, such as polyethyleneimine (PEI) [16,17]. Many research studies have reported the use of PEIs for enhancing the CO_2 adsorption capacity [16-19]. These polymers have higher CO_2 adsorption capacity due to the high number of amine groups present in the polymer structure. The primary as well as secondary amino groups present in PEI undergo reaction with CO_2 to generate carbamates. These carbamates react further to form bicarbonates in the presence of water, thereby, enhancing the CO_2 adsorption capacity. As expected, long polymer chains with amine groups grafted on each repeat unit exhibit higher CO_2 adsorption capacity than the short polymer chains. This is due to the presence of large number of CO_2 anchoring sites available to form carbamate group. However, these should meet some requirements for maximizing their ability as a polymeric adsorbent and enhancing the CO_2 removal capacity.

Post combustion CO_2 removal by using adsorbents consisting of basic nitrogen functional groups has also proved to be very efficient, as the addition of nitrogen groups to the carbon structure enhances the specific interaction between the CO_2 molecules and carbon surface [20]. In nano-casting approach, an inorganic removable template such as silica is employed to prepare porous adsorbent polymers having high surface area. Here, high temperature activation is not required for preparing adsorbents, and nitrogen is included inside the polymer matrix. On dissolution of inorganic template, the polymer remains as a mirror image and inherits the template porosity.

1.2.1 Carbon Adsorbents

Carbon based materials are effective for gas sorption as well as storage due to abundance, tunable surface area, strong pore structure, chemical stability and easy preparation at industrial scale. Different types of carbon based materials have been studied for CO_2 removal such as activated carbons [21], carbon molecular sieves [22], carbon nanotubes [23] and graphene [24]. Activated carbons are suitable for CO_2 capture because of advantages like high adsorption capacity at high pressure, low cost and are easy regeneration. However, the use of naturally occurring precursor for the activated carbon preparation limits the strength, purity and physical form of end products. This limitation can be overcome by the application of polymeric precursors. Here, the purity and reproducibility can be controlled, and the physical form of end products can be tailored during. The surface properties of activated carbons can be optimized by controlling the synthetic process of polymers, thereby, increasing the CO_2 adsorption capacity. Recently, preparation of activated carbons from synthetic materials has acquired immense research attention. Recently, few studies have reported increased adsorption capacity of activated carbon for H_2S [25] and CO_2 [26] by synthesizing adsorbents using nitrogen containing compounds. Carbon adsorbents consisting of nitrogen can be generated using nitrogen accommodating compounds like melamine, aniline, acetonitrile, etc., along with templates such as zeoliteY, SBA-45 and MCM-41 [27]. The cost of such systems is dependent on the cost of the precursor as well as the carbon yield during the activation and carbonization stages. Phenolic resin (PF), prepared from the phenol and formaldehyde, is one of the most commonly used types of precursors. Phenolic resin based activated carbons provide benefits such as good physical strength and very low level of impurities. In one such study, Martin *et al.* [28] employed different types of phenolic resins as the precursor materials, and ultimate CO_2 adsorption capacities till 10.8 wt% were obtained. It was observed that the narrow micropores were extremely active towards CO_2 separation at atmospheric pressure. It was concluded that for the pre-combustion CO_2 removal, the PF resin derived activated carbons exhibited higher potential as adsorbents. In another work by Martin *et al.* [29], the pre-combustion CO_2 adsorption capacity of two PF resin based activated carbons (Resol PFNA and Novolac PFCLA) were analyzed under dynamic and static conditions. Both adsorbents exhibited similar CO_2 adsorption under equilibrium at 298 K and 15 bar. For a mixture of 60% H_2 and 40% CO_2, at the aforementioned pressure and temperature conditions, PFNA exhibited higher CO_2/H_2 selectivity than PFCLA, due to the narrower average pore width. At 298 K, the breakthrough tests were performed at a total pressure of 15 bar in a fixed bed column and the CO_2 adsorption capacity was observed to be 6.5 mmol.g^{-1} and 6.4 mmol.g^{-1} for PFCLA and PFNA, respectively. However, at 313K, a decrease in the CO_2 adsorption capacity was observed (5.3 mmol.g^{-1} for PFCLA, 5.8 mmol.g^{-1} for PFNA). Further, both activated carbon adsorbents displayed excellent cyclability as well as regenerability over successive pressure swing adsorption (PSA) cycles. Overall, it was concluded that the adsorbents exhibited higher potential for the application of pre-combustion CO_2 adsorption by PSA process.

Long *et al.* [30] generated activated carbons through physical activation process from phenol-melamine-formaldehyde gels. The activated carbons had a specific area of about 525–685 m^2.g^{-1}. In a similar study by Ru-Ling *et al.* [31] for the CO_2 adsorption, highly porous activated carbon was prepared from melamine modified phenol formaldehyde (MPF) resin by steam activation at different ranges of activation temperature (700 °C - 950 °C) at atmospheric pressure. The modification of PF resins helped to provide high nitrogen content for activated carbon. The vessel containing MPF resin beads was placed in an oven and the temperature was increased at a rate of 5 °C.min^{-1} until the activation temperatures were attained (700 °C, 750 °C, 800 °C, 850 °C, 900 °C and 950 °C). Subsequently, activation was carried out using deionized water. It was observed that on enhancing the activation temperature from 700 °C to 950 °C, the specific surface area developed from 382 m^2.g^{-1} to 1439 m^2.g^{-1}. Further, the elemental and FTIR analysis confirmed on increasing the activation temperature, nitrogen atom content present in the MPF resin decreased. The CO_2 adsorbed amount on MPF850 at 1 atm and 0 °C was observed to be 6.71 mol.kg^{-1}. The adsorption capacity of MPF 850 remained unaltered even after five adsorption/desorption cycles at atmospheric pressure.

Pevida *et al.* [32] generated a range of highly porous melamine-formaldehyde adsorbents, without any high temperature activation for CO_2 removal. Adsorbents were prepared by template synthesis of MF resin and carbonization at varying temperatures, using silica as template material. The adsorbents were characterized in terms of their elemental composition, textural properties as well as surface chemistry. The materials contained 42 wt% nitrogen and had a surface area of 880 m^2.g^{-1}. The synthesis method was observed to be very efficient as the adsorbents exhibited CO_2 capture ability of upto 2.25 mmol.g^{-1} at 25 °C, thereby exceeding the adsorption capacity of several commercial activated carbons. The adsorption capacity was observed to be highest at 25°C and eventually decreased with increasing the temperature. The behavior was different from the performance of PEI based adsorbents, where the adsorption capacity remained constant until 90°C [33]. This was due to the difference in chemistry of the nitrogen present in PEI based adsorbents and MF resin. In another study, Pevida *et al.* [34] also developed nitrogen enriched activated carbon adsorbents by silica templating technique using melamine formaldehyde resin. Four different silica templated samples were generated, and the carbonization was carried out by heating the samples up to 400 °C, 500 °C, 600 °C and 700 °C for 1 h. It was observed that both texture as well as chemistry of the adsorbents played a remarkable role in determining the CO_2 removal capacity. It was confirmed that associating a convenient textural advancement with an agreeable chemistry resulted in an improvement in CO_2 adsorbing capacity of the adsorbents. The 600 °C carbonized adsorbents had the greatest adsorption capacity of 2.25 mmol.g^{-1} under pure CO_2 atmosphere at a temperature of 25 °C. The observed value was remarkably higher when compared to the previous values attained for MF polymer based sorbents without silica template (ultimate CO_2 removal capacity of 1.02 mmol.g^{-1} at 25 °C [20]). The addition of silica template enhanced the textural advancement of MF based adsorbents which enhanced the CO_2 adsorption.

Hao *et al.* [35] pyrolyzed copolymer of formaldehyde, lysine and resorcinol for the preparation of nitrogen doped porous carbon monolith. The CO_2 adsorption capacity of the generated material was observed to be 3.13 mmol.g^{-1} for 100% CO_2 at 1 atm and 25 °C. Drage *et al.* [36] prepared nitrogen enriched carbons from UF and MF resins, which were polymerized in the presence of K_2CO_3, the chemical activation agent. In this system, activation was performed over a range of temperatures. Further, carbons generated by chemical activation were observed to be more efficient for the CO_2 removal, when compared to the carbons prepared by CO_2 physical activation. The adsorbents, generated by the chemical activation of UF resin at 500°C, were able to capture 8 wt% CO_2 at a temperature of 25 °C. This exhibited the usefulness of surface chemistry of melamine functionalized carbon for the preparation of effective CO_2 adsorbents. In another study, Goel *et al.* [37] prepared a series of mesoporous carbon adsorbents with high nitrogen content, through nano-casting method. The mesoporous silica was employed as the template while melamine formaldehyde resin served as the precursor. A varying carbonization temperature range of 400-700 °C was used for the preparation of the adsorbents. The highest CO_2 adsorption capacity of about 0.83 mmol.g^{-1} was observed at 12.5% CO_2 (rest nitrogen) atmosphere at a temperature of 30 °C. Exothermic process was observed to take place and the adsorption capacity decreased with increasing temperature.

Wang *et al.* [38] reported the development of functional adsorbents by using commercially available carbon materials, such as activated carbons and carbon blacks, to prepare carbon-based "molecular basket" sorbent (CB-MBS) by modifying the surface with CO_2-philic PEI. Molecular basket sorbent (MBS) are solid amine materials which consists of mesoporous or nanoporous materials along with functional polymers. CO_2 sorption capacity of 154 mg-CO_2/g-sorb was observed for carbon black based MBS with 65 wt% PEI. The carbon black-supported PEI adsorbents were observed to have high sorption capacity due to high pore volume and large pore size. Figure 1.1 also exhibits CO_2 sorption capacity of carbon black based MBS as a function of surface area. In an earlier study to gain enhanced adsorption capacity of the PEI modified adsorbents, Ma *et al.* [39,40]

also used molecular basket system using silica support as basket, which helped in retaining large quantity of PEI. Figure 1.2 also exhibits its sorption breakthrough curves for CO_2 and H_2S in both single-stage and two-stage sorption processes using the as-generated MBS, when a model gas containing 0.40 v% H_2S, 2.40 v% CO_2, and 20 v% H_2 in N_2

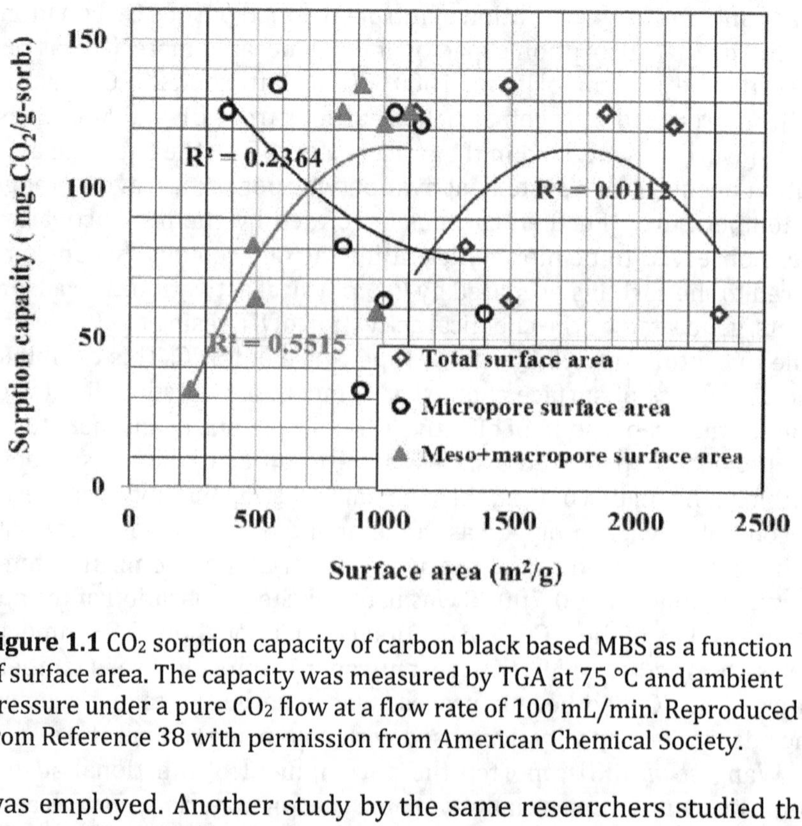

Figure 1.1 CO_2 sorption capacity of carbon black based MBS as a function of surface area. The capacity was measured by TGA at 75 °C and ambient pressure under a pure CO_2 flow at a flow rate of 100 mL/min. Reproduced from Reference 38 with permission from American Chemical Society.

was employed. Another study by the same researchers studied the CO_2 capture capacity of a series of molecular basket sorbents polyethylenimine-SBA-15 (mesoporous silica) containing different amount of PEI [41]. The PEI-SBA-15 sorbents were generated using the wet-impregnation techniques. It was observed that the CO_2 sorption by PEI-SBA-15 occurred through chemisorption even under high pressure, while in the system with SBA-15 alone, physisorption occurred at high pressure. Further, it was also confirmed that the best loading for CO_2 adsorption ability depended on the sorption temperature as well as on the pore structure of the supports. The CO_2 dissipation in the PEI bulk played a remarkable role in determining the CO_2 sorption ability as well as the adsorption/desorption rate at lower temperatures. At 75 °C, the highest CO_2 removal ability of 154.0 mg/g was obtained over PEI-60/SBA-15 sorbents. However, the highest ratio of CO_2 adsorption to the possible amine sites was exhibited by PEI-30/SBA-15, which was due to the improved dispersion inside the pores with lesser CO_2 adsorption barricade in the PEI-30/SBA-15 adsorbent.

Jitong *et al.* [42] reported CO_2 removal by PEI loaded mesoporous carbon sorbents. The sorbents were observed to combine the advantages of impregnated PEI as well as excellent porous characteristics of mesoporous carbons. It was observed that the PEI loading of 65 wt% was ideal due to less mass transfer resistance as well as the highest utilization ratio of amine compound. As the loading was further increased, the benefit of mesoporous structure and large pore volume of the support material was reduced. The carbon based sorbents offered better CO_2 sorption ability, as compared to mesoporous silica supports. Overall, it was concluded that the PEI loaded carbon sorbents provided benefits like high CO_2 removal capacity, good regeneration ability, low regeneration temperature at 100°C and high utilization ratio of the amine groups.

Carbon nanotubes (CNT) possess good ability to separate CO_2 from natural gas because of their high thermal stability, chemical stability and unique physico-chemical properties. CO_2 adsorption in carbonaceous materials like multi-walled carbon nanotubes (MWCNT) occurs near the carbon surface solid because of the physical forces that carbon atoms exert on the CO_2 molecules. Ngoy *et al.* [43] reported about increasing the CO_2 adsorption behavior by using ethylenediamine for grafting onto polysuccinimide (PSI) to prepare polyaspartamide (PAA), which was covalently linked to MWCNT. An increase in CO_2 adsorption ability was observed due to the fact that primary amine reacted with CO_2 by chemisorption process to form carbamate group. The PAA modified MWNTs exhibited higher CO_2 adsorption capacity of 70 gCO_2/kg as compared to 46.17 gCO_2/kg for PAA, 26.90 gCO_2/kg PSI, and 15.20 gCO_2/kg MWNTs alone.

Nowadays, graphene, a novel two-dimensional carbon nanomaterial, has acquired immense importance as a solid adsorbent for CO_2 removal from flue gases [44]. This is because of its exclusive molecular structure and several interesting properties like large surface area, excellent chemical stability, tunable porosity, excellent thermal conductivity and good mechanical strength. As control on surface functionalization, doping, enhancement of the specific surface area as well as pore structure are of substantial importance to increase the gas adsorption capacity, thus, graphene offers tremendous opportunities for preparing tailor made carbon based adsorbents [45]. As discussed before, amine functionalized adsorbents have enhanced CO_2 adsorption capacity, which increases even further when nitrogen is effectively added to the support. Hence, the amino group surface density has a remarkable role in the CO_2 adsorption capacity. As a result, polyaniline (PANI), which is an abundant source of nitrogen containing groups, was used with graphene to produce graphene-polyaniline nanocomposites with good CO_2 adsorption capacity [46].

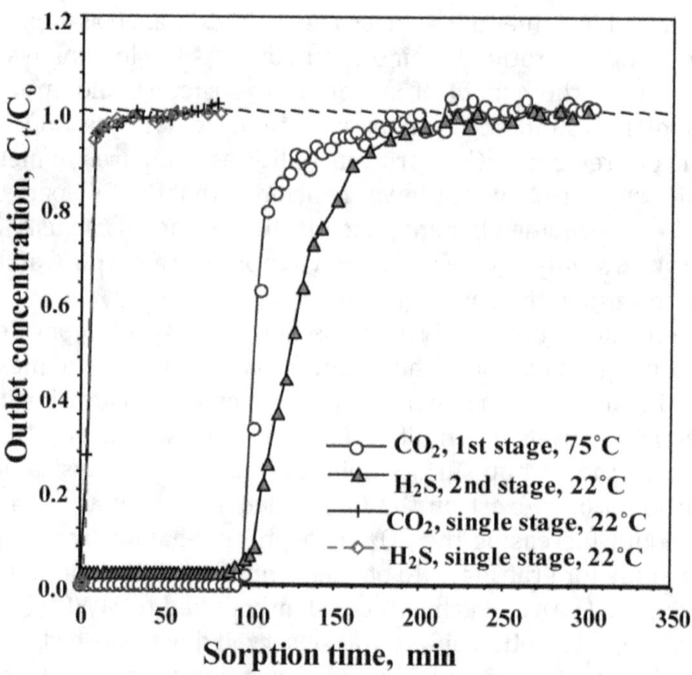

Figure 1.2 Sorption breakthrough curves for CO_2 and H_2S in both single-stage and two-stage sorption processes using the as-generated MBS, when a model gas containing 0.40 v% H_2S, 2.40 v% CO_2, and 20 v% H_2 in N_2 was employed. Reproduced from Reference 39 with permission from American Chemical Society.

In addition to graphene, its oxidized derivative, graphene oxide (GO) has also received significant research attention for development as a CO_2 adsorbent. The incorporation of GO with different polymers may lead to the preparation of CO_2 adsorbents with significantly superior performance. The structural as well as physical properties of

GO-PEI porous materials was examined by Sui *et al.* [47]. The authors used a facile approach for the generation of GO-PEI materials having three dimensional (3D) interconnected networks under moderate conditions. The materials possessed lower density along with high adsorption capacity for CO_2. The generated material was regarded as a unique adsorbent as it exhibited combined properties like excellent porous structure together with high amine density for increasing the CO_2 adsorption capacity. The samples had an extensive surface area (472 $m^2.g^{-1}$) as well as total pore volume (1.3 $cm^3.g^{-1}$). Figure 1.3 compares the carbon dioxide adsorption isotherms of GO, hydrothermal reduced graphene (HTG), and GO-PEI porous material (GEPM) samples, along with the adsorption capacity. The results confirmed enhanced level of CO_2 adsorption of the GEPM materials due to the presence of basic sites and large specific surface area.

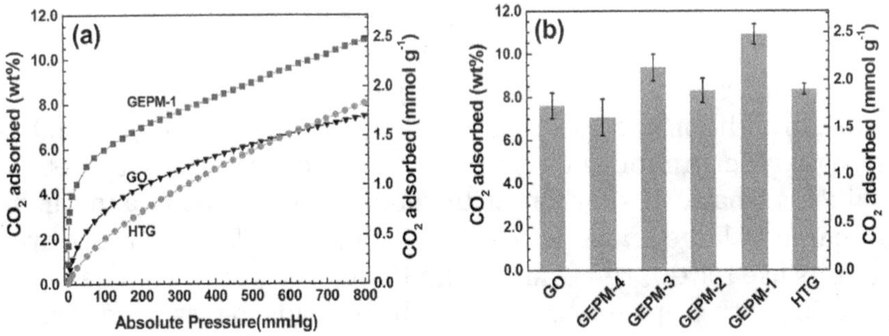

Figure 1.3 Comparison of the (a) carbon dioxide adsorption isotherms of GO, hydrothermal reduced graphene (HTG), and GO-PEI porous material (GEPM) samples and (b) comparison of the capacity of these materials. Reproduced from Reference 47 with permission from American Chemical Society.

Preparation of porous adsorbents for CO_2 capture from renewable sources has also been reported. Out of the different amino groups enriched natural polymers, chitosan has been utilized for various applications like waste water treatment, heterogeneous catalysis. This is because of its various advantages like presence of beneficial functional groups (-NH_2 and –OH) and environmental compatibility. Valechha *et al.* [48] reported a series of chitosan based adsorbents for CO_2 capture. The authors studies the interaction of free available amine with carbon dioxide molecules, along with studying the effect of crosslinking and functionalization. It was observed that chitosan glutaraldehyde exhibited superior adsorption than chitosan due to the formation of imine functionality.

In general, the mechanical as well as physical properties of chitosan are not as high as many synthetic polymers. Hence, nano-fillers like graphene, GO, CNT are generally incorporated in the polymer matrix in order to achieve these benefits. Alhwaige *et al.* [49] reported chitosan-GO hybrid aerogels as an interesting adsorbent for CO_2 capture. Aerogels, which are continuous pore materials with large surface area and ultra-low density, have been recently investigated for CO_2 removal from natural gas. Organic aerogels are pyrolyzed in an inert atmosphere for generating porous carbon materials with large surface area [50]. Examples include melamine-formaldehyde, resorcinol-formaldehyde, chitosan carbon gel, etc. [51]. In the studies by Alhwaige *et al.*, chitosan-GO aerogels were generated with different compositions using freeze drying method and the CO_2 adsorption was examined at various operating conditions. It was observed that the addition of 20 wt% GO led to the doubling of the quantity of CO_2 adsorbed on chitosan-GO hybrid aerogels as compared to chitosan alone. In addition, the adsorbents were observed to have good stability, easy regeneration and cost effectiveness over multiple cycles.

PPy is a basic CO_2 sorbent with good thermal stability and can be prepared at large scales. However, high cost and lower surface area limit its use for CO_2 removal. The limitations can be overcome by interfacing with graphene, which specifically leads to increased surface area. Chandra *et al.* [52] prepared N-doped porous carbons by chemical activation of graphene-polypyrrole (PPy) composites. Specifically, PPy functionalized graphene was prepared by chemical polymerization of pyrrole in GO utilizing ammonium persulfate and subsequent consecutive reduction employing hydrazine. The CO_2 removal capacity was observed to be 4.3 mmol.g^{-1} for the samples activated at a higher temperature of 500-600 °C, due to the enhanced adsorbent-adsorbate interaction on the micro-porous surface. It was confirmed that the graphene-PPy composites had high adsorption capacity as well as selectivity for CO_2 removal, along with a relatively inexpensive and easy synthesis process for the purpose of industrial scaling. Similar to the aforementioned work, a series of S- or N- doped graphene based adsorbents using various kinds of polymer-graphene nanocomposites (r-GO /polythiophene [53], r-GO/polyaniline [54], GO/ polyindole [55]) have been prepared by chemical activation at 400-800 °C. The materials were observed to have large pore volume, deep micro-porosity and high surface area. These doped samples exhibited higher CO_2 adsorption under ambient conditions (>4 mmol.g^{-1}), which was higher than several other solid adsorbents like activated carbon, MOFs and amine functionalized silica. In addition, the materials also exhibited excellent selectivity for CO_2 over N_2 (post combustion CO_2 removal), CH_4 and H_2 (pre-combustion CO_2 removal) and could also be regenerated without any difficulty for repeated use.

In another study by Mishra and Ramaprabhu [56], polyaniline-graphene nanocomposite was confirmed to be a promising adsorbent for CO_2 removal. An ultimate adsorption capacity of 75 mmol.g^{-1}, 47 mmol.g^{-1} and 31 mmol.g^{-1} was observed at 11 bar and temperatures of 25, 50, and 100 °C, respectively. The adsorption capacities were significantly higher as compared to pure graphene and the adsorbents exhibited higher degree of recyclability. A decrease in the adsorption capacity was observed with increase in temperature, which was opined to be due to the enhancement in the kinetic energy of gas molecules at higher temperature. Yang *et al.* [57] prepared a graphene based organic-inorganic ternary (PEI-G-silica) CO_2 solid adsorbent by nano-casting method. A unique and effective structure was generated by strong collaborative influence of each component. For instance, graphene offered large surface area and high thermal conductivity, while flexible morphology and uniform porosity was provided by mesoporous silica. With the presence of PEI, large CO_2 adsorption capacity was attained. Thus, the ternary adsorbent offered excellent CO_2 removal capacity and long term cycling stability.

Recently, Tsoufis *et al.* [58] intercalated diaminobutane poly(propylene imine) dendrimers into the interlayer spacing between GO sheets without employing any cross linking agent. The CO_2 adsorption capacity of resulting hybrids was observed to exhibit faster kinetics and higher values under wet circumstances, as compared to dry conditions. Thus, it was clear that the presence of water had an important role on the adsorption capacity of the graphene material.

Overall, from the existing research studies, it is clear that the unique structure and interesting properties of polymer-graphene nanocomposites provide immense opportunities for designing adsorbents suited for industrial applications.

1.2.2 Metal Organic Frameworks (MOF) and Porous Polymer Networks (PPN)

MOF, also referred as porous co-ordination polymers (PCP), are considered as one of the most promising porous materials with high degree of crystallinity, very high surface area (1000–8000 m^2/g) and porosity (50-90% free volume) [59]. In these physical adsorbents, the metal clusters are bonded strongly by organic linkers. This combination creates wide variety of porous materials with exclusive properties like chemical and structural tenability [60]. These materials have been reported to have significant CO_2 uptake ability and CO_2/N_2 selectivity at room temperature [61]. However, majority of MOFs have trouble in meeting the strict industrial requirements [62]. On the other hand, purely organic porous polymers, a class of adsorbents with proportionate pore size as well as surface areas, have good physico-chemical properties and stability due to covalent bonding in the network structure [63]. In spite of the fact that most of these materials are amorphous, their stability is highly useful in industrial applications. For porous material to be industrially acceptable, in addition to the high physico-chemical stability, high CO_2/N_2 selectivity as well as high CO_2-uptake ability is equally important. Both of these factors are influenced by the addition of CO_2-philic moieties in the structure. Lu *et al.* [64] prepared polyamine tethered PPN for CO_2 separation from flue gas. The addition of polyamine groups to PPN resulted in significant CO_2 adsorption behavior at low pressure and 295 K. The results confirmed that the CO_2 adsorption capacity was related to the amine loading rather than the surface area, and the results were consistent with the results obtained in another study by Dawson *et al.* [65]. In a study by Lin *et al.* [66], it was observed that the PEI incorporated MOF adsorbents were capable to achieve good selectivity of CO_2 over nitrogen up to 1200 at 50 °C and 770 at 25 °C. In another study, the adsorption ability of PEI impregnated resins for flue gas containing CO_2 concentration of 400 ppm and 15 vol % was determined to be 99.3 mg/g and 181 mg/g respectively at 25°C [67] (Figure 1.4).

Lu *et al.* [68] reported the preparation of a porous polymer network PPN-6 with permanent porosity and grafted the PPN-6 with lithium sulfonate and sulfonic acid. The resulting products, i. e., PPN-6-SO_3Li and PPN-6-SO_3H exhibited significant enhancement in CO_2 uptake ability and significantly large CO_2/N_2 adsorption selectivities at ambient conditions. At zero loading, the materials exhibited heat of adsorption of 30.4 $kJ.mol^{-1}$ (PPN-6-SO_3H) and 35.7 $kJ.mol^{-1}$ (PPN-6-SO_3Li), which were significantly higher than non-grafted PPN-6 (17 $kJ.mol^{-1}$). Figure 1.5 demonstrates the increment in the isosteric heats of CO_2 adsorption and CO_2 uptake capacities of the developed materials as a function of pressure.

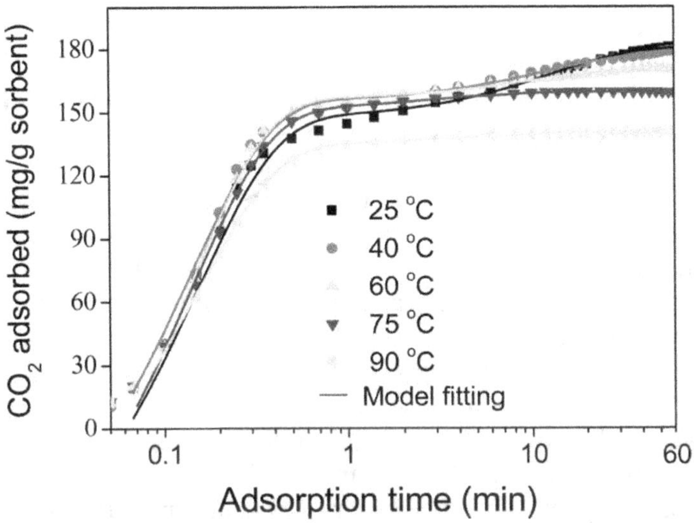

Figure 1.4 CO_2 adsorption (60 min) kinetics on PEI-impregnated resin (HP20/PEI-50) as a function of temperature. Reproduced from Reference 67 with permission from American Chemical Society.

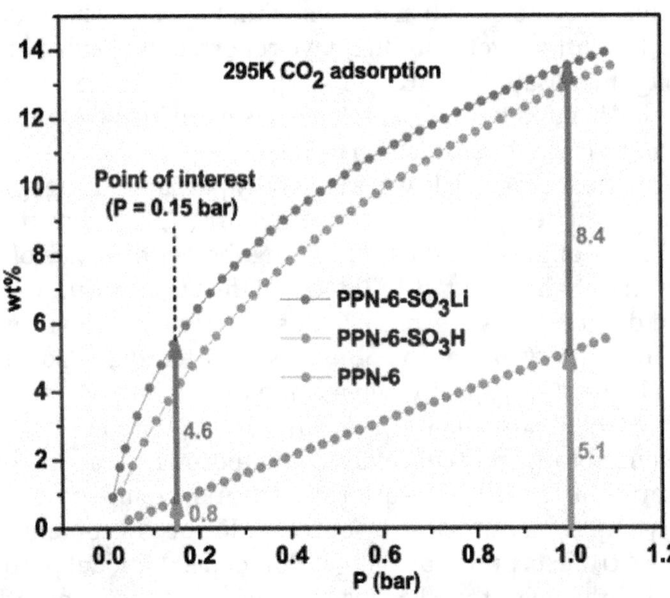

Figure 1.5 PPN grafted with sulfonic acid (PPN-6-SO_3H) and its lithium salt (PPN-6-SO_3Li). Reproduced from Reference 68 with permission from American Chemical Society.

1.2.3 Silica Based Adsorbents

Amine modified mesoporous silica has also been widely investigated for CO_2 capture. The unmodified silica materials have large pore size that can be tuned by proper selection of organic surfactants which act as template during the preparation, along with the surface hydroxyl groups, which can be chemically modified. For CO_2 capture, solid adsorbents are generated by the surface grafting of primary or secondary amines onto silica. This grafting can be achieved chemically (covalent bonding) as well as physically by wet impregnation. Amines are physically consolidated into silica materials by immersing the silica in a solution of a polymeric amine like polyethyleneimine. Thus, weak van der Waals forces generate between the amine and pore surface. Silica nano-particles that are modified by PEI are relatively easy to prepare, inexpensive and regenerable adsorbent with good capacity to capture CO_2 [69]. The main disadvantage of using amine grafted silica materials for the CO_2 removal is the hydrolysis of siloxane linkages which is promoted by either amine [70] or acidic conditions, thereby, decreasing the performance. Further, the ability of silica materials with physically attached amine may decrease over time due to the leaching process or redistribution within the material at the time of usage. Recently, several studies have reported PEI loaded mesoporous silica materials like SBA-12 [71], SBA-15 [72], SBA-16 [73], MCM-48 [74] and KIT-6 [75], and have observed high CO_2 capacity of these materials in the range of 1.5 to 3.5 mol.kg^{-1}. In these studies, the mesoporous silica materials were generally prepared using synthetic polymers as the template and tetra-ethyl-ortho-silicate as the silica source. In one of these studies comparing the performance of different silica materials, Son *et al.* [72] studied the CO_2 adsorption performance of different mesoporous silicas supported by PEI adsorbents. The authors observed that the SBA-15 supported PEI adsorbent exhibited a higher CO_2 capacity when compared to MCM-41 and MCM-48 supported PEI adsorbents. In another study, Witoon [76] studied the influence of PEI content and bimodal porous silica supports on CO_2 adsorption capacity at various adsorption temperatures. Rice husk was employed as the silica source. At lower PEI content (10-20 wt%), the CO_2 adsorption capacity of PEI loaded bimodal porous silica was slightly lower than PEI loaded unimodal silica. Recently, hyper-branched aminosilicas generated by in-situ aziridine polymerization on porous solids, were also used as adsorbents for the CO_2 removal from air [77]. Goeppert *et al.* [18] prepared adsorbents based on fumed silica (FS) with PEI impregnation. Samples with PEI loading of 33 wt % (FS-PEI-33) and 50 wt% (FS-PEI-50) were generated. The materials had superior adsorption capacity for the removal of CO_2 from air. The adsorbents were also able to adsorb CO_2 reversibly in repeated cycles under mild conditions. At lower loadings, the polymeric amine was observed to be better dispersed on the support surface. However, at higher loadings, a significant portion of the amino groups present in the PEI was not accessible because of the poor dispersion on the surface of the support. Further, the adsorbents could perform well under both humid as well as dry conditions. Figure 1.6 illustrates the results attained on FS-PEI-50 and FS-PEI-33 for the CO_2 adsorption under humid and dry conditions. These findings indicated that the generated materials were superior to zeolites, which lose CO_2 adsorption capacity under humid conditions. Thus, FS-PEI adsorbents can be used for CO_2 separation from gas streams in submarines as well as other closed environments.

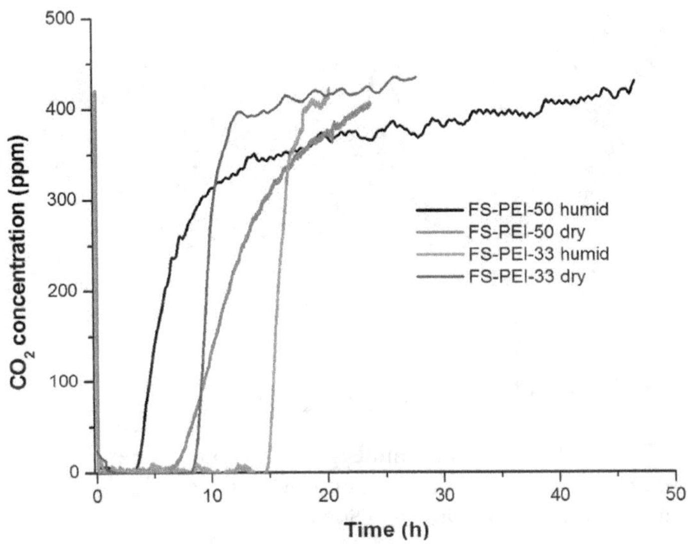

Figure 1.6 Adsorption of CO_2 from the air at 25 °C on FS-PEI-50 and FS-PEI-33 under humid and conditions. Reproduced from Reference 18 with permission from American Chemical Society.

Lively *et al.* [78,79] prepared a new type of CO_2 removal adsorbent system based on hollow fiber based solid adsorbent that operated in rapid temperature swing adsorption (TSA) mode. This prevented many deficiencies related to the cyclic sorption processes. The solid adsorbent particles were embedded in a polymeric porous hollow fiber matrix, which provided several advantages as compared to polymer fibers which are traditionally used for membrane applications. In the hollow fiber system, high volume of adsorbent materials could be incorporated, which allowed rapid mass transfer to the adsorbent due to large voids. Rezaei *et al.* [80] prepared silica/amine/polymer hollow fiber adsorbents using post spinning infusion method, as shown in Figure 1.7. The fibers exhibited good CO_2 removal capacity from the simulated flue gas. Thus, the post spinning infusion technique provided a new platform for preparing silica/amine/polymer hollow fibers for flue gas CO_2 capture applications.

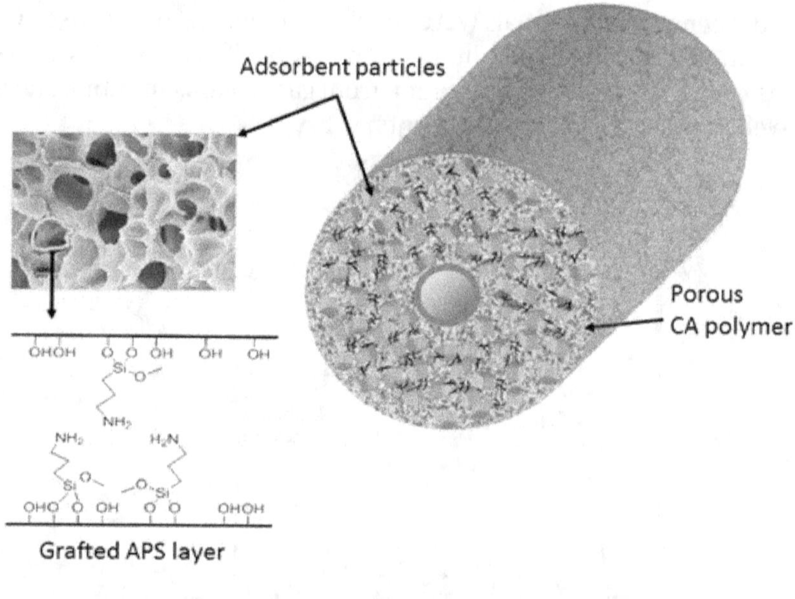

Figure 1.7 Generation of silica/amine/polymer hollow fiber adsorbents using post spinning infusion method. Reproduced from Reference 80 with permission from American Chemical Society.

1.3 Polymer Enhanced Adsorbents for H_2S Capture

H_2S is an immensely toxic gas and causes damage to the pipelines, along with deterioration of natural gas quality, thus, requiring its removal to ppm levels. As mentioned earlier, the amount of H_2S in natural gas can be minimized by reaction with amines. The reaction is highly exothermic and regardless of the amine structure, H_2S reacts with primary, secondary as well as tertiary amines. Several types of adsorbents like activated carbon [81], MOF [82], metal oxides [83] and graphite [84] have also been used for the effective H_2S removal. Other research studies have also reported different solid materials for H_2S separation at high temperature (>300 °C) from a series of industrial gas streams [85,86]. As mentioned earlier for CO2 removal, Ma *et al.* prepared molecular basket regenerable adsorbent (PEI-loaded MCM-41), which could adsorb CO_2 as well as H_2S from a model gas. Using the same molecular basket sorbent concept, Wang *et al.* [87] prepared sorbent by loading a PEI polymer on the mesoporous molecular sieve to remove H_2S by an acid-base reaction. The results indicated that the material could separate H_2S from gas streams to a minimum of 2 ppmv at ambient conditions. The amount of PEI on the molecular sieve was observed to have a strong effect on the sorption performance. The mesoporous molecular sieve with three dimensional channel structure as well as large pore size helped to enhance the kinetic ability of the PEI sorbent. The generated sorbents exhibited a saturation capacity of 3.02 mmol H_2S/g-sorb and breakthrough capacity of 0.79 mmol H_2S/g-sorb at 22 °C, gas hourly space velocity (GHSV) of 674 h^{-1} with a gas consisting of 4000 ppmv of H_2S. Further, it could also be regenerated readily under mild conditions (75-100 °C temperature range). Thus, considering the remarkable sorption performance, stability and regenerability of PEI/SBA-15 sorbent, it represented one of mist effective sorbents for H_2S separation from gas streams in more energy efficient and environment friendly operation. Chen *et al.* [88] generated PEI loaded porous silica monolith having high stability and high H_2S breakthrough capacity of about 1.27 mmol of H_2S/g sorbent at a temperature of 22 °C. The adsorbent could be regenerated at 75 °C. Further, it was also observed that H_2S adsorption performance was influenced by the amount of PEI added to the adsorbent as well as the interactions between the H_2S and amine functional group present in PEI. Figure 1.8 also shows the sorption results of the sorbents as a function of PEI molecular weight. It was observed that the sorbent with M_w of 600 exhibited the best sorption performance. Increasing the molecular weight of PEI led to reduction in the sorption due to steric hindrance and low fluidity, which hindered the internal diffusion of H_2S, thus, reducing the utilization efficiency of amine groups. PEI with lowest molecular weight also did not exhibit high sorption probably due to its composition as it was a mixture of linear monomers and branched polymers. In addition, the interaction of H_2S with the polymer chains possessing a varying range of molecular weights contributed to different performance of these PEI polymers.

In another study by Wang *et al.* [89], it was observed that the nano-porous composite adsorbent PEI/SBA-15 had high adsorption capacity for H_2S and the presence of moisture had a promoting effect on the H_2S separation from gas streams. Recently Jaiboon *et al.* [90] prepared PEI modified high porosity adsorbent silica xerogels for low temperature H_2S separation. Using fixed bed system, the H_2S adsorption as well as desorption capacities were characterized with respect

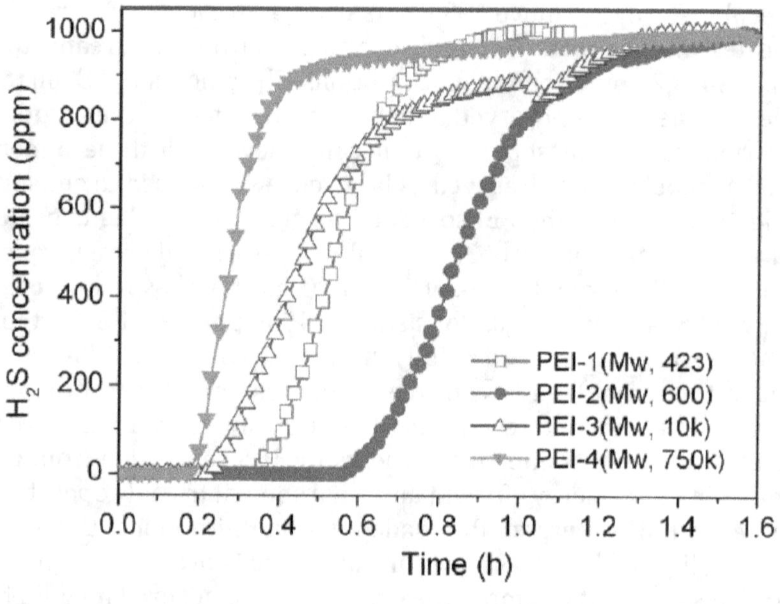

Figure 1.8 Breakthrough curves of H_2S for porous silica sorbents loaded with 65 wt % PEI as a function of PEI molecular weight. Reproduced from Reference 88 with permission from American Chemical Society.

to operating temperature, amine type, loading quantity and gas flow rate. On enhancing the amine loading level and reducing the temperature, the H_2S removal ability could be increased. At 30 °C, the xerogel containing 50 wt% PEI (fumed silica -PEI800-50) exhibited the highest saturation capacity and breakthrough time. Out of the different types, the optimal sorbent (fumed silica-PEI800-50) was observed to regenerate easily for minimum ten successive adsorption-desorption cycles, at a mild temperature, without any decrease in the regeneration ability or adsorption capacity.

1.4 Polymer Enhanced Adsorbents for SO_x and NO_x Capture

Adsorption of weak acidic gases like SO_2, CO_2, NO and NO_2 onto amine containing polymeric adsorbents has been examined [91,92]. The authors analyzed the influence of amine structure on the adsorption of acid gases as well as thermal reversibility of adsorption-desorption processes. It was observed that the thermal reversibility of gas removal process reduced in the order $CO_2 > SO_x > NO_x$. Poor CO_2 adsorption capacity was observed for the tertiary amine functionalized polymer. However, the adsorbent exhibited good affinity for the weak acidic gases. Further, it was also noticed that the SO_2 adsorption mechanism on amino-polymers was the same as CO_2 adsorption. Only few studies have dealt with the SO_x/NO_x activated degradation process at molecular level in supported amine adsorbents. The mechanism of reaction of amino-polymers with NO under ambient condition is as follows:

$$\mathbf{R_1R_2NH + NO \rightleftharpoons R_1R_2NHNO}$$

$$\mathbf{R_1R_2NHNO + NO \rightleftharpoons R_1R_2NHN_2O_2}$$

$$\mathbf{R_1R_2NHN_2O_2 + R_1R_2NH \rightleftharpoons R_1R_2NH_2\ R_1R_2N_2O_2^-}$$

Hallenbeck and Kitchin [93] conducted research on the polymeric CO_2 adsorbent functionalized using primary amines and the stability was examined in the presence of SO_2. There was a decrease in the CO_2 adsorption capacity of the polymer from 1.4 to 0.1 mmol/g on exposing it to 431 ppm SO_2 over nine cycles at 50 °C. Figure 1.9 demonstrates CO_2 capture capacity of resin in the the presence of SO_2. It was obvious that the CO_2 adsorption capacity of the adsorbent decreased in each cycle, till achieving a stage where the capacity was no longer available. More specifically, the activity decreased initially almost linearly, and stabilized afterwards at the minimum capacity of 0.2 mol/kg. Analysis in mass spectrometer also revealed no presence of SO_2 for the first six cycles, indicating that all of the SO_2 was captured by the adsorbent. The authors also confirmed that the reduction in CO2 adsorption capacity was not attributed to the reduction in surface area or pore volume, but predominantly resulted from the interaction of SO_2 with amine groups in the adsorbent. In another recent study, Fan *et al.* [94] analyzed the cyclic stability of polymeric fibers functionalized with amine. An optimal stability was observed after more than 120 cycles for the hollow fibers containing the secondary amines with 25% capacity loss for CO_2, after exposing to 200 ppm SO_2 at 35°C.

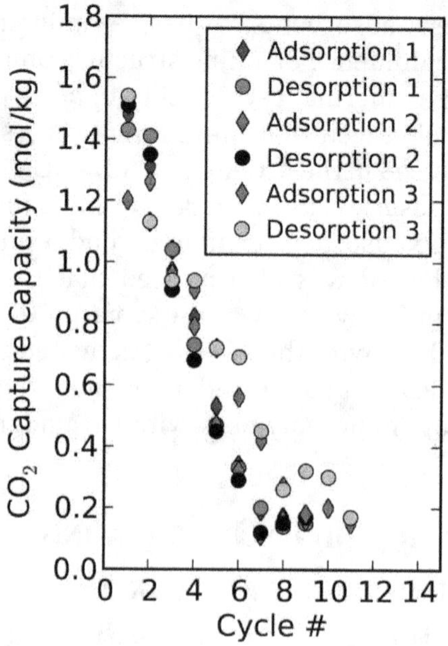

Figure 1.9 CO_2 capture capacity of resin in presence of SO_2. Reproduced from Reference 93 with permission from American Chemical Society.

1.5 Summary and Outlook

In summary, the chapter reviewed the performance of various polymers that are extensively employed for acid gas adsorption from the natural gas. The state of art in this direction was critically analyzed with a focus on the adsorbent material, mechanism of adsorption and challenges to be resolved.

To summarize, the new developments have resulted in the design and development of advanced materials with required chemical, physical and structural properties for effective acid gas removal when compared to the currently used technologies. Different types of synthetic polymers like polyethyleneimine, phenolic resin, melamine formaldehyde resin, urea formaldehyde resin, polyaniline, polypyrrole and natural polymers like chitosan have been reported to significantly contribute to the separation of CO_2 and H_2S from natural gas. Most of the polymer based adsorbents have been confirmed to have high efficiency towards the adsorption of acid gases. In addition, the concentration of amine groups available in the adsorbent as well as the interactions between the acid gases and amine groups are the two dominant factors that influence the amine efficiency and adsorption capacity. In addition, several other factors like the method of activation of carbons, length of polymer chains, degree of polymerization, pore structure and sorption temperature of the supports also play a vital role in determining the adsorption capacity. It was observed that the carbons prepared by chemical activation process were more efficient for the CO_2 separation than the carbons prepared by physical activation. Higher CO_2 adsorption ability was observed in long chain polymers due to the incorporation of more amine group in the adsorbent. Thus, a well developed and adequate porosity along with a favorable chemistry remarkably increased the acid gas adsorption capacity of the adsorbents. As a result, adsorbents need to be generated with sufficient basic sites to increase the affinity towards CO_2 and H_2S.

In spite of high adsorption capacity of the adsorbents, several economical and technical barriers hamper their large scale application for acid gas removal. The greatest obstacles for the usage of these adsorbents are absence of long term stability, limited range of convenient operating conditions, high preparation cost and inability for simultaneous multiple impurity gas removal. In addition to this, despite the fact that the new emerging materials like MOF exhibit good acid gas adsorption capacity, such adsorbents are unstable in wet conditions. Hence, more insights need to be gained for improving the stability under humid conditions. Significant research is still being carried out in order to investigate different classes of polymers so as to develop adsorbents with sufficient amount of nitrogen to enhance the adsorption as well as to overcome the aforementioned barriers. Despite several recent approaches, most of the potential remains unexplored and need to be translated into widespread use for the acid gas removal process through technology development.

References

1. Berg, S. V. (1998) Lessons in electricity market reform: regulatory processes and performance. *The Electricity Journal*, **11**, 13-20.
2. Rufford, T. E., Smart, S., Watson, G. C. Y., Graham, B. F., Boxall, J., Diniz da Costa, J. C., and May, E. F. (2012) The removal of CO_2 and N_2 from natural gas: a review of conventional and emerging process

technologies. *Journal of Petroleum Science and Engineering*, **94**, 123-154.

3. Iliuta, M. C., Larachi, F., and Grandjean, B. (2004) Solubility of hydrogen sulfide in aqueous solutions of Fe(II) complexes of trans-1,2-cyclohexanediaminetetraacetic acid. *Fluid Phase Equilibria*, **218**(2), 305-313.
4. Engineering Data Book, Gas Processors Suppliers Association (GPSA), USA (2004).
5. Abdulrahman, R. K., and Sebastine, I. M. (2013) Natural gas sweetening process simulation and optimization: a case study of Khurmala field in Iraqi Kurdistan region. *Journal of Natural Gas Science and Engineering*, **14**, 116-120.
6. Dortmundt, D., and Doshi, K. (1999) Recent developments in CO_2 removal membrane technology, UOP, USA. Online: http://www.membrane-guide.com/download/CO2-removal-membranes.pdf
7. Rochelle, G. T. (2009) Amine scrubbing for CO_2 capture. *Science*, **325**, 1652-1654.
8. Xie, H.-B., Zhou, Y., Zhang, Y., and Johnson, J. K. (2011) Reaction mechanism of monoethanolamine with CO_2 in aqueous solution from molecular modeling. *Journal of Physical Chemistry A*, **114**, 11844-11852.
9. Herzog, H. J., and Drake, E. M. (1993) Greenhouse gas R&D programme, IEA/93/OE6.
10. Khatri, R. A., Chuang, S. S. C., Soong, Y., and Gray, M. (2006) Thermal and chemical stability of regenerable solid amine sorbent for CO_2 capture. *Energy and Fuels*, **20**, 1514-1520.
11. Gray, M. L., Champagne, K. J., Fauth, D., Baltrus, J. P., and Pennline, H. (2008) Performance of immobilized tertiary amine solid sorbents for the capture of carbon dioxide. *International Journal of Greenhouse Gas Control*, **2**(1), 3-8.
12. Li, J. R., Sculley, J., and Zhou, H. C. (2012) Metal-organic frameworks for separations. *Chemical Reviews*, **112**(2), 869-932.
13. Lu, C., Bai, H., Wu, B., Su, F., and Hwang, J. F. (2008) Comparative study of CO_2 capture by carbon nanotubes, activated carbons, and zeolites. *Energy and Fuels*, **22**(5), 3050-3056.
14. Cavenati, S., Grande, C. A., and Rodrigues, A. E. (2004) Adsorption equilibrium of methane, carbon dioxide, and nitrogen on zeolite 13× at high pressures. *Journal of Chemical Engineering Data*, **49**(4), 1095-1101.
15. Sevilla, M., Valle-Vigón, P., and Fuertes, A. B. (2011) N-doped polypyrrole-based porous carbons for CO_2 capture. *Advanced Functional Materials*, **21**(14), 2781-2787.
16. Arenillas A., Smith K., Drage T. C., and Snape C. E. (2005) CO_2 capture using some fly ash-derived carbon materials. *Fuel*, **84**, 2204-2210.
17. Wang, X. R., Li, H. Q., Liu, H. T., and Hou, X. J. (2011) As-synthesized

mesoporous silica MSU-1 modified with tetraethylenepentamine for CO_2 adsorption. *Microporous & Mesoporous Materials*, **142**(2-3), 564-569.

18. Goeppert, A., Czaun, M., May, R. B., Surya Prakash, G. K., Olah, G. A., and Narayanan, S. R. (2011) Carbon dioxide capture from the air using a polyamine based regenerable solid adsorbent. *Journal of American Chemical Society*, **133**, 20164-20167.
19. Satyapal, S., Filburn, T., Trela, J., and Strange, J. (2001) Performance and properties of a solid amine sorbent for carbon dioxide removal in space life support applications. *Energy and Fuels*, **15**, 250-255.
20. Drage, T. C., Arenillas, A., Smith, A. M., Pevida, C., Piippo, S., and Snape, C. E. (2007) Preparation of carbon dioxide adsorbents from the chemical activation of urea–formaldehyde and melamine–formaldehyde resins. *Fuel*, **86**, 22-31.
21. Heuchel, M., Davies, G. M., Buss, E., and Seaton, N. A. (1999) Adsorption of carbondioxide and methane and their mixtures on an activated carbon: simulation and experiment. *Langmuir*, **15**, 8695-8705.
22. Okoye, I. P., Benham, M., and Thomas, K. M. (1997) Adsorption of gases and vapors on carbon molecular sieves. *Langmuir*, **13**, 4054-4059.
23. Yim, W. L., Byl, O., Yates, J. T., and Johnson, J. K. (2004) Vibrational behavior of adsorbed CO_2 on single-walled carbon nanotubes. *Journal of Chemical Physics*, **120**, 5377-5386.
24. Ghosh, A., Subrahmanyam, K. S., Krishna, K. S., Datta, S., Govindaraj, A., Pati, S. K., and Rao, C. N. R. (2008) Uptake of H_2 and CO_2 by graphene. *Journal of Physical Chemistry C*, **112**, 15704-15707.
25. Adib, F., Bagreev, A., and Bandosz, T. J. (2000) Adsorption/oxidation of hydrogen sulfide on nitrogen-containing activated carbons. *Langmuir*, **16**, 1980-1986.
26. Arenillas, A., Drage, T. C., Smith, K. M., and Snape, C. E. (2005) CO_2 removal potential of carbons prepared by co-pyrolysis of sugar and nitrogen containing compounds. *Journal of Analytical and Applied Pyrolysis*, **74**, 298-306.
27. Vinu, A., Anandan, S., Anand, C., Srinivasu, P., Ariga, K., and Mori, T. (2008) Fabrication of partially graphitic three-dimensional nitrogen-doped mesoporous carbon using polyaniline nanocomposite through nano-templating method. *Microporous Mesoporous Materials*, **109** (1-3), 398-404.
28. Martin, C. F., Plaza, M. G., Garcia, S., Pis, J. J., Rubiera, F., and Pevida, C. (2011) Microporous phenol-formaldehyde resin-based adsorbents for pre-combustion CO_2 capture. *Fuel*, **90**, 2064-2072.
29. Martin, C. F., Garcia, S., Beneroso, D., Pis, J. J., Rubiera, F., and Pevida, C. (2012) Pre-combustion CO_2 capture by means of phenol–formaldehyde resin-derived carbons: From equilibrium to dynamic cond-

itions. *Separation and Purification Technology*, **98**, 531-538.

30. Long, D., Zhang, J., Yang, J., Hu, Z., Cheng, G., Liu, X., Zhang, R., Zhan, L., Qiao, W., and Ling, L. (2008) Chemical state of nitrogen in carbon aerogels issued from phenol-melamine-formaldehyde gels. *Carbon*, **46**, 1253-1269.
31. Tseng, R.-L., Wu, F.-C., and Juang, R.-S. (2015) Adsorption of CO_2 at atmospheric pressure on activated carbons prepared from melamine-modified phenol–formaldehyde resins. *Separation and Purification Technology*, **140**, 53-60.
32. Pevida, C., Snape, C. E., and Drage, T. C. (2009) Templated polymeric materials as adsorbents for the post combustion capture of CO_2. *Energy Procedia*, **1**, 869-874.
33. Drage, T. C., Arenillas, A., Smith, K. M., and Snape, C. E. (2008) Thermal stability of polyethylenimine based carbon dioxide adsorbents and its influence on selection of regeneration strategies. *Microporous and Mesoporous Materials*, **116**(1), 504-512.
34. Pevida, C., Drage, T. C., and Snape, C. E. (2008) Silica-templated melamine-formaldehyde resin derived adsorbents for CO_2 capture. *Carbon*, **46**(11), 1464-1474.
35. Hao, G.-P., Li, W.-C., Qian, D., and Lu, A.-H. (2010) Rapid synthesis of nitrogen-doped porous carbon monolith for CO_2 capture. *Advanced Materials*, **22**(7), 853-857.
36. Drage, T. C., Arenillas, A., Smith, K. M., Pevida, C., Piippo, S., and Snape, C. E. (2007) Preparation of carbon dioxide adsorbents from the chemical activation of urea-formaldehyde and melamine formaldehyde resins. *Fuel*, **86**(1-2), 22-31.
37. Goel, G., Bhunia, H., and Bajpai, P. K. (2015) Mesoporous carbon adsorbents from melamine-formaldehyde resin using nano-casting technique for CO_2 adsorption. *Journal of Environmental Science*, **32**, 238-248.
38. Wang, D. X., Ma, X. L., Sentorun-Shalaby, C., and Song, C. S. (2012) Development of carbon-based "molecular basket" sorbent for CO_2 capture. *Industrial and Engineering Chemistry Research*, **51**(7), 3048-3057.
39. Ma, X. L., Wang, X. X., and Song, C. S. (2009) "Molecular basket" sorbents for separation of CO2 and H2S from various gas streams. *Journal of the American Chemical Society*, **131**(16), 5777-5783.
40. Wang, X. X., Schwartz, V., Clark, J. C., Ma, X. L., Overbury, S. H., Xu, X. C., and Song, C. S. (2009) Infrared study of CO_2 sorption over "molecular basket" sorbent consisting of polyethylenimine-modified mesoporous molecular sieve. *Journal of Physical Chemistry C*, **113**, 7260-7268.
41. Wang, X., Ma, X., Song, C., Locke, D. R., Siefert, S., Winans, R. E., Mollmer, J., Lange, M., Moller, A., and Glaser, R. (2013) Molecular basket sorbents polyethylenimine–SBA-15 for CO_2 capture from

flue gas: Characterization and sorption properties. *Journal of Microporous and Mesoporous Materials*, **169**, 103-111.

42. Wang, J., Chen, H., Zhou, H., Liu, X., Qiao, W., Long, D., and Ling, L. (2013) Carbon dioxide capture using polyethylenimine-loaded mesoporous carbons. *Journal of Environmental Science (China)*, **25**(1), 124-32.
43. Ngoy, J. M., Wagner, N., Riboldi, L., and Bolland, O. (2014) A CO_2 capture technology using multi-walled carbon nanotubes with polyaspartamide surfactant. *Energy Procedia*, **63**, 2230-2248.
44. Najafabadi, A. T. (2015) Emerging applications of graphene and its derivatives in carbon capture and conversion: Current status and future prospects. *Renewable Sustainable Energy Reviews*, **41**, 1515-1545.
45. Gadipelli, S., and Guo, Z. X. (2015). Graphene-based materials: Synthesis and gas sorption, storage and separation. *Progress in Materials Science*, **69**, 1-60.
46. Kemp, K. C., Chandra, V., Saleh, M., and Kim, K. S. (2013) Reversible CO_2 adsorption by an activated nitrogen doped graphene/polyaniline material. *Nanotechnology*, **24**, 235703.
47. Sui, Z.-Y., Cui, Y., Zhu, J.-H., and Han, B.-H. (2013) Preparation of three-dimensional graphene oxide-polyethylenimine porous materials as dye and gas adsorbents. *ACS Applied Materials and Interfaces*, **5**(18), 9172-9179.
48. Valechha, A., Thote, J., Labhsetwar, N., and Rayalu, S. (2012) Biopolymer based adsorbents for the post combustion CO_2 capture. *International Journal of Knowledge Engineering*, **3**(1), 103-106.
49. Alhwaige, A. A., Agag, T., Ishida, H., and Qutubuddin, S. (2013) Biobased chitosan hybrid aerogels with superior adsorption: Role of graphene oxide in CO_2 capture. *RSC Advances*, **3**, 16011-16020.
50. Tao, Y., Endo, M., and Kaneko, K. (2008) A review of synthesis and nanopore structures of organic polymer aerogels and carbon aerogels. *Recent Patents on Chemical Engineering*, **1**, 192-200.
51. Sevilla, M., and Fuertes, A. B. (2011) Sustainable porous carbons with a superior performance for CO_2 capture. *Energy and Environmental Science*, **4**, 1765-1771.
52. Chandra, V., Yu, S. U., Kim, S. H., Yoon, Y. S., Kim, D. Y., Kwon, A. H., Meyyappan, M., and Kim, K. S. (2012) Highly selective CO_2 capture on N-doped carbon produced by chemical activation of poly- pyrrole functionalized graphene sheets. *Chemical Communications*, **48**, 735-737.
53. Seema, H., Kemp, K. C., Le, N. K., Park, S.-W., Chandra, V., Lee, J. W., and Kim, K. S. (2014) Highly selective CO_2 capture by S-doped microporous carbon materials. *Carbon*, **66**, 320-326.
54. Kumar, N. A., Choi, H.-J., Shin, Y. R., Chang, D. W., Dai, L., and Baek, J.-B. (2012) Polyaniline-grafted reduced graphene oxide for efficient

electrochemical supercapacitors. *ACS Nano*, **6**(2), 1715-1723.

55. Saleh, M., Chandra, V., Kemp, K. C., and Kim, K. S., (2013) Synthesis of N-doped microporous carbon via chemical activation of polyindole-modified graphene oxide sheets for selective carbon dioxide adsorption. *Nanotechnology*, **24**, 255702.
56. Mishra, A. K., and Ramaprabhu, S. (2012) Nanostructured polyaniline decorated graphene sheets for reversible CO_2 capture. *Journal of Materials Chemistry*, **22**, 3708-3712.
57. Yang, S., Zhan, L., Xu, X., Wang, Y., Ling, L., and Feng, X. (2013) Graphene-based porous silica sheets impregnated with polyethyleneimine for superior CO_2 capture. *Advanced Materials*, **25**, 2130-2134.
58. Tsoufis, T., Katsaros, F., Sideratou, Z., Romanos, G., Ivashenko, O., Rudolf, P., Kooi, B. J., Papageorgiou, S., and Karakassides, M. A. (2014) Tailor-made graphite oxide–DAB poly(propylene imine) dendrimer intercalated hybrids and their potential for efficient CO_2 adsorption. *Chemical Communications*, **50**, 10967-10970.
59. Furukawa, H. , Muller, U., and Yaghi, O. M. (2015) "Heterogeneity within order" in metal-organic frameworks. *Angewandte Chemie International Edition*, **54**, 3417-3430.
60. Zhou, H.-C., and Kitagawa, S. (2014) Metal-organic frameworks (MOFs). *Chemical Society Reviews*, **43**, 5415-5418.
61. Mason, J. A., Sumida, K., Herm, Z. R., Krishna, R., and Long, J. R. (2011) Evaluating metal-organic frameworks for post-combustion carbon dioxide capture via temperature swing adsorption. *Energy and Environmental Science*, **4**, 3030-3040.
62. D'Alessandro, D. M., Smit, B., Long, J. R. (2010) Carbon dioxide capture: prospects for new materials. *Angewandte Chemie International Edition*, **49**, 6058-6082.
63. Rabbani, M. G., and El-Kaderi, H. M. (2011) Template-free synthesis of a highly porous benzimidazole-linked polymer for CO_2 capture and H_2 storage. *Chemistry of Materials*, **23**, 1650-1653.
64. Lu, W., Sculley, J. P., Yuan, D., Krishna, R., Wei, Z., and Zhou, H.-C. (2012) Polyamine-tethered porous polymer networks for carbon dioxide capture from flue gas. *Angewandte Chemie International Edition*, **51**, 7480-7484.
65. Dawson, R., Stockel, E., Holst, J. R., Adams, D. J., and Cooper, A. I. (2011) Microporous organic polymers for carbon dioxide capture. *Energy and Environmental Science*, **4**, 4239-4245.
66. Lin, Y., Yan, Q., Kong, C., and Chen, L. (2013) Polyethyleneimine incorporated metal-organic frame works adsorbent for highly selective CO_2 capture. *Scientific Reports*, **3**, 1859.
67. Chen, Z., Deng, S., Wei, H., Wang, B., Huang, J., and Yu, G. (2013) Polyethylenimine- impregnated resin for high CO_2 adsorption: an efficient adsorbent for CO_2 capture from simulated flue gas and ambient air. *ACS Applied Materials and Interfaces*, **5**, 6937-6945.

68. Lu, W., Yuan, D., Sculley, J., Zhao, D., Krishna, R., and Zhou, H.-C. (2011) Sulfonate-grafted porous polymer networks for preferential CO_2 adsorption at low pressure. *Journal of the American Chemical Society*, **133**, 18126-18129.
69. Meth, S., Goeppert, A., Prakash, G. K. S, and Olah, G. A. (2012) Silica nanoparticles as supports for regenerable CO_2 sorbents. *Energy and Fuels*, **26**, 3082-3090.
70. Smith, E. A., and Chen, W. (2008) How to prevent the loss of surface functionality derived from aminosilanes. *Langmuir*, **24**, 12405-12409.
71. Zelenak, V., Badanicova, M., Halamova, D., Cejka, J., Zukal, A., Murafa, N., Goerigk, G. (2008) Amine-modified ordered mesoporous silica: Effect of pore size on carbon dioxide capture. *Chemical Engineering Journal*, **144**, 336-342.
72. Son, W. J., Choi, J. S., and Ahn, W. S. (2008) Adsorptive removal of carbon dioxide using polyethyleneimine-loaded mesoporous silica materials. *Microporous and Mesoporous Materials*, **113**, 31-40.
73. Wei, J., Shi, J., Pan, H., Zhao, W., Ye, Q., and Shi, Y. (2008) Adsorption of carbon dioxide on organically functionalized SBA-16. *Microporous and Mesoporous Materials*, **116**, 394-399.
74. Jang, H. T., Park, Y. K., Ko, Y. S., Lee, J. Y., and Margandan, B. (2009) Highly siliceous MCM-48 from rice husk ash for CO2 adsorption. *International Journal of Greenhouse Gas Control*, **3**, 545-549.
75. Liu, Y., Shi, J., Chen, J., Ye, Q., Pan, H., Shao, Z. H., and Shi, Y. (2010) Dynamic performance of CO_2 adsorption with tetraethylene-pentamine loaded KIT-6. *Microporous and Mesoporous Materials*, **134**, 16-21.
76. Witoon, T. (2012) Polyethyleneimine-loaded bimodal porous silica as low-cost and high-capacity sorbent for CO_2 capture. *Materials Chemistry and Physics*, **137**, 235-245.
77. Nhut, J. M., Vieira, R., Pesant, L., Tessonnier, J. P., Keller, N., Ehret, G., Cuong, P. H., and Ledoux, M. J. (2002) Synthesis and catalytic uses of carbon and silicon carbide nanostructures. *Catalysis Today*, **76**, 11-32.
78. Lively, R. P., Leta, D. P., DeRites, B. A., Chance, R. R., and Koros, W. J. (2011) Hollow fiber adsorbents for CO_2 capture: Kinetic sorption performance. *Chemical Engineering Journal*, **171**, 801-810.
79. Lively, R. P., Chance, R. R., Mysona, J. A., Babu, V. P., Deckman, H. W., Leta, D. P., Thomann, H., and Koros, W. J. (2012) CO_2 sorption and desorption performance of thermally cycled hollow fiber sorbents. *International Journal of Greenhouse Gas Control*, **10**, 285-294.
80. Rezaei, F., Lively, R. P., Labreche, Y., Chen, G., Fan, Y., Koros, W. J., and Jones, C. W. (2013) Aminosilane-grafted polymer/silica hollow fiber adsorbents for CO_2 capture from flue gas. *ACS Applied Materials and Interfaces*, **5**, 3921-3931.

81. Choi, S., Drese, J. H., Chance, R. R., Eisenberger, P. M., and Jones, C. W. (2011) US Patent Application 2011/0179948A1.
82. Britt, D., Tranchemontagne, D., and Yaghi, O. M. (2008) Metal-organic frameworks with high capacity and selectivity for harmful gases. *PNAS*, **105**, 11623-11627.
83. Elseviers, W. F., and Vereist, H. (1999) Transition metal oxides for hot gas desulfurization. *Fuel*, **78**, 601-612.
84. Huang, P. H. (2014) Molecular dynamics investigation of separation of hydrogen sulfide from acidic gas mixtures inside metal-doped graphite micropores. *Physical Chemistry and Chemical Physics*, **17**, 22686-22698.
85. Novochinskii, I. I., Song, C. S., Ma, X. L., Liu, X., Shore, L., Lampert, J., and Farrauto, R. J. (2004) Low-temperature H_2S removal from steam-containing gas mixtures with ZnO for fuel cell application. 1. ZnO particles and extrudates. *Energy and Fuels*, **18**(2), 576-583.
86. Karayilan, D., Dogu, T., Yasyerli, S., and Dogu, G. (2005) Mn-Cu and Mn-Cu-V mixed oxide regenerable sorbents for hot gas desulfurization. *Industrial and Engineering Chemistry Research*, **44**, 5221-5226.
87. Wang, X., Ma, X., Xu, X., Sun, L., Song, C. (2008) Mesoporous-molecular-sieve-supported polymer sorbents for removing H_2S from hydrogen gas streams. *Topics in Catalysis*, **49**, 108-117.
88. Chen, Q., Fan, F., Long, D., Liu, X., Liang, X., Qiao, W., and Ling, L. (2010) Poly(ethyleneimine)-loaded silica monolith with a hierarchical pore structure for H_2S adsorptive removal. *Industrial and Engineering Chemistry Research*, **49**, 11408-11414.
89. Wang, X., Ma, X., Sun, L., and Song, C. (2007) A nanoporous polymeric sorbent for deep removal of H_2S from gas mixtures for hydrogen purification. *Green Chemistry*, **9**, 695-702.
90. Jaiboon, V., Yoosuk, B., and Prasassarakich, P. (2014) Amine modified silica xerogel for H_2S removal at low temperature. *Fuel Process Technology*, **128**, 276-282.
91. Diaf, A., Garcia, J. L., and Beckman, E. J. (1994) Thermally-reversible polymeric sorbents for acid gases: CO_2, SO_2, and NO_x. *Journal of Applied Polymer Science*, **53**, 857-875.
92. Diaf, A., and Beckman, E. J. (1995) Thermally reversible polymeric sorbents for acid gases, IV. Affinity tuning for the selective dry sorption of NO_x. *Reactive Polymers*, **25**, 89-96.
93. Hallenbeck, A. P., and Kitchin, J. R. (2013) Effects of O_2 and SO_2 on the capture capacity of a primary-amine based polymeric CO_2 sorbent. *Industrial and Engineering Chemistry Research*, **52**, 10788-10794.
94. Fan, Y., Rezaei, F., Labreche, Y., Lively, R. P., Koros, W. J., Jones, C. W. (2015) Stability of amine-based hollow fiber CO_2 adsorbents in the presence of NO and SO_2. *Fuel*, **160**, 153-164.

2

MOFs and Metal Oxides for Gas Adsorption and Environmental Applications

2.1 Introduction

Acid gases like carbon dioxide (CO_2) and hydrogen sulfide (H_2S) represent two most abundant harmful gases in the natural gas streams which need to be removed for achieving usefulness of natural gas. Similarly, environmental pollution due to the release of harmful gases from different industrial and household processes leads to severe degradation of the quality of air and significant efforts are required to minimize pollution level as well as to remove the pollutants from the environment. Some of the harmful materials that are present in our environment are H_2S, CO_x, SO_x, NO_x, nitrogen containing compounds (NCCs), sulfur containing compounds (SCCs), volatile organic compounds (VOCs), dyes, pharmaceuticals and personal care products (PPCPs).

2.1.1 Classification of Hazardous Materials

Depending on the sources, the most plentiful hazardous materials can be categorized into two classes, namely naturally originating hazardous materials and anthropogenic. A substantial quantity of naturally occurring materials is available underground as well as in water, soil, air and minerals. However, the anthropogenic materials generate from chemical reactions, combustion or from the effluent of toxic materials. The global energy need is satisfied by the naturally originating materials such as coal, crude oil and natural gas. As mentioned earlier, natural gas contains harmful gases like CO_2 and H_2S among others, which require removal. In addition, these energy sources when burnt generate high amounts of harmful gases to the atmosphere. The major toxic gases causing environmental air pollution are CO_2, H_2S, SO_x, NH_3, NO_x, VOCs and other hydrocarbons [1]. The gases like SO_2, CH_4, O_3 and N_2O are considered as greenhouse gases and the ejection of these gases enhances the ozone levels pre-

Haleema Saleem and Vikas Mittal, Khalifa University of Science and Technology, Abu Dhabi, UAE

sent in the troposphere. Vehicles related pollutants such as CH_4, CO, NO_2, SO_2 as well as carbon black also cause global warming. Currently, the carbon balance of the world is considered as one of the critical environmental problem, and hence the reduction of anthropogenic CO_2 release has now become a very significant issue. VOCs are chemicals having high vapor pressure, and are usually discharged from adhesives, resins, paints, solvents, etc. [2]. Benzene, toluene, xylene and phenolics are some of the common hazardous VOCs. Nitrogen or sulfur containing organic compounds are naturally originating species and are available in fossil fuels as well as oils like jet fuel, gasoline, crude oil, heating oil and diesel. The combustion of fossil fuels is a large source of toxic releases, which causes global warming, greenhouse effect, dangerous impact on living species and hazardous air pollution [3-5].

Further, due to the global industrial advancement, the severity of water pollution has become critical, thus, posing significant challenges to the health of living beings [6]. Dissolved heavy metal salts or organic pollutants are often present in the polluted water [7,8]. The organic pollutants existing in the natural water resources cause a great damage to the aquatic habitats and also to the human health. Some of the toxic pollutants existing in the waste water are benzene, phenols, nitro-benzenes, chlorinated benzenes, phthalates and chlorinated phenols. Dye materials are commonly used in plastics, textiles, paper and leather industries, and a large amount of these materials ends up in water as harmful pollutant [9,10]. Separation of these dyes from waste water is very crucial, but challenging. PPCPs are chemical contaminants that are present in water and are also unsafe for living organisms and have to be removed [11]. The disposal of heavy metal ions like chromium (Cr), lead (Pb), copper (Cu), arsenic (As), mercury (Hg), antimony (Sb), cadmium (Cd) and manganese (Mn) in processed water is also highly toxic to human beings as well as ecological systems [12,13].

2.1.2 Adsorption of Hazardous Materials

In the last few decades, the porous materials have evolved as good adsorbents that exhibit significant potential for the purification of air, gas and water streams by adsorptive separation of the contaminants, and hence a great deal of research focus has been devoted for the examination and analysis of the advanced porous materials [14]. For the decontamination, adsorption has been regarded as the preferable technique due to its low cost, no or harmless secondary products, design simplicity, ease of operation and efficient adsorbent regeneration. This method is based on the porous adsorbent ability to selectively adsorb specific materials from either refinery streams or from the atmosphere. The adsorbent materials can be structurally tuned to generate suitable pore shape and size, thus, providing efficient contact with the adsorbate materials, thus, leading to effective separation of pollutants during the adsorption process. Depending on the interaction between the porous sorbents and the adsorbates, the adsorbents can be classified as physical and chemical adsorbents [15]. With respect to physical adsorbents, the adsorbates are trapped inside the adsorbent pores through weak van der Waals forces. Hence, the adsorbent material can be easily regenerated by physical treatments or simple solvent exchange. However, in the chemical adsorbents, absorption takes place by the chemical bond formation between the adsorbent and the adsorbate. Therefore, chemical treatments are required for the regeneration of the spent adsorbents, which may also occasionally lead to permanent loss of adsorbent performance. The capability of the adsorption process depends on the selectivity for specific compounds, regenerability of adsorbents, durability and adsorption ability of the adsorbents. For the adsorptive removal of harmful compounds, different porous adsorbents like zeolites [16,17], activated carbons [18,19] , mesoporous materials [20,21] and metal organic frameworks (MOFs) [22,23] have been studied. For the effective adsorption, specific pore geometry, porosity and adsorption sites are needed [24]. In addition to this, certain active species such as different functional groups (basic or acidic), metal oxides, metal ions, polyoxometalates and metal salts are generally included into the adsorbents for the selective removal of harmful compounds by interactions such as hydrogen bonding, acid-base and π- π interactions.

In this chapter, the main focus is to elucidate the role of MOFs and metal oxides for the adsorptive removal of hazardous gases like H_2S, NH_3, CO_2, and other hydrocarbons present in the atmosphere as well as the dissolved heavy metal salts or the organic pollutants present in the polluted water. Beneficial characteristics of MOFs and metal oxides have been discussed that ensure their effective role in the adsorption of specific toxic compounds. The effect of linkers, open metal sites, MOF modification/functionalization, central metal ions and the adsorbent/adsorbate interactions are also explained. Further, the analysis of adsorption mechanisms such as electrostatic interactions, hydrogen bonding and acid-base interactions for the effective adsorption is also presented. Thus, this review presents an up-to-date study of theoretical predictions, experimental results as well as developing concepts on the adsorption ability of different MOFs and metal oxides, with specific emphasis on the removal of harmful gases like CO_2, H_2S, NH_3, and water contaminants like the heavy metals and organic contaminants, among other contaminants. In addition to this, observations in terms of benefits and disadvantages of employing different metal oxide frameworks as well as metal oxides have also been pointed out, along with discussing in detail the thermodynamic and kinetic analysis.

2.2 Metal Organic Frameworks

MOFs consists of two major components, i.e., an organic molecule (linker) and a metal ion or a group of metal ions. The linkers are generally di-, tri- or tetradendate ligands [25]. Examples of some MOFs are Cu-BTC, MOF-5 and CPO-27. The porous MOFs are very attractive because of their versatile applications like CO_2 storage/adsorption [26], vapor adsorption [27], hydrogen storage[28], chemical separation [29], drug delivery [30] and so on. These materials are preferred than other porous materials due to their pore functionality, high and tunable porosity, pore compositions and open metal sites. For the adsorption related applications, MOFs are promising materials due to their controlled tuning of the pore surface, which causes the selective adsorption of certain guest molecules which have specific functional groups. The properties of the MOFs can be tuned for obtaining the desired performance, by orderly and intentionally tuning the functionalities and structures. Hence, many research studies have been carried out on the adsorption of different liquid and gaseous components using MOF, due to their beneficial features of pore geometry and porosity [31]. Depending on the adsorbates, both modified as well as virgin MOFs can be used for the adsorptive removal of different toxic gases and liquids.

2.2.1 Adsorptive Removal of Harmful Gases

Using different MOFs, adsorption of various harmful gases/vapors like H_2S [32,33], NH_3 [4], CO_2 [34,35], SO_2 [36], benzene [37], NO [38] have been studied. Most of the aforementioned gases/vapors are either required to be removed from products or are ejected from different industries as waste byproducts. For a clean and safe environment, efficient removal of these harmful gases is very important. As mentioned earlier, MOFs have advantages like high porosity as well as easy tunability of pore shape and size, from mesoporous to microporous scale, by altering the nature of organic molecule and connectivity of inorganic component [39]. The characteristic examples include zirconium based MOFs namely UiO-68(Zr- terphenyldicarboxylate, Zr-TPDC), UiO-67 (Zr-BPDC, Zr- biphenyldicarboxylate) and UiO-66 (Zr- benzenedicarboxylate, Zr-BDC) having high surface areas 4170, 3000 and 1187 m^2/g respectively [40]. The large enhancement in the surface area is obtained by varying the organic linker. The following sections describe the specific use of MOFs for the removal of various harmful gases.

CO_2 Adsorption

The adsorption as well as desorption of CO_2 in MOFs are attributed to the variations happening in the framework structures, and have also been described using the "gate effect mechanism" or a "breathing type mechanism". In a further study, Walton *et al.* [41] added through the comparison of Monte Carlo simulation and experimental adsorption of CO_2 for MOF IRMOF-1 that attractive electrostatic interactions between the CO_2 molecules resulted in the unusual shape of the adsorption isotherms (Figure 2.1).

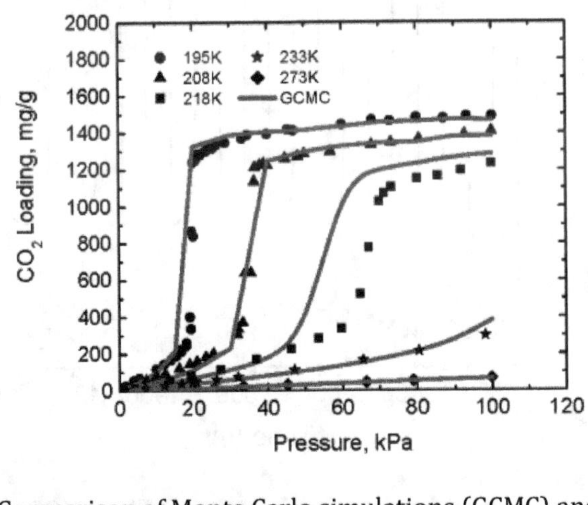

Figure 2.1 Comparison of Monte Carlo simulations (GCMC) and experimental adsorption isotherms for CO_2 in IRMOF-1. Reproduced from Reference 41 with permission from American Chemical Society.

For the application in flue streams, MOF adsorbents for the CO_2 capture should meet the requirements such as high CO_2 adsorption ability, corrosion resistance, high thermal stability and good selectivity for CO_2 over several other components present in flue gas. It should also be noted that most of these adsorbents exhibit little adsorption in the low-pressure conditions i.e., 0.1–0.2 bars, which is the realistic pressure range employed for the CO_2 adsorption from flue gas stream [42]. Further, a reduction in their adsorption capacity is seen when it exposed to gaseous mixtures under dynamic conditions. Hence, significant efforts are required for the enhancement in the CO_2 adsorption capacity under equilibrium conditions. In this study, the modifications in MOFs were achieved in three aspects namely (1) metal ions, (2) organic linkers and (3) novel consolidation of both. As shown in Figure 2.2, the functionalized MOF was subjected to cyclic gas treatments. A small 0.75 wt% change in the weight of the sample was observed over repeated cycles, which confirmed that the material was able to withstand cyclic exposure to the mixed gas stream. The adsorption performance was maintained over the cycles and the material could be regenerated effectively.

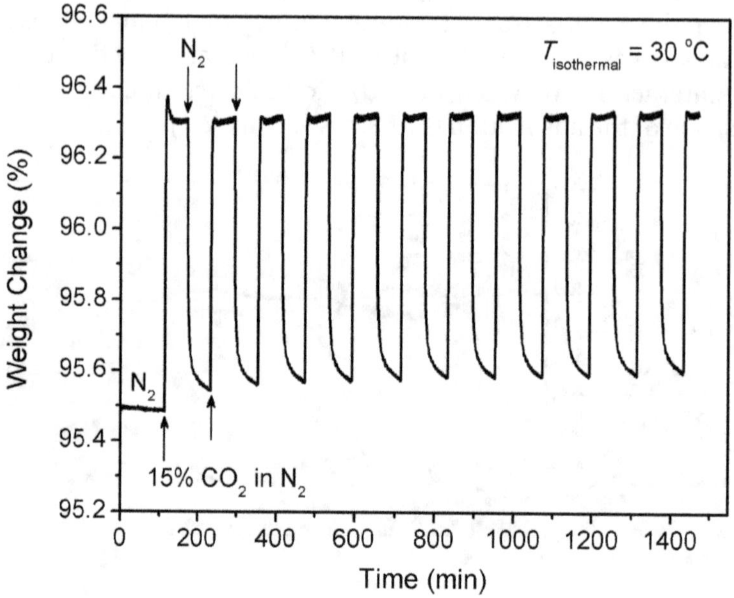

Figure 2.2 Gas cycling experiment for the functionalized MOF at 30°C, with 15% CO_2 in N_2 followed by a flow of pure N_2. Reproduced from Reference 42 with permission from American Chemical Society.

For the metal ions, generating open metal sites is an effective method, as the vacant coordination sites at metal ions act as primary sites for the guest molecules [43]. Britt *et al.* [43] stated that MOF loaded with open Mg sites, Mg-MOF-74 (Mg2(2,5-dioxidoterephthalate)) had high selectivity, easy regeneration and excellent performance for the CO_2 removal. Mg-MOF-74 was exposed to a gas stream consisting of 20% CO_2 in CH_4, and the adsorbent was observed to capture CO_2 alone, without CH_4. The adsorbent retained 89 g of CO_2/kg of material prior to breakthrough (2.23 mmol.g^{-1}). Dietzel *et al.* [44] studied the outcome of the metal center on the CO_2 adsorption ability and selectivity using a series of isostructural MOFs namely M-CPO-27s (M=Ni, Mn, Co, Mg, Zn). It was observed that at high pressure (50 bar) and 298 K, the CO_2 adsorption was 63 wt% for MOF CPO-27(Mg) and 51 wt% for the MOF CPO-27(Ni). In another research by Caskey *et al.* [45], it was observed that at 1 atm, the CPO-27(Co) exhibited CO_2 adsorption of 30.6 wt%, whereas CPO-27(Mg) showed an adsorption of 35.2 wt%. The CO_2 adsorption ability of CPO-27(Mg) was found to be superior as compared to the other members of the series, thereby representing a major development in the CO_2 adsorption ability of the MOFs. In addition to this, for enhancing the performance, the unsaturated metal sites could be functionalized using groups like amine. Couck *et al.* [46] proved the efficient separation of CH_4 and CO_2 gases using amino functionalized MIL-53(Al). Due to the presence of quadrupole moment in CO_2, these CO_2 molecules have a greater affinity for the NH_2 groups. Breakthrough experiments confirmed that the CH_4, which was weakly adsorbed, was able to transmit through the MOF packed column during the adsorption of CO_2. Thus, at ambient conditions, effective separation of CO_2 and CH_4 could be achieved using the aforementioned MOF adsorbent.

For the organic linkers modification, Bae *et al.* [47] analyzed the feasibility of utilizing MOFs which are carborane based or coordinated with ligand mixture. A higher selectivity for CO_2 over CH_4 was noticed in these MOF materials. Carboranes have several beneficial material properties like thermal stability, rigidity and chemical stability. In the pursuit of preparing novel MOFs having high CO_2 adsorption capacity as well as stability, the capability to generate MOFs having different organic linkers as well as metal joints is considered to contribute significant flexibility, along with particular physical behavior and chemical functionalities. Recently, Yaghi *et al.* [48] prepared a new class of MOFs called zeolitic imidazole frameworks (ZIFs), where metal atoms like Zn were bonded with nitrogen atoms by ditopic imidazolate or functionalized Im links to generate the neutral frameworks. The materials had good chemical as well as thermal stabilities (until 500 °C). In another study by Phan *et al.* [49], it was stated that at 0 °C, 1 L of ZIF-69 had the ability to store 82.6 L of CO_2. Further, ZIFs also exhibited higher selectivity than other MOF types for the adsorption of CO_2 from flue gases. It was found that only CO_2 could slip into the ZIF cages, whereas other gas molecules could just pass through them without any obstruction.

Hu *et al.* [50,51] studied the effect of addition of alkaline metal cations like Na^+, Li^+ and K^+ into a previously studied Zr-MOF having excellent CO_2 adsorption properties, by facile neutralization reactions. The CO_2 adsorption capacity of the MOFs was analyzed employing breakthrough experiments and sorption isotherms. From the binary mixed gas breakthrough experiment, it was seen that out of the different Zr-MOFs, the MOF UiO-66 (Zr)-$(COONa)_2$ exhibited high CO_2/N_2 selectivity value of 99.6. This was considered to be highest among all the previously obtained values under analogous testing conditions. The MOFs were reported to have good CO_2 adsorption capacity, high water stability and excellent CO_2/N_2 selectivity. Further, their preparation could be scaled up very efficiently by utilizing commercially accessible reagents with batch reactors as well as environmental friendly solvent like water. The results obtained in the study confirmed that the synthesized MOFs are very efficient adsorbents for the removal of CO_2 from flue gas. Yang *et al.* [52] observed enhanced CO_2 adsorption with adsorbents attained by sonication or microwave heating. At 298 K, the small sized and homogenous particles of CPO-27(Mg), generated by sonication method, adsorbed almost 350 $mg.g^{-1}$ of CO_2. In addition to this, the materials also exhibited large isosteric heat of adsorption. Table 2.1 also demonstrates the CO_2 adsorption capacities of some of the MOFs.

Table 2.1 Comparison of CO_2 adsorption capacities of some of the MOF adsorbents

Sl. No.	Adsorbent	Temperature	Pressure	CO_2 adsorption capacity (wt%)	Reference
1	CPO-27(Mg)	298 K	50 bar	63 wt%	131
2	CPO-27(Ni)	298 K	50 bar	51 wt%	131

3	CPO-27(Co)	Ambient temperature	1 atm	30.6 wt%	132
4	CPO-27(Mg)	Ambient temperature	1 atm	35.2 wt%	132

Adsorption of Sulfur Containing Compounds

The use of fossil resources, like petroleum, natural gas and coal, generates sulfur containing species like SO_2, H_2S, mercaptan, thiophene and thioether. These sulfur containing compounds are harmful to human health and the environment and hence the desulphurization processes are needed. As mentioned earlier, MOFs are also promising absorbents for the removal of sulfur containing compounds. In a study by Achmann *et al.* [53], it was noted that the adsorbent MOF-199 possessed high efficiency for the removal of sulfur from low sulfur model fuels and oil. The MOF also exhibited the maximal sorption rates as well as rapid sorption characteristics. The minor pores in the MOF-199 provided better stabilization for Cu-adsorbed sulfur species by the van der Waals force, and thereby caused an excellent sulfur adsorption performance. In another study, Chavan *et al.* [54] analyzed the H_2S adsorption performance on the adsorbent Ni-MOF-74. The adsorbent exhibited maximal uptake and higher H_2S adsorption enthalpy, when compared to other adsorbents, while preserving its original porosity and structure (Figure 2.3).

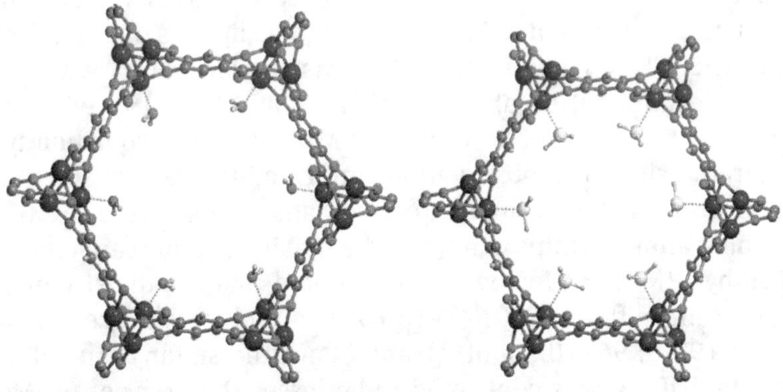

Figure 2.3 H_2O (left) and H_2S (right) molecules as arranged in the CPO-27-Ni channels. Reproduced from Reference 54 with permission from American Chemical Society.

In a study reported by de Voorde *et al.* [55], DFT analysis on the adsorption behavior of S-heterocyclic compounds on MOF-74 series (Mg, Cu, Ni, Zn and Co) was carried out. It was confirmed that the metal ion had a remarkable effect on the MOFs adsorption ability and S-heterocyclic species affinity. It was suggested that the adsorbent Ni-MOF-74 offered the best desulfurization behavior. Hamon *et al.* [56] examined the efficiency of MIL-series e. g. MIL-100(Cr), MIL-53(Cr, Fe and Al), MIL-101(Cr) and MIL-47 for the adsorption of H_2S at 303 K. The study suggested that the –OH groups present in the MIL-53 adsorbent promoted the H_2S adsorption. In a different study by Hamon *et al.*, [57] it was reported that the –OH groups existing in the MOFs performed as proton donor, thereby increasing the H_2S adsorption. The authors observed that on H_2S adsorption, the MIL-47(V) structure remained rigid up to a pressure of 1.8 MPa. On the other hand, the MIL-53(Cr) initially present in the large pore form (LP) switched to its narrow pore version (NP) at very low pressure, and exhibited a second structural transition from the NP to the LP at higher pressure. The authors observed that experimental and simulated adsorption enthalpies for H_2S decreased in the sequence: MIL-53(Cr) NP > MIL-47(V) > MIL- 53(Cr) LP, also depicted in Figure 2.4.

Several previous studies have showed that the efficiency of MOFs can be influenced by the factors such as metals ions, substituent groups and coordinatively unsaturated sites (CUS). Further, most of the studies were based on the adsorption of thiophene species from liquid fuel. Chen *et al.* [58] performed research on the desulfurization ability of MOFs for the H_2S, mercaptan and thioether species. A systematic DFT analysis on the influence of MOFs metal centre (structures as well as metal ions) and organic ligands on the adsorption of H_2S, CH_3SCH_3 and CH_3CH_2SH was carried out. It was noted that the desulfurization ability of MOF-74 and MOF-199 with open metal site, was superior to the other MOFs without coordinatively unsaturated sites. The observation confirmed the remarkable influence of CUS on the sulfur compounds adsorption. Here, IRMOF-3 containing amino groups exhibited the maximal sulfur capacity, succeeded by IRMOF-8. However, it was noted that the density of CUS sites present in the Zn-MOF-74 (6.16×10^{-3} mol/g) was greater than Cu-MOF-199 (4.96×10^{-3} mol/g), and hence the sulfur removal ability of Zn-MOF-74 was confirmed to be lesser than that of Cu-MOF-199. Fernandez *et al.* [59] studied the adsorptive removal of toxic acid gases by using fluorinated MOF (FMOF-2), which was prepared using zinc nitrate hexahydrate and 2,2'-bis(4-carboxyphenyl) hexafluoropropane. The MOF was observed to be stable for the H_2S and SO_2 adsorption. With FMOF-2, the estimated weight capacities for

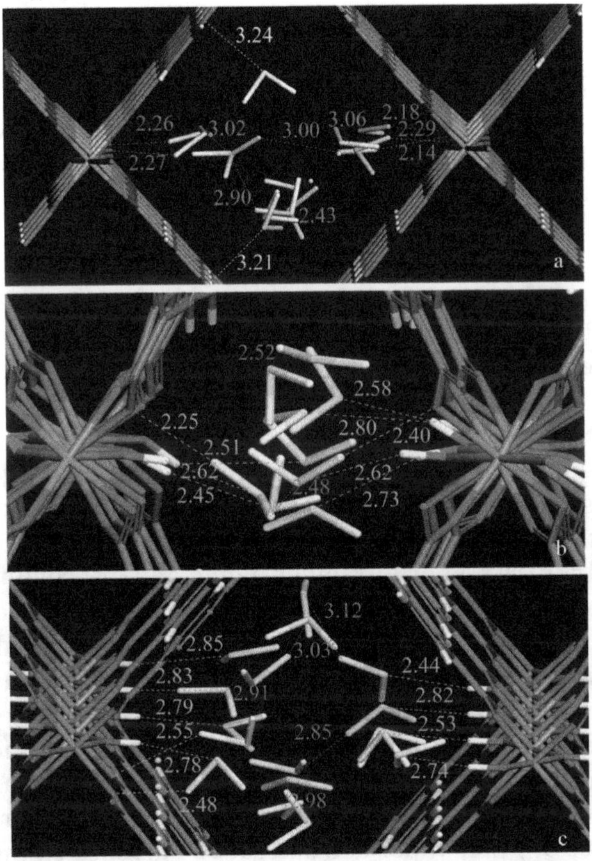

Figure 2.4 Illustrations of the preferential arrangements for H_2S simulated at 303 K in MIL-47(V) (a), MIL-53(Cr) NP (b) and MIL-53(Cr) LP (c) solids. Reproduced from Reference 57 with permission from American Chemical Society.

H_2S and SO_2 were 8.3% and 14% respectively, at 1 bar and room temperature. Dathe *et al.* [36] studied the SO_2 adsorption using $Ba(CH_3COO)_2$ (or $Ba(NO_3)_2$ and $BaCl_2$)-impregnated Cu-BTC. The impregnation accelerated the formation of small micro-crystals of barium salts in Cu-BTC pores. Also, the sectional destruction of the host structure took place in $BaCl_2$. It was also noticed that at greater temperature, the SO_2 adsorption surpassed the stoichiometric ability based on Ba^{2+} concentration. Accordingly, the excess SO_2 adsorption was because of the chemical bonding between the SO_2 and metal cations, thereby resulting in the generation of Cu-sulfates. Hence, at lower temperature, Cu-BTC performed as an excellent host material for obtaining very well dispersed barium salts. However, at higher temperature, the adsorbent Cu-BTC decomposed and generated unique Cu species that performed as SO_x storage sites, thus producing Cu-sulfates finally. Thus, the main limitation of these barium salts impregnated Cu-BTC adsorbent is the irreversible SO_x storage in their structure.

Adsorption of Other Harmful Gases

For the adsorption of various toxic gases, the open metal sites of MOFs have been stated repeatedly as the active sites. Further, a large number of studies stated that the properties of MOFs could be controlled by the loading of active metals (Pd, Cu or Ag) or active species (sulfated zirconia or polyoxymetalates) or by generating composites [60]. Glover *et al.* [61] demonstrated the adsorption of various toxic gases like SO_2, NH_3, octane vapor and CNCl using M-CPO-27 (M=Zn, Ni, Mg or Co) under humid and dry conditions. The breakthrough results confirmed that only under the dry conditions, the adsorbents with open metal sites were able to adsorb the aforementioned toxic gases. In humid conditions, the adsorptive ability decreased significantly as the aggressive adsorption of water started dominating. In the case of NH_3, the reduction in the adsorption capacity was not identified, as it was adsorbed similarly under both humid and dry conditions. Hence, it could be confirmed that the presence of water vapor had a negative influence on the adsorption of different toxic gases using the CPO-27 type adsorbents.

Petit and Bandosz [62] studied different composites of MOFs (MIL-100(Fe), MOF-5 or Cu-BTC) and graphitic compound (GO, graphite or graphite oxide), under ambient conditions, for the adsorption of NO_2, H_2S and NH_3. It was observed that the open metal sites of MOFs coordinated with oxide groups of GO, which caused the generation of a new pore space in the interface between the MOF units and carbon layers, to form composites having definite properties. With the GO/Cu-BTC composites, more than 50% (for H_2S), 12% (for NH_3) and 4% (for NO_2) enhancement in the adsorption ability was observed. The increase in the adsorption ability for the toxic gases seen in the composites was attributed to the porosity generated in the interface where the dispersive forces were substantial. Britt *et al.* [63] studied the possible suitableness of six MOFs, consisting of 2-amino terephthalate (IRMOF-3), diacetylene-1,4-bis(4-benzoic acid) (IRMOF-62), $Zn_4O(CO_2)_6$ group linked by terephthalate (MOF-5), benzene-1,3,5-tris(4- benzoate) (MOF-177), Cu-BTC and $Zn_2O_2(CO_2)_2$ chains linked by 2,5-dihydroxyterephthalate (CPO-27), for the adsorption of different hazardous gases/vapors including NH_3, chlorine, SO_2, benzene, tetrahydrothiophene and ethylene oxide. The obtained results were correlated with BPL carbon. It was seen that different factors like the active adsorption site having specific functional group and open metal sites of Cu-BTC or CPO-27 played a critical role in deciding the dynamic adsorption ability of these MOFs. Compared to the other MOFs, CPO-27 exhibited excellent performance, and the adsorption ability was almost 6 times higher than the BPL carbon. The improved adsorption was because of the existence of an immensely reactive 5 coordinate Zn species, together with the reactive oxo-group present in CPO-27. However, in the case of all other gases except Cl_2, the MOF Cu-BTC exhibited higher efficiency than BPL carbon. For NH_3 adsorption in IRMOF-3, the existence of NH_2 in the MOF greatly increased the adsorption ability, when compared to MOF-5 or IRMOF-1. Prior to the breakthrough, the IRMOF-3 adsorbed NH_3 approximately 71 times, when compared to the BPL carbon, due to the ability of NH_3 to easily form the hydrogen bonds.

CO, another toxic gas in the environment, has good binding capacity with the metal sites and MOFs with CUS can easily adsorb CO from gaseous mixture. Karra and Walton [64] conducted molecular simulation analysis and observed that at 298 K, Cu-BTC was a good adsorbent for CO over N_2 and H_2. It was proposed that the electrostatic interaction between the CO dipole and partial charge of CUS of Cu-BTC controlled the adsorptive behavior. Different MOF materials can also be used successfully to adsorb NO by adsorption processes [65]. The gravimetric NO adsorption on the open metal sites of Cu-BTC at 196 K and 1 bar was examined by Xiao *et al.* [65]. It was observed that by using 1 g of Cu-BTC almost 9 mmol of NO was adsorbed, which was remarkably greater than any other porous solids for the NO adsorption (Figure 2.5). The adsorption of benzene (both liquid and vapor phase) was also analyzed by Jhung *et al.* [66] using MIL-101(Cr), chromium terephthalate adsorbent. 16.7 $mmol.g^{-1}$ of benzene (vapor phase) was adsorbed by the adsorbent MIL-101(Cr). The obtained value was approximately 8.7, 5.5 and two times higher than the H-ZSM-5, SBA-15 and a commercial activated carbon, respectively. Due to the high porosity of MIL-101(Cr), it displayed higher benzene adsorption capacity than the pitch based activated carbon (12.4 mmol.g^{-1}) with a higher surface area of 2600–3600 m^2 g^{-1}. The potential utilization of MIL-101(Cr) in the vapor phase adsorption of VOCs like p-xylene and ethyl acetate was also studied by Shi *et al.* [67].

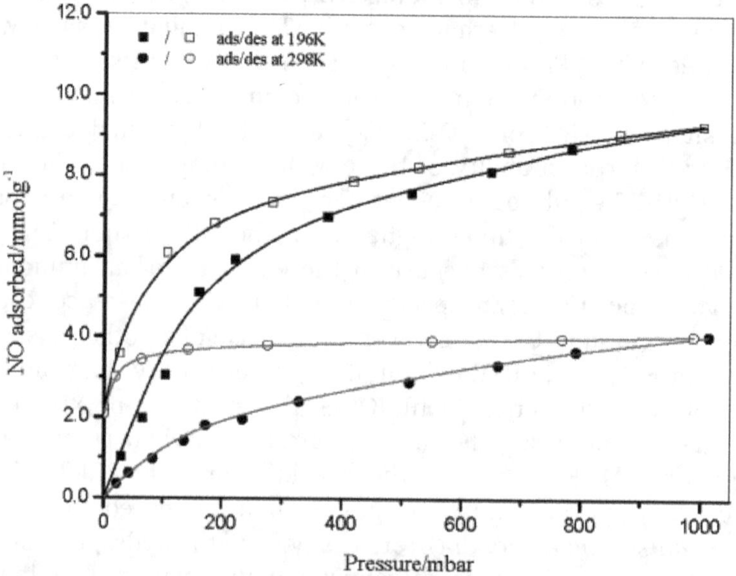

Figure 2.5 Adsorption (filled symbols) and desorption (open symbols) isotherms at 196 K (squares) and 298 K (circles) for nitric oxide on HKUST-1. Reproduced from Reference 65 with permission from American Chemical Society.

2.2.2 Adsorptive Elimination of Contaminants from Wastewater

Various MOFs have been employed for the adsorption of different hazardous contaminants from water. The adsorption occurs through different mechanisms like the electrostatic interactions, hydrogen bonding, acid-base interactions, hydrophobic interactions and π-π interactions. One of the substantial barrier for the application of several types of MOFs is the water instability (both in vapor and liquid phase). This instability of MOFs in water is a critical problem and many research studies have been carried out to increase their water stability [68]. Computational as well as experimental studies showed that several MOFs like MIL-101-V, MOF-5 experienced degradation by ligand displacement, when these materials were exposed to water [69,70]. On the other hand, MOFs like UiO-66 [71], ZIF-8 [72] MIL-101-Cr [73], MIL-100 (Fe, Al, Cr) [74,75] had good water stability and were suitable for the adsorption application in water or water purification. Several methods like incorporation of water repellent functional groups [76], composite formation [77] and fluorination [78] have been achieved to improve the water stability [79].

Haque *et al.* [80] reported the adsorptive separation of dye using MOFs. The authors utilized two types of porous Cr based MOFs, namely MIL-53-Cr and MIL-101-Cr, for the adsorption of methyl orange (MO), an anionic hazardous dye, from the aqueous solution. It was noticed that the performance of the MOFs was superior to the activated carbon. The adsorption kinetic constant and the adsorption capacity of MIL-101-Cr were higher than that of MIL-53-Cr, thereby portraying the significance of pore size as well as porosity for the adsorption. On the other hand, a better adsorption was seen after the functionalization of MIL-101-Cr with ethylenediamine (ED) and protonated ED (PED). Generally, MO exists in the sulfate form, and hence it exhibits greater electrostatic interaction with an adsorbent having positive charge. It was observed that the positive charge dispersion enhanced in the order MIL-101-Cr < ED-MIL-101-Cr < PED-MIL-101-Cr. On the other hand, the positive charge on PED-MIL-101-Cr reduced with enhancing the solution pH, due to the deprotonation of the protonated adsorbent. In another study, Haque *et al.* [81] analyzed the adsorption of methylene blue (MB-cationic) and MO from aqueous solution, using NH_2-MIL-101-Al, an NH_2-functionalised MOF. The ultimate adsorption ability of NH_2 -MIL-101-Al, for the removal of MB was observed to be 762±12 mg/g at 30 °C, which was higher than most other porous materials and MOFs. However, by using non-amine functionalized framework (MIL-101-Al), lower adsorption capacity was noticed (195 mg/g). Thus, it suggested that the electrostatic interactions between the cationic MB and the amino groups of the MOF favored the adsorption. Haque *et al.* [82] also reported that the iron terephtahalate MOF (MOF-235) could adsorb both the cationic (MB) as well as anionic (MO) dyes from the waste water. Here, the adsorption kinetic constant and adsorption capacity of MOF-235 were higher as compared to that of the activated carbon. The higher adsorption of MB or MO was explained on the basis of the electrostatic interactions between the adsorbents and dyes. MB existed in positive form and MO existed in negative form, and hence electrostatic interactions were observed with the adsorbents having negative (charge balancing anion) charges as well as positive framework respectively. After the equilibration with MOF-235, MB and MO adsorptions at varying pH values were determined. By increasing the pH of the MO solution, it was observed that the quantity of adsorbed MO reduced, as the positive charge density reduced with increasing pH. However, the quantity of adsorbed MB increased with pH increase, as the negative charge concentration increased with pH increase. In another study, Leng *et al.* [83] analyzed the adsorption behavior of uranine using MIL-101-Cr and observed the aforementioned type of interactions. The adsorptive removal ability of uranine using MIL-101-Cr was found to be 126.9 mg/g, which was much higher than the activated carbon (17.5 mg/g). This was due to the electrostatic interactions, which were explained by the zeta potential analysis. The zeta potential of MIL-101-Cr was observed to be around 21.1 mV, and zeta potential reduced to 12.0 mV after the uranine interaction. Chen *et al.* [84] studied the xylenol orange (XO) adsorption on MIL-101-Cr and observed significant adsorption ability of the material. It was observed that the enhancement in the pH of the solution decreased the amount of adsorbed XO. No adsorption was observed, when the pH of the solution attained 12. Thus, after the analysis of zeta potential, the authors concluded that the interactions between the adsorbent and dye played a key role in the adsorption process.

Hydrogen bonding also has a major role in adsorption of harmful materials using MOFs. Liu *et al.* [85] examined the adsorption of phenol and para-nitrophenol (PNP) from aqueous solution, using different MOFs (NH_2-MIL-101-Al and MIL-100-Fe, Cr). During the phenol adsorption, limited as well as similar adsorption ability was exhibited by all the three MOFs. On the other hand, NH_2-MIL-101-Al exhibited very high adsorption of PNP, which was about 1.9 times and 4.3 times greater than the MIL-100-Cr and MIL-100-Fe, respectively. The significant adsorption ability of NH_2-MIL-101-Al was because of the hydrogen bonding between the PNP and NH_2 groups in NH_2-MIL-101-Al, as illustrated in Figure 2.6. Xie *et al.* [86] examined nitrobenzene adsorption from waste water using two types of aluminum based MOFs, namely MIL-68-Al (1130 ± 10 mg/g) and CAU-1 (970 ± 10 mg/g). The adsorption values were greater than the experimental values reported for other porous materials. The two MOFs had μ-OH groups in Al-O-Al units, which played a crucial role in attaining higher adsorption by the generation of hydrogen bonding between the nitrogen atom in nitrobenzene and μ-OH groups of MOFs.

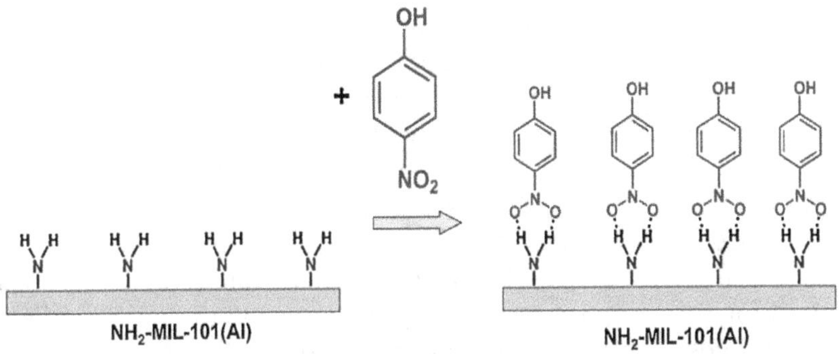

Figure 2.6 Mechanism of PNP adsorption on NH2-MIL-101-Al by hydrogen bonding. Reproduced from Reference 85 with permission from American Chemical Society.

Khan *et al.* [87] studied the adsorption of phthalic acid from water using ZIF-8 and other MOFs like NH_2-UiO-66, MIL-53-Cr, UiO-66, NH_2-MIL-100-Cr, MIL-100-Cr and MIL-100-Fe. When compared to other MOFs, the ZIF-8 exhibited higher adsorption at high pH. This might be due to the electrostatic interactions between the ZIF-8 (positive charged surface) and PA^{2-} as well as $H\text{-}PA^{-}$ ions. Further, it was also observed that the amine functionalized MOFs exhibited higher adsorption of phthalic acid at lower pH, than the MOFs without NH_2 group. At low pH, the phthalic acid did not deprotonate, and the acid-base interaction between the basic sites and phthalic acid exhibited dominancy. Tong *et al.* [88] studied the effect of framework metal ions in the adsorption of dyes using MIL-100-Fe/Cr. The adsorbents MIL-100-Cr and MIL-100-Fe had identical pore volumes (0.75 cm^3/g and 0.76 cm^3/g) as well as surface areas (1760 m^2/g and 1770 m^2/g). On the other hand, it was noticed that both the MOFs exhibited different behavior for the adsorption of MB and MO from wastewater. For the adsorption of MO, the absorbent MIL-100-Fe displayed excellent adsorption (1045.2 mg/g), when compared to the adsorbent MIL-100-Cr (211.8 mg/g). Due to the greater binding energy of water molecules with the metal sites of MIL-100-Cr, the aggressive adsorption of water and MO was more outstanding at the MIL-100-Cr surface. The adsorption of water molecules occurred at the apertures of hexagonal and pentagonal windows, which limited the accessibility of MIL-100-Cr cages for MO to a higher degree. Size selective adsorption is also noticed during adsorption using MOFs. Wang *et al.* [89] analyzed the size selective adsorption of dyes having different sizes, where neutral MOFs were used for the adsorption. The MOFs had face centered cubic topologies with 3-dimensional frameworks and were used for the adsorption of three different dyes, i.e., MB, MO and rhodamine (RB). It was noticed that MB and MO were adsorbed, however, RB was not adsorbed due to its larger size that blocked the transmission through the smaller windows of the MOFs.

Presence of uranium in water is one of the most serious means of hazard to human health [90]. Adsorption is considered as the best method for the uranium removal from water due to its design flexibility, low cost and toxic pollutant insensitivity. Feng *et al.* [91] examined uranium adsorption capacity of the MOF HKUST-1 from aqueous solution. The most advantageous adsorption conditions were fixed along with the adsorbent dosage (0.03g), solution pH (at 6) as well as the agitation time (2 h). It was noticed that when the initial concentration of uranium was less than 200 mg/g, the temperature had no visible effect on the adsorption. However, when the initial concentration of uranium exceeded 200 mg/g, a greater temperature favored the adsorption process. At 318 K, the adsorbent HKUST-1 displayed an excellent adsorption capacity, at initial uranium concentration of 800 mg/L and solution pH value of 6. Further, the endothermic and impulsive nature of the adsorption process were indicated by the thermodynamics of HKUST-1/uranium system.

2.3 Metal Oxides

Transitional metal oxide nanoparticles have been adequately researched because of the attractive behavior as advanced nanomaterials in environmental as well as energy areas due to their excellent adsorption performance [92,93]. As mentioned earlier, harmful gases pose serious challenges to the process industries as well as environment. Wastewater from food processing, pharmaceutical and textile industries has constantly been a crucial environmental issue, as the contaminants have adverse effects on aquatic life as well as human health. Several dyes as organic contaminants are stable and difficult to degrade in conventional wastewater treatment methods. Thus, establishing effective and beneficial adsorption methods are needed for removing the organic contaminants present in the wastewater [94]. Metal oxide based adsorbents confirmed to be an efficient solution for the removal of a wide variety of contaminants. These oxides are available in various shapes and sizes, in the form of micro-particles, nanoparticles, nanocomposites and granules, thereby demonstrating distinct characteristics which are critical to their sorption properties. Several research studies have been carried out during the last two decades for developing various optimum methods of preparation of metal oxide adsorbents. Out of the different methods, thermal decomposition and reduction, sol–gel processes and hydrothermal preparation [95] are the most promising ones.

2.3.1 Adsorption of CO_2

In addition to the removal of CO_2 from natural gas, the removal of CO_2 from the large CO_2 emission sources like thermal power plants has also a significant role for preventing the global warming [96]. For the efficient removal of CO_2 using solid adsorbents, higher selectivity of CO_2 against other gases like nitrogen is very important. Chemical adsorption of CO_2 is considered as one of the methods to enhance the selectivity of the adsorbents. The metal oxides display improved selectivity due to the chemical adsorption between the CO_2 molecules and their surfaces. Chemical adsorption also allows the metal oxides for CO_2 adsorption in the existence of water vapor. H_2O undergoes physical adsorption and hence it has limited effect on the CO_2, which is adsorbed chemically. This behavior is different from zeolites, which are noted to have higher loss of CO_2 in the presence of water vapor. Together with the high selectivity for CO_2, the high surface area of the adsorbent is also very important for the efficient adsorption. The advancements in nanostructure control have facilitated the preparation of oxides with higher surface area by generating nanoparticles and porous structures. The utilization of these methods allows added enhancement in the CO_2 adsorption ability of the metal oxides.

Cai *et al.* [97] explored the synthesis of mesoporous alumina (MA) and MA-supported metal oxides. Aluminum isopropoxide, aluminum chloride, and aluminum nitrate nonahydrate were used as aluminum precursors, whereas nickel, magnesium, iron, chromium, copper, cerium, lanthanum, yttrium, calcium, tin chlorides, or nitrates were metal precursors. The authors observed that the use of aluminum nitrate nonahydrate for the synthesis led to large mesopores with size ranging from ~7 nm to 16 nm, improved ordering of the oxides as well as improved adsorption affinity toward CO_2. Figure 2.7 shows the temperature-programmed desorption (TPD) profiles of the <A and MA supported metal oxide materials for CO_2. The profiles exhibited a broad peak, which indicated the presence of a wide range of basic sites on the surface of these materials.

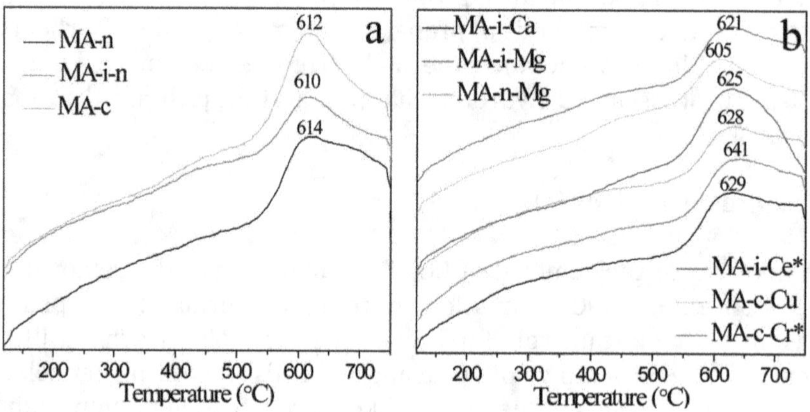

Figure 2.7 CO_2-TPD profiles for (a) MAs and (b) MA supported-metal oxides. Reproduced from Reference 97 with permission from American Chemical Society.

In another study, Leon *et al.* [98] studied magnesium-aluminum double oxides derived from the thermal treatment of layered hydroxides for CO_2 adsorption. The authors analyzed the effects of various factors, such as the incorporated cation (K or Na), mode of addition of precursors, sonication, and calcination temperature, and related these with the adsorption capacity by the use of thermogravimetry and calorimetry. As observed in Figure 2.8, the equilibrium adsorption isotherms at 6.7 kPa and 323 K of the mixed oxides, recorded using microcalorimetry, exhibited the presence of combined chemisorption and physisorption behavior. It was observed that Langmuir transformation fit of the curves was achieved at low coverages, however, did not fit the data at higher extent of coverage, thus, indicating multilayer adsorption.

Yoshikawa *et al.* [99] also studied the CO_2 adsorption behavior of single metal oxides. CeO_2, Al_2O_3, SiO_2 and ZrO_2 were generated, and the chemically adsorbed CO_2 amount was measured using CO_2 pulse injection. From the adsorption analysis, it was observed that the

CeO_2 based adsorbents exhibited the highest CO_2 adsorption capacity. The temperature programmed desorption investigation of CO_2 confirmed that the CO_2, adsorbed on the CeO_2, got desorbed mostly at temperatures from 323K to 473 K, with the peak at 393 K. Further, a very small amount of desorption remained till the temperature reached 723 K. For analyzing the CO_2 adsorption as well as desorption mechanisms, the amount of CO_2 adsorbed on CeO_2 and the temperature dependence were examined with the help of FTIR

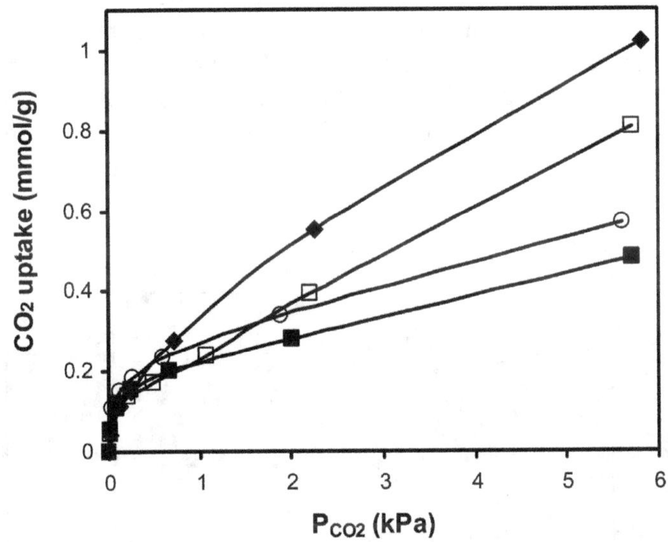

Figure 2.8 CO_2 adsorption isotherms at 323 K and 6.7 kPa, measured by microcalorimetry for (♦) 1K723F, (□) 2K723F, (■) 1KUS723F, and (○) 2KUS723F. Reproduced from Reference 98 with permission from American Chemical Society.

spectroscopy. The FTIR analysis results confirmed that the hydrogen carbonate and bidentate carbonate species were essentially formed on the CeO_2 adsorbent surface. The lowest decomposition temperature was exhibited by the hydrogen carbonate species, which revealed that this species could be utilized to reduce the energy consumption for the capture of CO_2. It was also noted that for increasing the hydrogen carbonate, treatment with water vapor proved to be efficient. As the exhaust gases from the thermal power plants consist of CO_2 and water vapor, it has the ability to help the generation of hydrogen carbonate species. Hence, CeO_2 can be considered as an efficient adsorbent for the CO_2 adsorption in thermal power plants. Wang *et al.* [100] also studied high-temperature adsorption of CO_2 on mixed oxides, which were derived from hydrotalcite-like compounds. The oxides were characterized with various physicochemical techniques such as x-ray diffraction analysis, Fourier transform infrared spectroscopy, thermogravimetric analysis and BET. It was observed that the generated oxides were of either periclase or spinel phase, with an interparticle pore diameter of 9.6-15.4 nm. The oxides were observed to exhibit high CO_2 adsorption capability at 350 °C. For instance, mixed oxide CaCoAlO was observed to adsorb 1.39 mmol/g of CO_2 (6.12 wt%) from a gas mixture (8% CO_2 in N2) at 350 °C and 1 atm. The range of adsorption for other mixed oxides was 0.87–1.28 mmol/g (3.83–5.63 wt%) of CO_2. Figure 2.9 also demonstrates the CO2 adsorption data in the fixed bed reaction, in comparison with the simulated breakthrough curves.

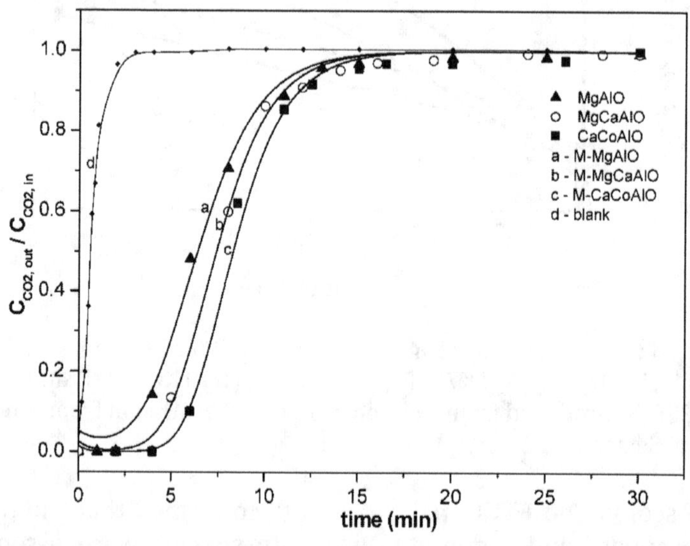

Figure 2.9 CO_2 sorption data in the fixed-bed reactor and simulated breakthrough curves. "M-" represents predicated curves. Reproduced from Reference 100 with permission from American Chemical Society.

In other studies for CO_2 adsorption, Zhao *et al.* [101] synthesized MgO nano/micro-particles with various morphologies as well as porous structures and studied the CO_2 adsorption ability of these porous materials. Song *et al.* [102] analyzed the CO_2 adsorption kinetics of MgO adsorbent under various CO_2 partial pressures and adsorption temperatures. The adsorption behaviors of MgO were successfully predicted by the pseudo-second order models. The material exhibited two stage process of adsorption, having a fast initial step succeeded by a sluggish second step. The first step was due to the diffusion of film from the bulk gas phase to the MgO exterior, whereas the second slow step was attributed to the strong interparticle diffusion resistance due to the large number of narrow pores, thus, reducing the diffusion rate of CO_2. From the pore size and the pore area distribution, it was noted that the surface area of the generated MgO primarily located in small pores under 3 nm, where the CO_2 molecules underwent strong diffusion resistance. The study confirmed that the slow diffusion of CO_2 in the narrow pores reduced the MgO adsorption kinetics in spite of the fact that numerous pores contributed to larger surface area. Thus, it was revealed that an outstanding chemical adsorbent should have greater surface area for chemisorption as well as suitable pore size distribution for promoting the diffusion of CO_2 in the adsorbent porous structure. Kumar *et al.* [103] analyzed the CO_2 adsorption ability of mesoporous MgO, which acted as a promising pre-combustion CO_2 adsorbent. It was observed that at 350 °C temperature and CO_2 pressure of 10 bar, almost 96.96% of MgO changed to $MgCO_3$. In another study by Vu *et al.* [104], mesoporous MgO adsorbent was synthesized using aerogel technique and analysis of the CO_2 adsorption behavior at intermediate temperature range of 250–400 °C was carried out. At 325 °C and 120 min, the MgO sample exhibited higher adsorption capacity of 13.9 wt%. Thus, it was confirmed that MgO exhibited higher CO_2 adsorption ability after modification and had the potential of being an excellent adsorbent for CO_2 removal.

Yu-Dong *et al.* [105] examined the CO_2 adsorption efficiency of MgO adsorbent in the presence of water vapor in a fixed bed. Initially, the analysis of CO_2 adsorption capacity of MgO adsorbent under dry condition was carried out. At 100 mL/min flow rate, it took almost 14 min for the CO_2 concentration of the outlet to reach 14.5%. A decrease in the time to reach the equilibrium was seen with the increased flow rate and it took just 4 min for that of 300 mL/min to obtain equilibrium. It was found that at 50 °C, with the increase in relative humidity from 30% to 70%, the CO_2 adsorption ability of MgO adsorbent enhanced from 0.82 mol/kg to 3.42 mol/kg. At a concentration of water vapor of 8.547% v/v and at an adsorption temperature of 75 °C, the CO_2 adsorption capacity reached a maximum value of 3.54 mol/kg. It was reported that the CO_2 adsorption efficiency of MgO was greatly dependent on the presence of water vapor. However, the time to obtain equilibrium increased because of the controlled diffusion of CO_2 arising due to the condensation of water vapor. The results obtained from the cycle experiments confirmed that the condensed water increased the aggregation of MgO particles. Further, it was also noted that at higher temperature, i.e., 100 °C, no water vapor condensation took place and the porous structure remained stable during the cycling adsorption. Elimination of the porous structure degradation and enhancement in the CO_2 adsorption capacity was observed at adsorption temperature between 100 °C and 110 °C. The study confirmed that MgO adsorbent could be developed as a promising material for CO_2 adsorption from wet flue gas.

2.3.2 Adsorption of H_2S

As mentioned earlier, H_2S is a toxic, corrosive gas and is one of the commonly seen contaminants present in several industrially essential feedstocks like coal gas, natural gas, liquefied petroleum gas, jet fuel and naphtha. H_2S is also a dominant source of acid rain, as the oxidation of H_2S causes the formation of sulfur oxide. Further, it can also poison several industrial catalysts like the supported nickel catalyst, which is employed in hydrocarbon steam reforming. Hence, the separation of H_2S from these liquid or gaseous streams is very essential. H_2S adsorption using different metal oxides or their mixtures has been explored in great details to overcome these challenges.

Metal oxides such as iron oxides are widely used for H_2S adsorption. These oxides separate H_2S by forming iron sulfides which are insoluble in nature. The different chemical reactions occurring in this process are as described as follows:

$$Fe_2O_3 + 3\ H_2S \rightarrow Fe_2S_3 + 3\ H_2O$$
$$Fe_2S_3 + 3/2\ O_2 \rightarrow Fe_2O_3 + 3\ S$$

For the adsorption process, iron oxide is commonly employed in a form known as "iron sponge". The iron sponge consists of iron oxide (Fe_2O_3 and Fe_3O_4) impregnated wood chips and can be used in batch as well as continuous systems. After the saturation, the sponge can be regenerated. However, after each regeneration cycle, the activity generally gets decreased by one-third [106]. In continuous system operation, the air is progressively added to the gaseous stream, and the regeneration of iron sponge occurs simultaneously. The H_2S removal rate is generally very high i.e., 2500 mg H_2S/g Fe_2O_3. However, the iron sponge has some disadvantages like high operating cost, chemically intensive process and production of continuous waste stream that might be either disposed of as dangerous waste or regenerated expensively. Some commercially generated iron oxide systems like Sulfatreat 410 HPR are available, which can produce non-hazardous waste. The adsorption capacity of Sulfatreat 410 HPR was reported to be 150 mg H_2S/g adsorbent [106]. Carnes *et al.* [107] reported that the metal oxide reactivities depend on intrinsic crystallite reactivity, crystallite size and the surface area. Further, it was also seen that when compared to the micro-crystalline structures, the nano-crystalline structures exhibited enhanced reactivity with H_2S. This is because of the higher surface area that promotes adsorption. Further, the presence of Fe_2O_3 enhanced the reaction due to the formation of iron sulfides which acted as the catalyst. Rodriguez and Maiti [108] reported that the H_2S adsorption ability of a metal oxide depended on the electronic band gap energy. More H_2S adsorption took place, when electronic band gap energy was less. This is due to the fact that the electronic bad gap is negatively related to oxide chemical reactivity. However, the reactivity depends on the mixing of oxide's bands with H_2S orbitals. In the case of better mixing, the oxides react well with sulfur containing molecules, thus forming metal sulfides. This causes the dissociation of H_2S molecules and sulfur immobilization in metal sulfides. In spite of this, the application of metal oxides for H_2S removal has some drawbacks like high cost, low selectivity, low separation effectiveness and lesser sorption/desorption rate.

At higher temperatures (200 °C - 400 °C), zinc oxide (ZnO) based materials are employed for removing H_2S traces. This is because, ZnO have better selectivity for sulfides than the iron oxides [109]. Davidson *et al.* [109] conducted research on the H_2S removal capacity of ZnO, and it was observed that the ZnO surface reacted with H_2S to form a zinc sulfide insoluble layer, thus, separating the H_2S from the gas stream. Almost 40% of available H_2S was converted and the reaction depicted in the equation below led to the H_2S removal.

$$ZnO + H_2S \rightarrow ZnS + H_2O$$

Many commercial products have been developed by employing ZnO, and the ultimate sulfur loading on these oxides materials has been observed in the range 300-400 mg S/g sorbent. Majority of research studies carried out till date for the advancement of solid H_2S adsorbents focused on the materials which are acceptable for higher temperatures (> 300 °C). Ca- and Fe-containing materials are among the most researched solid materials for H_2S removal from coal gas. This is because of their low cost and higher reactivity [110]. However, due to the presence of large amount of H_2 and large CO/CO_2 ratio in the coal gas, these materials have some disadvantages. For example, the Fe_3O_4 is reduced to FeO and Fe, however, the Fe_3C formed during the process decreases the sulfur uptake of the solid [111]. Lew *et al.* [112] also observed that ZnO was superior to iron oxide due to its agreeable sulfidation thermodynamics. However, the sulfidation kinetics of ZnO was slower than iron oxide. The major disadvantage of using ZnO based solid adsorbents for the application of hot gas cleaning is the damage which occurs because of the reduction of ZnO to elemental Zn which is volatile. Thus, a number of metal oxides have the potential for efficient use as gas adsorbents, however, many challenges are required to be resolved.

Several mixed metal oxides like Cu-Fe-O, Zn-Fe-V-O, Zn-V-O, Zn-Fe-Ti-O, Zn-Ti-O, Cu-Mo-O and Cu-Mn-O have been examined for the H_2S adsorption applications [113-116]. It was observed that the regeneration ability as well as reactivity got enhanced when these materials were deposited on an appropriate support. The mixed metal oxide $ZnFe_2O_4$ also exhibited good H_2S adsorption ability at higher temperature range of 500 °C – 700 °C [117]. For high temperature fuel gas desulfurization, Kobayshi *et al.* [118] examined the zinc ferrite silicon dioxide composites. Zinc ferrite was observed to undergo sulfidation in the reducing environment to form iron sulfides, zinc blende and wurtzite in the presence of 500 ppm of H_2S.

In another study, Huang *et al.* [119] studied the adsorption and reaction of H_2S on TiO_2 rutile (110) and anatase (101) surfaces using periodic density functional theory (DFT). The authors observed that H2S, HS, S, and H preferentially adsorbed at the Ti_{5c}, O_{2c}, $(Ti_{5c})_2$, and O_{2c} sites, respectively, on the rutile surface, whereas these sites were Ti_{5c}, $(Ti_{5c})_2$, $(\text{-}O_{2c})(\text{-}Ti_{5c})$, and O_{2c}, respectively, on the anatase surface. Figure 2.10 also shows the adsorbed H_2S and the fragments HS, S and H on the TiO_2 rutile surface. Novochinskii *et al.* [120] also reported a functional ZnO based adsorbent for low-temperature H_2S removal from steam-containing gas mixtures [120]. The authors observed very low H_2S outlet concentrations (20 parts per billion by volume) over the modified ZnO adsorbent. The authors observed the sulfur-trap capacity (which is the amount of H_2S trapped before breakthrough) to be dependent on factors such as space velocity, temperature, steam concentration, CO_2 concentration, and particle size. Figure 2.11 also demonstrates the performance of a ZnO-based trap at different concentrations of water vapor and H_2S. The authors observed higher trap capacity when the H_2S inlet concentration was higher. The trap capacity was also observed to decrease as a function of temperature. Presence of steam was noticed to inhibit the capture of H_2S by the adsorbent due to the shifting of the equilibrium of the reaction. Furthermore, increasing the CO_2 concentration in the feed up to 12 vol % was also observed to decreases the capacity of ZnO for H_2S adsorption.

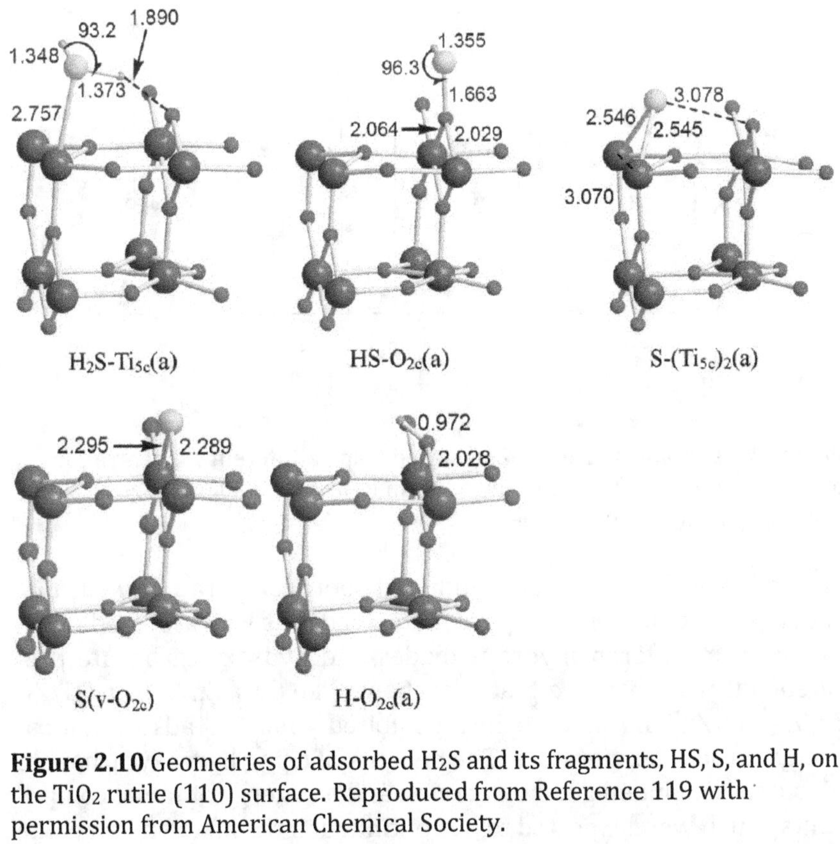

Figure 2.10 Geometries of adsorbed H_2S and its fragments, HS, S, and H, on the TiO_2 rutile (110) surface. Reproduced from Reference 119 with permission from American Chemical Society.

For low temperature H_2S removal applications, several other studies have also been reported [107,109,121,122]. Carnes and Klabunde [107] examined the nano-crystalline metal oxides which were generated using sol gel technique. It was confirmed that at lesser temperature range (25 °C - 100 °C), the activity order was observed to be ZnO > CaO > Al_2O_3 >> MgO. Davidson *et al.* [109] examined the adsorption ability of H_2S using doped and high surface area ZnO at lower temperatures of 25 °C - 45 °C and observed about 40% conversion of H_2S. Baird *et al.* [123] examined the ZnO doped with different oxides of Fe, Co and Cu, mixed Co–Zn oxides, Co-Zn-Al oxides, Co–Fe oxides to generate H_2S adsorbents at lower temperature. Out of the mixed metal oxides, Co-Zn oxide was reported to

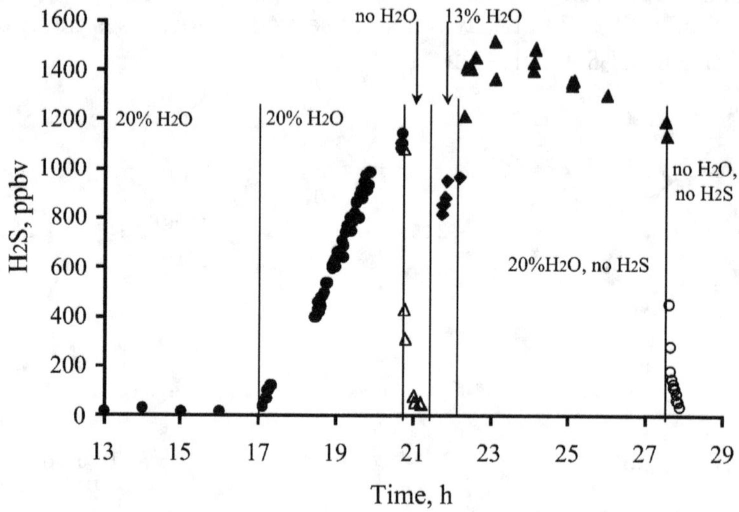

Figure 2.11 Performance of a ZnO-based trap at different concentrations of water vapor and H_2S. Reproduced from Reference 120 with permission from American Chemical Society.

have the best H_2S adsorption ability at room temperature. Xue *et al.* [124] generated a series of pure metal oxides as well as mixed metal oxides from different hydrous oxides and hydroxy-carbonate precursors. It was observed that adsorbents like CuO, Zn/Co, Zn/Ti/Zr, Zn/Al and Zn/Mn mixed oxides exhibited good H_2S adsorption capacity at room temperature. The benefits of using hydroxycarbonate precursors included good dispersion and a synergy of metal components which were expected to be continued in the mixed metal oxide succeeding calcination [125]. The high degree of dispersion of metal oxides is one of the most attractive features of metal oxide based adsorbents for the low temperature performance. Some studies have also focused on the advancement of the mixed metal oxide adsorbents like Cu-Zn and Cu-Zn-Al for the H_2S adsorption at lower temperatures. These Cu based mixed metal oxides are considered to effective adsorbents for low temperature applications because of their low cost than the Zn–Co mixed metal-oxide adsorbents. However, the Cu based mixed oxides, which are generally reduced by hydrogen before being used as adsorbents, were also analyzed for the H_2S adsorption and some other organic sulfur compounds for middle or higher temperatures also [126,127]. Jiang *et al.* [128] generated a series of Cu-Zn hydroxy-carbonate precursors by co-precipitation method, with changing Cu/Zn molar ratios. In addition to this, two series of Cu-Zn-Al hydroxy-carbonate precursors were prepared with varying metal molar ratio, by co-precipitation as well as multi-precipitation methods. The high surface area Cu-Zn-Al metal oxides displayed remarkable sulfur removal capacities at the temperature range 25 °C - 100 °C. It was observed that for H_2S adsorption at low temperature, the Cu rich adsorbents were more suited than the Zn rich adsorbents. The probable reason for this behavior could be due to the fast sulfidation rate of CuO than ZnO because of the lesser rearrangement of anion lattice.

Polychronopoulou *et al.* [129] also studied the efficiency of mixed metal oxides Fe-Mn-Zn-Ti-O of differing composition, generated using sol gel technique, for the removal of H_2S from a gaseous mixture consisting of 7.5 vol % CO_2, 1-3 vol % H_2O, 25 vol % H_2 and 0.06 vol % H_2S, analyzed at temperature range 25 °C - 100 °C. The chemical interaction between the mixed metal oxide and H_2S was studied, and it was observed that the molar ratio of Fe/Mn employed in the preparation of aforementioned metal oxides mostly determined the size and morphology of the generated solid particles. Comparison of the temperature effect on the H_2S adsorption for the sol gel generated 5Fe-15 Mn-40 Zn-40 Ti-O solid and the commercial nickel based solid was made in the presence of 1 vol % and 3 vol % H_2O in the feed stream. In the 1 vol % H_2O system, the commercial solid appeared to be better than the sol gel generated mixed metal oxide at 25 °C - 100 °C temperature range. For the commercial solid, the H_2S adsorption enhanced by 141% by the increase in adsorption temperature from 25 °C - 100 °C. The presence of 1-3 vol % H_2O in the 7.5% CO_2/25% H_2/0.06% H_2S/He gas mixture remarkably increased the H_2S adsorption at 25 °C for 5 Fe-15 Mn-40 Zn-40 Ti-O solid. In a recent study, in-situ formed graphene/ZnO nanostructured composites were employed for low temperature hydrogen sulfide removal from natural gas [130]. Nanostructured composites of graphene and highly dispersed sub-20 nm sized ZnO nanoparticles (TRGZ) were successfully prepared via by combining freeze-drying and thermal annealing processes. A series of compositions with different weight ratios of ZnO nanoparticles were prepared and used as a reactive sorbent in low temperature hydrogen sulfide (H_2S) removal from natural gas. The composite sorbent having a ZnO mass ratio of 45.1 wt% exhibited a significantly greater H_2S adsorption capacity (3.46 mmol.g^{-1}) than that of pure ZnO (1.06 mmol.g^{-1}), indicating that hybridization of ZnO with graphene significantly improved the H_2S removal ability. Figure 2.12 also shows the transmission elecrom micrographs (TEM) of (a) pristine ZnO nanoparticles, (b) TRGZ-1, (c) TRGZ-2, and (d) TRGZ-3 graphene/ZnO nanohybrids. The authors suggested that the possible mechanism involved initial physisorption of H_2S molecules with oxygenated functional groups of graphene which later can reach to the finely dispersed active ZnO nanoparticles and get chemisorbed through reactive adsorption. The EDX spectra of H_2S treated samples exhibited a strong peak for sulfur element indicating the reactive adsorption of H_2S on the sorbent.

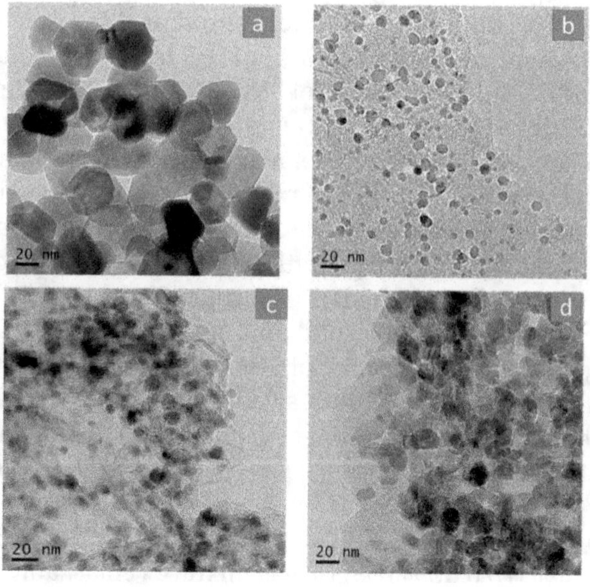

Figure 2.12 TEM images of (a) pristine ZnO nanoparticles, (b) TRGZ-1, (c) TRGZ-2, and (d) TRGZ-3 graphene/ZnO nanohybrids. Reproduced from Reference 130 with permission from Royal Society of Chemistry.

2.3.3 Adsorption of Organic Contaminants

Among the different physical methods employed for dye removal, adsorption is widely used method and considerable attention has been paid for the application of nano-sized adsorbent materials because of their higher surface area. In recent years, the utilization of magnetic particle technology for solving environmental issues has also gained attention. Different types of iron oxides have been studied for the dye adsorption, due to their reactive surface. These materials enable faster magnetic separation just after the adsorption. The magnetic particles can be employed for adsorbing the contaminants from gaseous or aqueous effluents and can subsequently be easily removed from the medium.

Luiz *et al.* [131] generated magnetic adsorbents by combining the magnetic properties of iron oxides and adsorption properties of the activated carbon. The iron oxide/activated carbon composites were used for the removal of volatile organic compounds like phenol, chloroform, drimaren red dye and chlorobenzene from the aqueous solution. It was noted that the existence of iron oxide did not influence the adsorption efficiency of activated carbon. In another study by Khosravi *et al.* [132], iron oxide nano-spheres were prepared using solvo-thermal method and were employed as an efficient adsorbent for anionic dye separation from the aqueous solution. Here, efficiency of iron oxide nano-spheres for the adsorption of anionic dyes namely reactive orange (RO) and reactive yellow (RY) was analyzed. The maximal adsorption capacity of RY and RO was observed to be 25.0 mg/g and 32.5 mg/g, respectively. The kinetic analysis confirmed rapid adsorption rate onto nano-spheres, which took place within 10 min. The adsorption process was exothermic, and the optimum pH was about 4, i.e., acidic solution. At the final stage, due to the magnetic nature of the iron oxide nano-spheres, smooth separation from solution by a magnet could be achieved.

2.3.4 Heavy Metal Removal from Water/Wastewater

Heavy metal exposure, even at a very small level, is considered to be a danger for the living beings [133]. For the heavy metal separation from the water/wastewater, adsorption methods have become prominent techniques. Out of the various adsorbents available, nano-sized metal oxides (NMOs) are regarded as promising ones for the heavy metal removal from aqueous systems [134]. This is due to their greater surface areas as well as higher activities generated by the size quantization effect. The NMOs offer numerous advantages like preferable sorption, higher capacities and faster kinetics towards the heavy metals present in the water as well as wastewater. The examples for NMOs include nano-sized ferric oxides, aluminum oxides, magnesium oxides, manganese oxides, titanium oxides and cerium oxides. These materials exist in different forms such as tubes, particles, etc. On the other hand, when the metal oxide size decreases from the micrometer level to nanometer level, the enhancement in the surface energy causes poor stability. Hence, NMOs are prone to agglomeration because of the presence of van der Waals forces and other interactions, and thereby the high adsorption capacity as well as selectivity of NMOs can significantly reduce or even disappear [135]. Thus, the shape and size of these adsorbents are critical factors influencing their adsorption behavior.

The utilization of iron oxide based nanomaterials is considered to be very effective for the separation of metal contaminants from water, due to their remarkable features such as small size, magnetic property, reusability and high surface area [136]. For instance, iron oxide nanomaterials exhibit excellent adsorption ability for the separation of arsenic from water. Feng *et al.* [137] generated super magnetic Fe_3O_4 coated with ascorbic acid, by hydrothermal process. The material had a surface area of 179 m^2/g and diameter < 10 nm. At room temperature, the material exhibited super paramagnetic behavior and the saturation magnetization attained 40 $emug^{-1}$, which indicated that these materials were suitable for use as an adsorbent for the separation of arsenic from waste water. The ultimate adsorption capacity of As(III) and As(V) was found to be 46.06 mg/g and 16.56 mg/g respectively.

The excessive application of copper in the industries leads to the accumulation in the environment, thereby polluting the water. Hao *et al.* [138] generated a magnetic nano-adsorbent (MNP-NH_2) by covalent binding of 1,6-hexadiamine on the Fe_3O_4 nanoparticles surface and analyzed its adsorption capacity for the removal of copper ions from the aqueous solution. Different factors influencing the adsorption performance such as salinity, initial Cu^{2+} concentration, temperature, amount of MNP-NH_2, contact time and pH were examined. The magnetic nanoparticles were able to separate 98% of copper from tap water as well as the polluted water. It was noticed that the adsorption occurred faster, and the equilibrium was attained within 5 min. The chemical sorption was noticed to be the rate limiting step of the sorption mechanism, and the ultimate adsorption ability was observed to be 25.77 $mg.g^{-1}$ at 298 K and pH value 6. The adsorbent MNP-NH_2 exhibited good resuability as well as higher stability under the experimental conditions. Complete desorption of the Cu^{2+} ions was achieved by using 0.1 $mol.L^{-1}$ HCl solution for < 1 min and the regenerated MNP-NH_2 was able to retain the original metal discharge level. It was observed that the adsorption ability of the nano-adsorbent MNP-NH_2 remained constant, and no change in the desorption ability occurred during 15 adsorption - desorption cycles. In another study by Nashaat *et al.* [139], the adsorption mechanics, equilibria, thermodynamics and kinetics of Pb(II) ions onto Fe_3O_4 nano-adsorbents were analyzed. It was found that Fe_3O_4 nano-adsorbents had greater removal efficiency as Pb(II) was adsorbed from waste water in a very brief time and the equilibrium was accomplished within 30 min. Further, the adsorption was dependent on temperature, pH and the Pb(II) initial concentration. At pH value 5.5, the maximal removal of Pb(II) was observed. As the temperature and Pb(II) initial concentration was increased, the adsorption was also observed to enhance. The adsorption isotherms were determined using the Freundlich and Langmuir models. The thermodynamics of the adsorption of Pb(II) revealed the impulse nature, endothermic behavior and physisorption process. The regeneration as well as the desorption analysis confirmed the repeated use of nano-adsorbents without influencing the adsorption capacity. This confirmed the usefulness of Fe_3O_4 nano-adsorbents as inexpensive and effective adsorbents for the fast removal and recovery of the metal ions from wastewater.

The adsorptive behavior of nano-sized manganese oxides (NMnOs) is considered to be superior when compared to its bulk counterpart due to its greater surface area as well as polymorphic structures [140]. In the past several decades, NMnOs have been used for the sorption of anionic and cationic contaminants like arsenate, phosphate and heavy metal ions from natural water. Alumina (Al_2O_3) is also a conventional adsorbent for the removal of heavy metals and γ- $Al_2O_3$is regarded to be superior to α- Al_2O_3 in adsorption performance [141]. Nano-sized γ- Al_2O_3 adsorbents are generated using sol-gel method, and the materials have been used as solid phase extraction materials for the pre-concentration/separation of the trace metal ions. In the case of TiO_2, it was observed that the bulk and nano-TiO_2 differed in surface acidity, catalytic reactivity and chemical behavior, based on their various surface planes. From the BET analysis, it was seen that the nominal particle sizes were 8.3 and 329.8 nm, for nano-sized and bulk particles respectively [142]. The specific area of bulk particles was found to be 9.5 m^2/g, whereas the nano-particles' specific area was 185.5 m^2/g. In a research study by Liang *et al.* [143], the nano-TiO_2 adsorbent, with BET surface area 208 m^2/g and diameter 10–50 nm exhibited adsorption capacity of Cd and Zn as 7.9 mg/g and 15.3 mg/g respectively at pH value 9. The effects of contact time, pH, interfering ions and elution solution on the adsorption performance of nano-TiO_2 for the removal of Zn and Cd were examined. The presence of common anions and cations (in the concentration range of 100-5000 mg/L) was found to have no significant effect on the metal adsorption (Cd^{2+} ions and Zn^{2+} of 1.0 mg/mL) under the specified conditions. Many research studies have also confirmed the role of nano-structured ZnO to effectively remove the heavy metals from various media [144]. In a related study, Lee *et al.* [145] generated NZnOs powder using the solution combustion method. The authors observed that as compared with two TiO_2 powders, P25 and the other one generated by homogenous precipitation at lower temperature, the NZnO powder displayed greater Cu^{2+} adsorption from the solution.

Table 2.2 summarizes the heavy metals adsorption capacities of different metal oxides reviewed in this study.

Table 2.2 Heavy metal adsorption capacities of some of the metal oxide adsorbents

Sl. No.	Adsorbent	Temperature	pH value	Heavy metal	Adsorption capacity (mg/g)	Reference
1	Super-magnetic Fe_3O_4 coated with ascorbic acid	room temperature	-	As(III)	46.06 mg/g	115
2	Super-magnetic Fe_3O_4 coated with ascorbic acid	room temperature	-	As(V)	16.56 mg/g	115
3	Magnetic nano-adsorbent (MNP-NH_2)	298 K	6	Cu	25.77 mg/g	116
4	Nano-TiO_2 adsorbent		9	Cd	7.9 mg/g	103
5	Nano-TiO_2 adsorbent		9	Zn	15.3 mg/g	103

2.4 Summary and Outlook

Adsorption represents an immensely useful technology for the removal of harmful gases like H_2S, CO_2, SO_2, NH_3 present in the gas stocks and air as well as separation of organic contaminants and heavy metals from the wastewater. It has attracted remarkable consideration in both scientific exploration as well as commercial utilizations. Further, the development of state-of-the-art porous materials for the adsorption process is required to enhance the adsorption capacity or to overcome the existing disadvantages such as small pore volume or pore size of adsorbents. In this chapter, a comprehensive assessment on the use of MOFs and metal oxides for gas/vapor and liquid phase adsorptions has been presented. Further, focus has also been devoted to the separation of a variety of contaminants like dyes and various heavy metals present in the water using the MOFs and metal oxides. As mentioned before, MOFs are promising adsorbents for the aforementioned applications due to their unique properties like higher surface area as well as tunable porosities. In addition to these properties, MOFs are able to consolidate functionalities by means of grafting or loading active or functional species and thereby increasing the adsorption capacity. Several modifications in MOFs, which were done in three aspects namely (1) metal ions, (2) organic liners and (3) novel consolidation of both, for enhancing the adsorption ability were also analyzed. The metal oxides exhibit enhanced selectivity because of the presence of chemical adsorption between the gas molecules and their surfaces. Both the surface area as well as pore size of the adsorbent greatly affects the gas diffusion in the adsorbent porous structure. Their reactivity depends on the crystallite size, intrinsic crystallite reactivity as well as the surface area. Nano sized metal oxides (NMOs) are considered to be an excellent adsorbent for the heavy metal removal from aqueous systems. In spite of the high adsorption capacity of the MOFs and metal oxides, there are several economical and technical barriers to promote the usage of these materials for gas adsorption and environmental applications. Most of these challenges are associated with material stability, remigration, high temperature performance, durability, loss of adsorption and selectivity, etc. However, with the employment of advanced nanotechnologies, it is envisaged that many of the currently existing limitations would be overcome in due course of time, which will pave the way for the large-scale application of these MOFs and metal oxide materials.

References

1. Pawelec, B., Navarro, R. M., Campos-Martin, J. M., and Fierro, J. L. G. (2011) Towards near zero-sulfur liquid fuels: a perspective review. *Catalysis Science and Technology*, **1**, 23-42.
2. Vandenbroucke, A. M., Morent, R., Geyter, N. D., and Leys, C. (2011) Non-thermal plasmas for non-catalytic and catalytic VOC abatement. *Journal of Hazardous Materials*, **195**, 30-54.
3. Garces, H. F., Galindo, H. M., Garces, L. J., Hunt, J., Morey, A., and Suib, S. L. (2010) Low temperature H_2S dry-desulfurization with zinc oxide. *Microporous and Mesoporous Materials*, **127**, 190-197.
4. Petit, C., and Bandosz, T. J. (2010) Enhanced adsorption of ammonia on metal-organic framework/graphite oxide composites: analysis of surface interactions. *Advanced Functional Materials*, **20**, 111-118.
5. Colvile, R. N., Hutchinson, E. J., Mindell, J. S., and Warren, R. F. (2001) The transport sector as a source of air pollution. *Atmospheric Environment*, **30**, 1537-1565.
6. Bao, C., and Fang, C.-L. (2012) Water resources flows related to urbanization in China: challenges and perspectives for water management and urban development. *Water Resources Management*, **26**, 531-552.
7. Gupta, V. K., and Suhas (2009) Application of low-cost adsorbents for dye removal - a review. *Journal of Environmental Management*, **90**, 2313-2342.
8. Gong, Y., Zhao, X., Cai, Z., O'Reilly, S. E., Hao, X., and Zhao, D. (2014) A review of oil, dispersed oil and sediment interactions in the aquatic environment: influence on the fate, transport and remediation of oil spills. *Marine Pollution Bulletin*, **79**, 16-33.
9. Crini, G. (2006) Non-conventional low-cost adsorbents for dye removal: a review. *Bioresource Technology*, **97**, 1061-1085.
10. Chen, S., Zhang, J., Zhang, C., Yue, Q., Li, Y., and Li, C. (2010) Equilibrium and kinetic studies of methyl orange and methyl violet adsorption on activated carbon derived from phragmites australis. *Desalination*, **252**, 149-156.
11. Hasan, Z., Jeon, J., and Jhung, S. H. (2012) Adsorptive removal of naproxen and clofibric acid from water using metal-organic frameworks. *Journal of Hazardous Materials*, **209-210**, 151-157.
12. Ke, F., Qiu, L.-G., Yuan, Y.-P., Peng, F.-M., Jiang, X., Xie, A.-J., Shen, Y.-H., and Zhu, J.-F. (2011) Thiol-functionalization of metal-organic framework by a facile coordination based post-synthetic strategy and enhanced removal of Hg^{2+} from water. *Journal of Hazardous Materials*, **196**, 36-43.
13. Zhu, B.-J., Yu, X.-Y., Jia, Y., Peng, F.-M., Sun, B., Zhang, M.-Y., Luo, T., Liu, J.-H., and Huang, X.-J. (2012) Iron 1,3,5-benzenetricarboxylic

metal-organic coordination polymers prepared by solvothermal method and their application in effcient As(V) removal from aqueous solutions. *Journal of Physical Chemistry C*, **116**, 8601–8607.

14. Li, J.-R., Sculley, J., and Zhou, H.-C. (2012) Metal- organic frame works for separations. *Chemical Reviews*, **112**, 869-932.
15. Babich, I. V., and Moulijn, J. A. (2003) Science and technology of novel processes for deep desulfurization of oil refinery streams: a review. *Fuel*, **82**, 607-631.
16. Velu, S., Ma, X., and Song, C. (2003) Selective adsorption for removing sulfur from jet fuel over zeolite-based adsorbents. *Industrial and Engineering Chemistry Research*, **42**, 5293-5304.
17. Choi, S., Drese, J. H., and Jones, C. W. (2009) Adsorbent materials for carbon dioxide capture from large anthropogenic point sources. *ChemSusChem*, **2**, 796-854.
18. Zhou, A., Ma, X., and Song, C. S. (2006) Liquid-phase adsorption of multi-ring thiophenic sulfur compounds on carbon materials with different surface properties. *Journal of Physical Chemistry B*, **110**, 4699-4707.
19. Deliyanni, E., Seredych, M., and Bandosz, T. J. (2009) Interactions of 4,6-dimethyldibenzothiophene with the surface of activated carbons. *Langmuir*, **25**, 9302-9312.
20. Haque, E., Jun, J. W., Talapaneni, S. N., Vinu, A., and Jhung, S. H. (2010) Superior adsorption capacity of mesoporous carbon nitride with basic CN framework for phenol. *Journal of Materials Chemistry*, **20**, 10801-10803.
21. Wang, Y., Yang, R. T., and Heinzel, J. M. (2008) Desulfurization of jet fuel by π-complexation adsorption with metal halides supported on MCM-41 and SBA-15 mesoporous materials. *Chemical Engineering Science*, **63**, 356-365.
22. Li, J.-R., Kuppler, R. J., and Zhou, H.-C. (2009) Selective gas adsorption and separation in metal-organic frameworks. *Chemical Society Reviews*, **38**, 1477-1504.
23. Jiang, H.-L., and Xu, Q. (2011) Porous metal-organic frameworks as platforms for functional applications. *Chemical Communications*, **47**, 3351-3370.
24. Seredych, M., and Bandosz, T. J. (2011) Removal of dibenzothiophenes from model diesel fuel on sulfur rich activated carbons. *Applied Catalysis, B: Environmental*, **106**, 133-141.
25. Ferey, G. (2008) Hybrid porous solids: past, present, future. *Chemical Society Reviews*, **37**, 191-214.
26. Sumida, K., Rogow, D. L., Mason, J. A., McDonald, T. M., Bloch, E. D., Herm, Z. R., Bae, T.-H., Long, J. R. (2012) Carbon dioxide capture in metal-organic frameworks. *Chemical Reviews*, **112**, 724-781.
27. Wu, H., Gong, Q., Olson, D. H., and Li, J. (2012) Commensurate adsorption of hydrocarbons and alcohols in microporous metal orga-

nic frameworks. *Chemical Reviews*, **112**, 836-868.

28. Suh, M. P., Park, H. J., Prasad, T. K., and Lim, D.-W. (2012) Hydrogen storage in metal-organic frameworks. *Chemical Reviews*, **112**, 782-835.
29. Li, J.-R., Sculley, J., and Zhou, H.-C. (2012) Metal-organic frameworks for separations. *Chemical Reviews*, **112**, 869-932.
30. Horcajada, P., Gref, R., Baati, T., Allan, P. K., Maurin, G., Couvreur, P., Ferey, G., Morris, R. E., and Serre, C. (2012) Metal-organic frameworks in biomedicine. *Chemical Reviews*, **112**, 1232-1268.
31. Furukawa, H., Ko, N., Go, Y. B., Aratani, N., Choi, S. B., Choi, E., Yazaydin, A. O., Snurr, R. Q., O'Keeffe, M., Kim, J., and Yaghi, O. M. (2010) Ultrahigh porosity in metal-organic frameworks. *Science*, **329**, 424-428.
32. Hamon, L., Serre, C., Devic, T., Loiseau, T., Millange, F., Ferey, G., and Weireld, G. D. (2009) Comparative study of hydrogen sulfide adsorption in the MIL-53(Al, Cr, Fe),MIL-47(V), MIL-100(Cr), and MIL-101(Cr) metal organic frameworks at room temperature, *Journal of American Chemical Society*, **131**, 8775-8777.
33. Hamon, L., Leclerc, H., Ghoufi, A., Oliviero, L., Travert, A., Lavalley, J.-C., Devic, T., Serre, C., Ferey, G., Weireld, G. D., Vimont, A., and Maurin, G. (2011) Molecular insight into the adsorption of H_2S in the flexible MIL-53(Cr) and rigid MIL-47(V) MOFs:infrared spectroscopy combined to molecular simulations. *Journal of Physical Chemistry C*, **115**, 2047-2056.
34. Liu, J., Thallapally, P. K., McGrail, B. P., and Brown, D. R. (2012) Progress in adsorption-based CO_2 capture by metal-organic frameworks. *Chemical Society Reviews*, **41**, 2308-2322.
35. Sumida, K., Rogow, D. L., Mason, J. A., McDonald, T. M., Bloch, E. D., Herm, Z. R., Bae, T.-H., and Long, J. R. (2012) Carbon dioxide capture in metal-organic frameworks. *Chemical Reviews*, **112**, 724-781.
36. Dathe, H., Peringer, E., Roberts, V., Jentys, A., and Lercher, J. A. (2005) Metal organic frameworks based on Cu^{2+} and benzene-1,3,5-tricarboxylate as host for SO_2 trapping agents. *C. R. Chimie*, **8**, 753-763.
37. Planchais, A., Devautour-Vinot, S., Giret, S., Salles, F., Trens, P., Fateeva, A., Devic, T., Yot, P., Serre, C., Ramsahye, N,m and Maurin, G. (2013) Adsorption of benzene in the cation-containing MOFs MIL-141. *Journal of Physical Chemistry C*, **117**(38), 19393-19401.
38. Huxford, R. C., Rocca, J. D., and Lin W. (2010) Metal-organic frameworks as potential drug carriers. *Current Opinion in Chemical Biology*, **14(2)**, 262-268.
39. Kitagawa, S., Kitaura, R., and Noro, S.-I. (2004) Functional porous coordination polymers. *Angewandte Chemie International Edition*, **43**, 2334-2375.

40. Cavka, J. H., Jakobsen, S., Olsbye, U., Guillou, N., Lamberti, C., Bordiga, S., and Lillerud, K. P. (2008) A new zirconium inorganic building brick forming metal organic frameworks with exceptional stability. *Journal of American Chemical Society*, **130**, 13850-13851.
41. Walton, K. S., Millward, A. R., Dubbeldam, D., Frost, H., Low, J. J., Yaghi, O. M., and Snurr, R. Q. (2008) Understanding inflections and steps in carbon dioxide adsorption isotherms in metal-organic frameworks. *Journal of American Chemical Society*, **130**, 406-407.
42. Demessence, A., D'Alessandro, D. M., Foo, M. L., and Long, J. R. (2009) Strong CO_2 binding in a water-stable, triazolate-bridged metal-organic framework functionalized with ethylenediamine. *Journal of American Chemical Society*, **131**, 8784-8786.
43. Britt, D., Furukawa, H., Wang, B., Glover, T. G., and Yaghi, O. M. (2009) Highly efficient separation of carbon dioxide by a metal-organic framework replete with open metal sites. *Proceedings of the National Academy of Sciences of the USA*, **106**, 20637-20640.
44. Dietzel, P. D. C., Besikiotis, V., and Blom, R. (2009) Application of metal-organic frameworks with coordinatively unsaturated metal sites in storage and separation of methane and carbon dioxide. *Journal of Materials Chemistry*, **19**, 7362-7370.
45. Caskey, S. R., Wong-Foy, A. G., and Matzger, A. J. (2008) Dramatic tuning of carbon dioxide uptake via metal substitution in a coordination polymer with cylindrical pores. *Journal of American Chemical Society*, **130**, 10870-10871.
46. Couck, S., Denayer, J. F. M., Baron, G. V., Remy, T., Gascon, J., and Kapteijn, F. (2009) An amine-functionalized MIL-53 metal-organic framework with large separation power for CO_2 and CH_4. *Journal of American Chemical Society*, **131**, 6326-6327.
47. Bae, Y. S., Farha, O. K., Spokoyny, A. M., Mirkin, C. A., Hupp, J. T., and Snurr, R. Q. (2008) Carborane-based metal-organic frameworks as highly selective sorbents for CO_2 over methane. *Chemical Communications*, **37**, 4135-4137.
48. Hayashi, H., Cote, A. P., Furukawa, H., O'Keeffe, M., and Yaghi, O. M. (2007) Zeolite A imidazolate frameworks. *Nature Materials*, **6**, 501-506.
49. Phan, A., Doonan, C. J., Uribe-romo, F. J., Knobler, C. B., O'Keeffe, M., and Yaghi, O. M. (2010) Synthesis, structure, and carbon dioxide capture properties of zeolitic imidazolate frameworks. *Accounts of Chemical Research*, **43**, 58-67.
50. Hu, Z., Khurana, M., Seah, Y. H., Zhang, M., Guo, Z., and Zhao, D. (2015) Ionized Zr-MOFs for highly efficient post-combustion CO_2 capture, *Chemical Engineering Science*, **124**, 61-69.
51. Hu, Z. G., Zhang, K., Zhang, M., Guo, Z. G., Jiang, J. W., and Zhao, D. (2014) A combinatorial approach towards water-stable metal-organic frameworks for highly efficient carbon dioxide separation.

ChemSusChem, **7**, 2791-2795.

52. Yang, D.-A., Cho, H.-Y., Kim, J., Yang, S.-T., and Ahn, W.-S. (2012) CO_2 capture and conversion using Mg-MOF-74 prepared by a sonochemical method. *Energy & Environmental Science*, **5**, 6465-6473.
53. Achmann, S., Hagen, G., Hammerle, M., Malkowsky, I. M., Kiener, C., and Moos, R. (2010) Sulfur removal from low-sulfur gasoline and diesel fuel by metal-organic frameworks. *Chemical Engineering & Technology*, **33**, 275-280.
54. Chavan, S., Bonino, F., Valenzano, L., Civalleri, B., Lamberti, C., Acerbi, N., Cavka, J. H., Leistner, M., and Bordiga, S. (2013) Fundamental aspects of H_2S adsorption on CPO-27-Ni. *Journal of Physical Chemistry C*, **117**, 15615-15622.
55. van de Voorde, B., Hezinova, M., Lannoeye, J., Vandekerkhove, A., Marszalek, B., Gil, B., Beurroies, I., Nachtigall, P., and De Vos, D. (2015) Adsorptive desulfurization with CPO-27/MOF-74: an experimental and computational investigation. *Physical Chemistry and Chemical Physics*, **17**, 10759-10766.
56. Hamon, L., Serre, C., Devic, T., Loiseau, T., Millange, F., Frrey, G., and De Weireld, G. (2009) Comparative study of hydrogen sulfide adsorption in the MIL-53 (Al, Cr Fe), MIL-47 (V), MIL-100 (Cr), and MIL-101 (Cr) metal-organic frameworks at room temperature. *Journal of American Chemical Society*, **131**, 8775-8777.
57. Hamon, L., Leclerc, H., Ghoufi, A., Oliviero, L., Travert, A., Lavalley, J. C., Devic, T., Serre, C., Ferey, G., De Weireld, G., Vimont, A., and Maurin, G. (2011) Molecular insight into the adsorption of H_2S in the flexible MIL-53 (Cr) and rigid MIL-47 (V) MOFs: infrared spectroscopy combined to molecular simulations. *Journal of Physical Chemistry C*, **115**, 2047-2056.
58. Chen, Z., Ling, L., Wang, B., Fan, H., Shangguan, J., and Mi, J. (2016) Adsorptive desulfurization with metal-organic frameworks: A density functional theory investigation. *Applied Surface Science*, **387**, 483-490.
59. Fernandez, C. A., Thallapally, P. K., Motkuri, R. K., Nune, S. K., Sumrak, J. C., Tian, J., and Liu, J. (2010) Gas-induced expansion and contraction of a fluorinated metal-organic framework. *Crystal Growth & Design*, **10**, 1037-1039.
60. Ahmed, I., and Jhung, S. H. (2014) Composites of metal-organic frameworks: preparation and application in adsorption. *Materials Today*, **17**, 136-146.
61. Glover, T. G., Peterson, G. W., Schindler, B. J., Britt, D., and Yaghi, O. M. (2011) MOF-74 building unit has a direct impact on toxic gas adsorption. *Chemical Engineering Science*, **66**, 163-170.
62. Petit, C., and Bandosz, T. J. (2012) Exploring the coordination chemistry of MOF-graphite oxide composites and their applications as adsorbents. *Dalton Transactions*, **41**, 4027-4035.

63. Britt, D., Tranchemontagne, D., and Yaghi, O. M. (2008) Metal-organic frameworks with high capacity and selectivity for harmful gases. *Proceedings of the National Academy of Sciences of the USA*, **105**, 11623–11627.
64. Karra, J. R., and Walton, K. S. (2008) Effect of open metal sites on adsorption of polar and nonpolar molecules in metal organic framework Cu-BTC. *Langmuir*, **24**, 8620-8626.
65. Xiao, B., Wheatley, P. S., Zhao, X., Fletcher, A. J., Fox, S., Rossi, A. G., Megson, I. L., Bordiga, S., Regli, L., Thomas, K. M., and Morris, R. E. (2007) High-capacity hydrogen and nitric oxide adsorption and storage in a metal organic framework. *Journal of American Chemical Society*, **129**, 1203-1209.
66. Jhung, S. H., Lee, J.-H., Yoon, J. W., Serre, C., Ferey, G., Chang, J.-S. (2007) Microwave synthesis of chromium terephthalate MIL-101 and its benzene sorption ability. *Advanced Materials*, **19**, 121-124.
67. Shi, J., Zhao, Z., Xia, Q., Li, Y., Li, Z. (2011) Adsorption and diffusion of ethyl acetate on the chromium-based metal-organic framework MIL-101. *Journal of Chemical Engineering Data*, **56**, 3419-3425.
68. Taylor, J. M., Vaidhyanathan, R., Iremonger, S. S., and Shimizu, G. K. H. (2012) Enhancing water stability of metal- organic frameworks via phosphonatemonoester linkers. *Journal of American Chemical Society*, **134**, 14338–14340.
69. Greathouse, J. A., and Allendorf, M. D. (2006) The interaction of water with MOF-5 simulated by molecular dynamics. *Journal of American Chemical Society*, **128**, 10678-10679.
70. Der Voort, P. V., Leus, K., Liu, Y.-Y., Vandichel, M., Speybroeck, V. V., Waroquier, M., and Biswas, S. (2014) Vanadium metal-organic frameworks: structures and applications. *New Journal of Chemistry*, **34**, 1853-1867.
71. Valenzano, L., Civalleri, B., Chavan, S., Bordiga, S., Nilsen, M. H., Jakobsen, S., Lillerud, K. P., and Lamberti, C. (2011) Disclosing the complex structure of UiO-66 metal organic framework: A synergic combination of experiment and theory. *Chemistry of Materials*, **23**(7), 1700-1718.
72. Park, K. S., Ni, Z., Cote, A. P., Choi, J. Y., Huang, R., Uribe-Romo, F. J., Chae, H. K., O'Keeffe, M., and Yaghi, O. M. (2006) Exceptional chemical and thermal stability of zeolitic imidazolate frameworks. *Proceedings of the National Academy of Sciences of the USA*, **103**, 10186-10191.
73. Ehrenmann, J., Henninger, S. K., and Janiak, C. (2011) Water adsorption characteristics of mil-101 for heat-transformation applications of MOFs. European Journal of Inorganic Chemistry, 2011(4), DOI: doi:10.1002/ejic.201190006.
74. Jeremias, F., Khutia, A., Henninger, S. K., and Janiak, C. (2012) MIL-100(Al, Fe) as water adsorbents for heat transformation purposes

- a promising application. *Journal of Materials Chemistry*, **22**, 10148-10151.

75. Cychosz, K. A., and Matzger, A. J. (2010) Water stability of microporous coordination polymers and the adsorption of pharmaceuticals from water. *Langmuir*, **26**, 17198-17202.
76. Wu, T., Shen, L., Luebbers, M., Hu, C., Chen, Q., Ni, Z., and Masel, R. I. (2010) Enhancing the stability of metal-organic frameworks in humid air by incorporating water repellent functional groups. *Chemical Communications*, **46**, 6120-6122.
77. Zu, D.-D., Lu, L., Liu, X.-Q., Zhang, D.-Y., and Sun, L.-B. (2014) Improving hydrothermal stability and catalytic activity of metal-organic frameworks by graphite oxide incorporation. *The Journal of Physical Chemistry C*, **118**, 19910-19917.
78. Yang, C., Kaipa, U., Mather, Q. Z., Wang, X., Nesterov, V., Venero, A. F., and Omary, M. A. (2011) Fluorous metalorganic frameworks with superior adsorption and hydrophobic properties toward oil spill cleanup and hydrocarbon storage. *Journal of the American Chemical Society*, **133**, 18094-18097.
79. Hasan, Z., and Jhung, S. H. (2015) Removal of hazardous organics from water using metal-organic frameworks (MOFs): Plausible mechanisms for selective adsorptions. *Journal of Hazardous Materials*, **283**, 329-339.
80. Haque, E., Lee, J. E., Jang, I. T., Hwang, Y. K., Chang, J.-S., Jegal, J., and Jhung, S. H. (2010) Adsorptive removal of methyl orange from aqueous solution with metal-organic frameworks, porous chromium-benzenedicarboxylates. *Journal of Hazardous Materials*, **181**, 535-542.
81. Haque, E., Lo, V., Minett, A. I., Harris, A. T., and Church, T. L. (2014) Dichotomous adsorption behaviour of dyes on an amino-functionalised metal-organic framework, amino-MIL-101(Al). *Journal of Materials Chemistry A*, **2**, 193-203.
82. Haque, E., Jun, J. W., and Jhung, S. H. (2011) Adsorptive removal of methyl orange and methylene blue from aqueous solution with a metal-organic frame- work material, iron terephthalate (MOF-235). *Journal of Hazardous Materials*, **185**, 507-511.
83. Leng, F., Wang, W., Zhao, X. J., Hu, X. L., and Li, Y. F. (2014) Adsorption interaction between a metal-organic framework of chromium–benzenedicarboxylates and uranine in aqueous solution. *Colloids & Surfaces A*, **441**, 164-169.
84. Chen, C., Zhang, M., Guan, Q., and Li, W. (2012) Kinetic and thermodynamic studies on the adsorption of xylenol orange onto, MIL-101(Cr). *Chemical Engineering Journal*, **183**, 60-67.
85. Liu, B., Yang, F., Zou, Y., and Peng, Y. (2014) Adsorption of phenol and p-nitrophenol from aqueous solutions on, metal-organic frameworks: effect of hydrogen bonding. *Journal of Chemical Engi-*

neering Data, **59**, 1476-1482.

86. Xie, L., Liu, D., Huang, H., Yang, Q., and Zhong, C. (2014) Efficient capture of nitrobenzene from waste water using metal-organic frameworks. *Chemical Engineering Journal*, **246**, 142-149.
87. Khan, N. A., Jung, B. K., Hasan, Z., and Jhung, S. H. (2015) Adsorption and removal of phthalic acid and diethylphthalate from water with zeolitic imidazolate and metal- organic frameworks. *Journal of Hazardous Materials*, **282**, 194-200.
88. Tong, M., Liu, D., Yang, Q., Devautour-Vinot, S., Maurin, G., and Zhong, C. (2013) Influence of framework metal ions on the dye capture behavior of MIL-100 (Fe, Cr) MOF type solids. *Journal of Materials Chemistry A*, **1**, 8534-8537.
89. Wang, H.-N., Liu, F.-H., Wang, X.-L., Shao, K.-Z., and Su, Z.-M. (2013) Three neutral metal-organic frameworks with micro and mesopores for adsorption and separation of dyes. *Journal of Materials Chemistry A*, **1**, 13060-13063.
90. Han, R., Zou, W., Wang, Y., and Zhu, L. (2007) Removal of uranium (VI) from aqueous solutions by manganese oxide coated zeolite: discussion of adsorption isotherms and pH effect. *Journal of Environmental Radioactivity*, **93**, 127-143.
91. Feng, Y., Jiang, H., Li, S., Wang, J., Jing, X., Wang, Y., and Chen, M. (2013) Metal–organic frameworks HKUST-1 for liquid-phase adsorption of Uranium. *Colloids and Surfaces A*, **431**, 87-92.
92. Batzill, M. (2012) The surface science of graphene: Metal interfaces, CVD synthesis, nanoribbons, chemical modifications, and defects. *Surface Science Reports*, **67**, 83-115.
93. Furukawa, H., Cordova, K. E., O'Keeffe, M., and Yaghi, O. M. (2013) The chemistry and applications of metal-organic frameworks. *Science*, **341**, 1230444-1230456.
94. Tao, X., Ma, W., Li, J., Huang, Y., Zhao, J., and Yu, J. C. (2003) Efficient degradation of organic pollutants mediated by immobilized iron tetrasulfophthalocyanine under visible light irradiation. *Chemical Communications*, 80-81.
95. Kyzas, G. Z., and Matis, K. A. (2015) Nanoadsorbents for pollutants removal: a review. *Journal of Molecular Liquids*, **203**, 159-168.
96. House, K. Z., Harvey, C. F., Aziz, M. J., and Schrag, D. P. (2009) The energy penalty of post-combustion CO_2 capture & storage and its implications for retrofitting the U.S. installed base. *Energy & Environmental Science*, **2**, 193-205.
97. Cai, W., Yu, J., Anand, C., Vinu, A., and Jaroniec, M. (2011) Facile synthesis of ordered mesoporous alumina and alumina-supported metal oxides with tailored adsorption and framework properties. *Chemistry of Materials*, **23**, 1147-1157.
98. Leon, M., Diaz, E., Bennici, S., Vega, A., Ordonez, S., and Auroux, A. (2010) Adsorption of CO_2 on hydrotalcite-derived mixed oxides:

Sorption mechanisms and consequences for adsorption irreversibility. *Industrial and Engineering Chemistry Research*, **49**, 3663-3671.

99. Yoshikawa, K., Sato, H., Kaneeda, M., and Kondo, J. N. (2014) Synthesis and analysis of CO_2 adsorbents based on cerium oxide. *Journal of CO_2 Utilization*, **8**, 34-38.
100. Wang, X. P., Yu, J. J., Cheng, J., Hao, Z. P. and Xu, Z. P. (2008) High-temperature adsorption of carbon dioxide on mixed oxides derived from hydrotalcite-like compounds. *Environmental Science and Technology*, **42**, 614-618.
101. Zhao, Z., Dai, H., Du, Y., Deng, J., Zhang, L., and Shi, F. (2011) Solvo- or hydrothermal fabrication and excellent carbon dioxide adsorption behaviors of magnesium oxides with multiple morphologies and porous structures. *Materials Chemistry and Physics*, **128**, 348-356.
102. Song, G., Zhu, X., Chen, R., Liao, Q., Ding, Y.-D., and Chen, L. (2016) An investigation of CO_2 adsorption kinetics on porous magnesium oxide. *Chemical Engineering Journal*, **283**, 175-183.
103. Kumar, S., Saxena, S. K., Drozd, V., and Durygin, A. (2015) An experimental investigation of mesoporous MgO as a potential pre-combustion CO_2 sorbent. *Materials for Renewable and Sustainable Energy*, **4**:8, DOI: 10.1007/s40243-015-0050-0.
104. Vu, A. T., Park, Y., Jeon, P. R., and Lee, C. H. (2014) Mesoporous MgO sorbent promoted with KNO_3 for CO_2 capture at intermediate temperatures. *Chemical Engineering Journal*, **258**, 254-264.
105. Ding, Y.-D., Song, G., Liao, Q., Zhu, X., and Chen, R. (2016) Bench scale study of CO_2 adsorption performance of MgO in the presence of water vapor. *Energy*, **112**, 101-110.
106. Abatzoglou, N., and Boivin, S. (2009) A review of biogas purification processes. *Biofuels, Bioproducts and Biorefining*, **3**, 42-71.
107. Carnes, C. L., Klabunde, K. J. (2002) Unique chemical reactivities of nanocrystalline metal oxides toward hydrogen sulfide. *Chemistry of Materials*, **14**, 1806-1811.
108. Rodriguez, J. A., and Maiti, A. (2000) Adsorption and decomposition of H_2S on MgO(100), NiMgO(100), and ZnO(0001) surfaces: A first-principles density functional study. *The Journal of Physical Chemistry B*, **104**, 3630-3638.
109. Davidson, J. M., Lawrie, C. H., and Sohail, K. (1995) Kinetics of the absorption of hydrogen sulfide by high purity and doped surface area zinc oxide. *Industrial and Engineering Chemistry Research*, **34**, 2981-2989.
110. Ozdemir, S., and Bardakci, T. (1999) Hydrogen sulfide removal from coal gas by zinc titanate sorbent. *Separation and Purification Technology*, **16**, 225-234.
111. Ayala, R. E., and Marsh, D. W. (1991) Characterization and long-

range reactivity of zinc ferrite in high-temperature desulfurization processes. *Industrial and Engineering Chemistry Research*, **30**, 55-60.

112. Lew, S., Jothimurugesan, K., and Stephanopoulos, M. F. (1992) The reaction of zinc titanate and zinc oxides solids. *Chemical Engineering Science*, **47**, 1421-1431.
113. Woods, M. C., Gangwal, S. K., Harrison, D. P., and Jothimurugesan, K. (1991) Kinetic of the reactions of a zinc ferrite sorbent in high-temperature coal gas desulfurization. *Industrial and Engineering Chemistry Research*, **30**, 100-107.
114. Akyurtlu, J. F., and Akyurtlu, A. (1995) Hot gas desulfurization with vanadium-promoted zinc ferrite sorbents. *Gas Separation and Purification*, **9**, 17-25.
115. Tamhankar, S. S., Bagajewicz, M., Gavalas, G. R., Sharma, P. K., and Flytzani-Stephanopoulos, M. (1986) Mixed-oxide sorbents for high- temperature removal of hydrogen sulfide. *Industrial and Engineering Chemistry Process Design and Development*, **25**, 429-437.
116. Flytzani-Stephanopoulos, M., Gavalas, G. R., and Tamhankar, S. S. (1998) High Temperature Regenerative H_2S Sorbents, US Patent 4729889.
117. Focht, G. D., Ranade, P. V., and Harrison, D. P. (1988) High-temperature desulfurization using zinc ferrite: Reduction and sulfidation kinetics. *Chemical Engineering Science*, **48**(11), 3005-3013.
118. Kobayashi, H., Shirai, M., and Nunokawa, M. (2002) High-temperature sulfidation behavior of reduced zinc ferrite in simulated coal gas revealed by in situ X-ray diffraction analysis and Mossbauer spectroscopy. *Energy & Fuels*, **16**, 601-607.
119. Huang, W.-F., Chen, H.-T., and Lin, M. C. (2009) Density functional theory study of the adsorption and reaction of H_2S on TiO_2 Rutile (110) and anatase (101) surfaces. *The Journal of Physical Chemistry C*, **113**, 20411-20420.
120. Novochinskii, I. I., Song, C., Ma, X., Liu, X., Shore, L., Lampert, J., and Farruato, R. J. (2004) Low-temperature H_2S removal from steam-containing gas mixtures with ZnO for fuel cell application. 1. ZnO particles and extrudates. *Energy & Fuels*, **18**, 576-583.
121. Bagreev, A., Rahman, H., and Bandosz, T. J. (2001) Thermal regeneration of activated carbon previously used as hydrogen sulfide adsorbent. *Carbon*, **39**, 1319-1326.
122. Polychronopoulou, K., Fierro, J. L. G., and Efstathiou, A. M. (2004) Novel Zn-Ti-based mixed metal oxides for low-temperature adsorption of H_2S from industrial gas streams. *Applied Catalysis B: Environmental*, **57**, 125-137.
123. Baird, T., Campbell, K. C., Holliman, P. J., Hoyle, R., Noble, G., Stirling, D., and Williams, B. P. (2003) Mixed cobalt–iron oxide absor-

bents for low-temperature gas desulfurization. *Journal of Materials Chemistry*, **13**, 2341-2347.

124. Xue, M., Chitrakar, R., Sakane, K., and Ooi, K. (2003) Screening of adsorbents for removal of H_2S at room temperature. *Green Chemistry*, **5**, 529-534.
125. Bems, B., Schur, M., Dassenoy, A., Junkes, H., Herein, D., and Schlogl, R. (2003) Relations between synthesis and microstructural properties of copper/zinc hydroxycarbonates. *Chemistry - A European Journal*, **9**, 2039-2052.
126. Karvan, O., Sirkecioglu, A., and Atakul, H. (2009) Investigation of nano-CuO/mesoporous SiO_2 materials as hot gas desulphurization sorbents. *Fuel Processing Technology*, **90**, 1452-1458.
127. Bae, J. W., Kang, S. H., Dhar, G. M., and Jun, K. W. (2009) Effect of Al_2O_3 content on the adsorptive properties of $Cu/ZnO/Al_2O_3$ for removal of odorant sulfur compounds. *International Journal of Hydrogen Energy*, **34**, 8733-8740.
128. Jiang, D., Su, L., Ma, L., Yao, N., Xu, X., Tang, H., and Li, X. (2010) Cu–Zn–Al mixed metal oxides derived from hydroxycarbonate precursors for H_2S removal at low temperature. *Applied Surface Science*, **256**, 3216-3223.
129. Polychronopoulou, K., Galisteo, F. C., Granados, M. L., Fierro, J. L. G., Bakas, T., and Efstathiou, A. M. (2005) Novel Fe–Mn–Zn–Ti–O mixed-metal oxides for the low-temperature removal of H2S from gas streams in the presence of H_2, CO_2, and H_2O. *Journal of Catalysis*, **236**, 205-220.
130. Lonkar, S. P., Pillai, V., Abdala, A., and Mittal, V. (2016) In situ formed graphene/ZnO nanostructured composites for low temperature hydrogen sulfide removal from natural gas. *RSC Advances*, **6**, 81142-81150.
131. Oliveira, L. C. A., Rios, R. V. R. A., Fabris, J. D., Garg, V., Sapag, K., and Lago, R. M. (2002) Activated carbon/ iron oxide magnetic composites for the adsorption of contaminants in water. *Carbon*, **40**, 2177-2183.
132. Khosravi, M., and Azizian, S. (2014) Adsorption of anionic dyes from aqueous solution by iron oxide nanospheres. *Journal of Industrial and Engineering Chemistry*, **20**, 2561-2567.
133. Tchounwou, P. B., Yedjou, C. G., Patlolla, A. K., and Sutton, D. J. (2012) Heavy metals toxicity and the environment. *EXS*, **101**, 133-164.
134. Agrawal, A., and Sahu, K. K. (2006) Kinetic and isotherm studies of cadmium adsorption on manganese nodule residue. *Journal of Hazardous Materials*, **137**, 915-924.
135. Pradeep, T., and Anshup (2009) Noble metal nanoparticles for water purification: A critical review. *Thin Solid Films*, **517**, 6441-6478.

136. Warner, C. L., Chouyyok, W., Mackie, K. E., Neiner, D., Saraf, L. V., Droubay, T. C., Warner, M. G., and Addleman, R. S. (2012) Manganese doping of magnetic iron oxide nanoparticles: tailoring surface reactivity for a regenerable heavy metal sorbent. *Langmuir*, **28**(8), 3931-3937.
137. Feng, L., Cao, M., Ma, X., Zhu, Y., and Hu, C. (2012) Superparamagnetic high-surface-area Fe_3O_4 nanoparticles as adsorbents for arsenic removal. *Journal of Hazardous Materials*, **217-218**, 439-446.
138. Hao, Y.-M., Man, C., and Hu, Z.-B. (2010) Effective removal of Cu (II) ions from aqueous solution by amino-functionalized magnetic nanoparticles. *Journal of Hazardous Materials*, **184**(1-3), 392-399.
139. Nassar, N. N. (2010) Rapid removal and recovery of Pb(II) from wastewater by magnetic nanoadsorbents. *Journal of Hazardous Materials*, **184**(1-3), 538-546.
140. Wang, H. Q., Yang, G. F., Li, Q. Y., Zhong, X. X., Wang, F. P., Li, Z. S., and Li, Y. H. (2011) Porous nano- MnO_2: large scale synthesis via a facile quick-redox procedure and application in a supercapacitor. *New Journal of Chemistry*, **35**, 469-475.
141. Li, J. D., Shi, Y. L., Cai, Y. Q., Mou, S. F., and Jiang, G. B. (2008) Adsorption of di-ethyl-phthalate from aqueous solutions with surfactant-coated nano/microsized alumina. *Chemical Engineering Journal*, **140**, 214-220.
142. Engates, K. E., and Shipley, H. J. (2011) Adsorption of Pb, Cd, Cu, Zn, and Ni to titanium diox-ide nanoparticles: effect of particle size, solid concentration, and exhaustion. *Environmental Science and Pollution Research*, **18**, 386-395.
143. Liang, P., Shi, T. Q., and Li, J. (2004) Nanometer-size titanium dioxide separation/preconcentration and FAAS determination of trace Zn and Cd in water sample. *International Journal of Environmental Analytical Chemistry*, **84**, 315-321.
144. Wang, X. B., Cai, W. P., Lin, Y. X., Wang, G. Z., and Liang, C. H. (2010) Mass production of micro/nanostructured porous ZnO plates and their strong structurally enhanced and selective adsorption performance for environmental remediation. *Journal of Materials Chemistry*, **20**, 8582-8590.
145. Lee, J. H., Kim, B. S., Lee, J. C., and Park, S. (2005) Removal of Cu++ ions from aqueous Cu-EDTA solution using ZnO nanopowder. In: *Eco-Materials Processing & Design VI*, Kim, H. S., Park, S.-Y., Hur, B. Y., and Lee, S. W. (eds.), Trans Tech Publications, Korea, pp. 510-513.

3

A Novel Method for the Synthesis of Aerogels using Vacuum Drying

3.1 Introduction

The very first aerogel was produced by Kistler [1] in 1931. This startling discovery in materials science led to the development of a multitude of organic materials like nitrocellulose, gelatin or rubber, which can all be obtained by replacing the lyogel using a supercritical fluid. In spite of the interesting properties of the silicon based aerogels, their industrial applications are very limited due to the technical difficulties faced during the drying procedures, as the supercritical drying of methanol is not very optimal and cost effective [2,3].

The application of the sol-gel chemistry to silica aerogel synthesis was achieved by late 1970s, based on the acid catalyzed polymerization of alkoxysilanes. The immense progress in the field of organic polymer chemistry in the last few decades has prompted renewed interest in the aerogel research and application [4]. Thus, organic aerogels have been obtained from a multitude of organic precursors, such as resorcinol/formaldehyde, phenol/furfural and polyurethanes, by sol-gel processes and subsequent supercritical drying [4-7]. Organic aerogels can be further converted into carbon aerogels by subsequent pyrolysis at 600-2000 °C. The exceptionally multifaceted properties of all three classes of aerogels, viz., silica, organic and carbon aerogels, include high nanoscale porosity, large specific surface area, extremely low density, sound propagation, thermal expansion, heat transmission, large infrared optical adsorption coefficient and good electrical conductivity. Thus, these new materials are immensely attractive for many industrial applications [8-13].

In this chapter, novel aerogels were prepared using nanofibrillated cellulose (NFC). N-Methylmorpholine-N-oxide (NMMO) was used as the solvent to achieve complete dissolution of cellulose, and vac-

Hariharan Krithivasan and Vikas Mittal, Khalifa University of Science and Technology, Abu Dhabi, UAE*

**Current address: University of Waterloo, Waterloo, Canada* uum drying technique was used to obtain the final aerogel product. Specific properties of these aerogels were subsequently studied, which include density, porosity and pore size characteristics, along with mechanical properties. The prime reason for the preparation these aerogels was to optimize their characteristics in order to enable their coating with graphene. Afterwards, these aerogels can be converted into hydrophobic materials for effective oil spill clean-up applications.

3.2 Materials and Synthesis

Cellulose was obtained from Sigma-Aldrich, in the form of nanofibrillated cellulose (NFC). The solvent used in the preparation of the aerocellulose was N-methylmorpholine-N-oxide (50 wt% in H_2O) and was procured from Sigma-Aldrich. Poly(propyl gallate) (PGA) was used as the stabilizer. Absolute analytical grade ethanol was used for the regeneration and solvent exchange procedure.

Solution with 7.5 wt% cellulose in NMMO-H_2O was prepared by dissolving the NFC in the solution, along with 0.05 g of PGA stabilizer (Table 3.1). The solution was stirred at room temperature for 60 min. Subsequently, the cellulose/NMMO solution was charged into a rotary evaporator. The rotary evaporator was used to control the temperature and pressure while maintaining constant stirring. The temperature was increased to 60 °C, and the solution was stirred for 2 h to achieve complete dissolution. Afterwards, the pressure was reduced inside the rotation chamber, thus, resulting in the evaporation of the excess water in the solution, thereby leaving behind NMMO and cellulose mixture. The same procedure was followed for different concentrations of cellulose and NMMO. After removal of the excess water, the solution was poured in a casting mold for drying at room temperature for 72 h.

Table 3.1 Concentrations of materials for sample preparation

Material	Concentration (wt%)
NMMO (50 wt% in H_2O)	70-75
Cellulose (NFC)	7.5-12.5
Distilled water	10-20
PGA (stabilizer)	0.05 (g)

The regeneration was performed by transferring the solid cellulose/NMMO mold in analytical grade ethanol baths. The first step of regeneration was performed for 4-6 h. Afterwards, the second and third step of solvent exchange was performed by transferring the mold into fresh ethanol baths until NMMO was completely replaced by ethanol. The whole procedure was performed under room temperature and atmospheric pressure. The final stage of the aerogel preparation to replace ethanol by air was achieved by using vacuum dryer. The sample (cellulose/ethanol mold) was placed in the vacuum dryer and high vacuum was obtained (50 Psi). At these conditions, ethanol evaporated slowly without disturbing the cellulose structure. This formed the delicate cellulose aerogel structure, with ethanol totally replaced by air, thus, leaving pores in the structure of the final product. After 24 h, the sample was removed from the vacuum dryer, and the final aerogel was obtained. Figure 3.1 shows the three stages of aerogel preparation, whereas Figure 3.2 demonstrates the steps needed for the preparation of the aerogels.

Figure 3.1 Three stages of aerogel preparation.

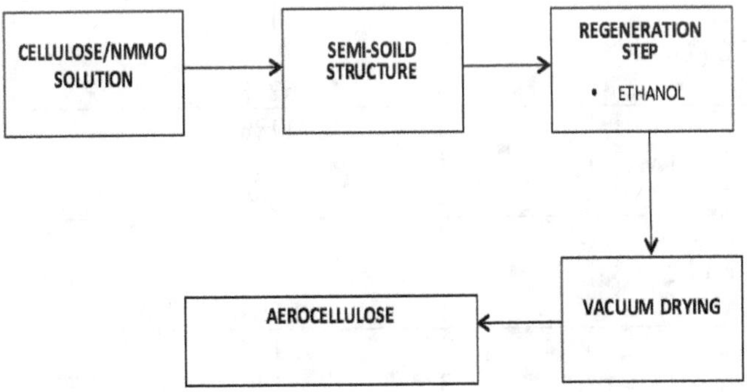

Figure 3.2 Aerogel preparation steps using vacuum dryer.

3.3 Results and Discussion

3.3.1 Density

The basic property to be characterized for the aerogels is their density. The density generally varies depending on various factors, e.g. the concentration (in wt%) of cellulose is the major contributor to the density of an aerogel. The number of regeneration baths used before the drying process also influence the density of the aerogels (Table 3.2). With the increase in the number of regeneration baths,

Table 3.2 Density and shrinkage measurement of aerogels at different cellulose concentrations

Cellulose Concentration	Number of Baths	Shrinkage (%)	Density (kg/m^3)
NFC (7.5 wt%)	2	32 %	153
	3	40%	
	4	52%	
NFC (8 wt%)	2	33%	168
	3	44%	
	4	57%	
NFC (8.5 wt%)	2	35%	177
	3	41%	
	4	52%	
NFC (9 wt%)	2	33%	192
	3	49%	
	4	58%	
NFC (9.5 wt%)	2	35%	197
	3	45%	
	4	59%	
NFC (10 wt%)	2	41%	210
	3	49%	
	4	62%	
NFC (11 wt%)	2	41%	223
	3	52%	
	4	68%	
NFC (12 wt%)	2	48%	241
	3	55%	
	4	62%	
NFC (12.5 wt%)	2	51%	251
	3	61%	
	4	69%	

the shrinkage factor of the aerogels increases. This leads to a much denser aerogel structure, as cellulose is packed in a tighter structure. Also, the drying time seems to have an impact on the density of the aerogels, with longer drying time leading to a reduction in the density of the aerogels after threshold time. Before threshold time, some ethanol is still left in the aerogel structure.

Overall, it can be seen from Table 3.2 that an increase in the concentration of cellulose enhanced the density of the aerogels. Also, increasing the number of regeneration baths resulted in enhanced shrinkage.

3.3.2 Thermal Conductivity

The thermal conductivity of the monolithic aerogels can be simply described as the addition of the solid thermal conductivity, gaseous thermal conductivity and radiative thermal conductivity. For this, the measurements were performed using a hot wire device. The platinum wire was squeezed between a pair of identical aerogels. The effective embedding of the wires was possible due to the elasticity of the aerogels. The specimens were enclosed inside a glass tube to reduce the noise from the environment. The platinum wire served as the heating element. The temperature increase in the aerogel during a heat pulse was recorded. It was clear that the aerogel density affected their thermal conductivity (Table 3.3). Cellulose is a poor conductor of heat, so the major heat conductor was the gas available

Table 3.3 Measurement of thermal conductivity for different cellulose concentrations

Cellulose Concentration	Thermal Conductivity $Wm^{-1}K^{-1}$
NFC (7.5 wt%)	0.05209
NFC (8 wt%)	0.05202
NFC (8.5 wt%)	0.05134
NFC (9 wt%)	0.05120
NFC (9.5 wt%)	0.05114
NFC (10 wt%)	0.05115
NFC (10.5 wt%)	0.05091
NFC (11 wt%)	0.04997
NFC (11.5 wt%)	0.04988
NFC (12 wt%)	0.04984
NFC (12.5 wt%)	0.04977

in the pores. When there is an increase in the density, the gas present inside the aerogels reduces, thus, leading to a reduction in the thermal conductivity.

3.3.3 Mechanical Properties

The stress-strain behaviors of the aerogels were studied by applying compressive stress to further understand the mechanical characteristics. The samples were compressed from 15 mm to 8 mm in thickness without any decrease in the mechanical load, as usually practiced for the stress tests in foamed plastics [14]. The compressive stress-strain curves were studied to evaluate the Young's modulus, compressive yield stress and yield strain (Table 3.4). From the data,

Table 3.4 Mechanical properties of cellulose based aerogels

Cellulose Concentration	E (Nmm^{-2})	$\sigma_y(Nmm^{-2})$	$\varepsilon_y(\%)$
NFC (7.5 wt%)	12.8	0.16	4.14
NFC (8 wt%)	13.7	0.18	4.11
NFC (8.5 wt%)	15.1	0.19	4.07
NFC (9 wt%)	16.3	0.22	4.09
NFC (9.5 wt%)	65.8	0.91	2.87
NFC (10 wt%)	66.7	0.96	2.81
NFC (10.5 wt%)	68.3	0.99	2.79
NFC (11 wt%)	68.9	1.02	2.75
NFC (11.5 wt%)	69.7	1.07	2.73
NFC (12 wt%)	70.4	1.09	2.77
NFC (12.5 wt%)	70.6	1.13	2.70

it can be seen that the initial cellulose concentration affected the mechanical properties of the aerogel significantly. The mechanical properties were poor until the cellulose concentration reached the threshold value (9.5 wt%). After this point, the mechanical properties of the aerogels improved significantly. The observed behavior resulted due to the relation between the pore characteristics and mechanical behavior. Before the threshold point, the pore structure collapsed, and the aerogel was damaged. After the threshold value was obtained, the aerogel did not collapse under compressive stress. Figure 3.3 also shows the scanning electron microscopy (SEM) image of the aerogel after tensile test indicating collapsed aerogel structure.

Figure 3.3 SEM image of the aerogel after tensile test.

3.3.4 Pore Characteristics

The pore characteristics of the aerogels were analyzed using nitrogen adsorption test, performed under constant temperature. The surface area, pore diameter and pore volume were obtained from the nitrogen adsorption test (Table 3.5). The isotherms for all aerogel samples were of type IV, thus, the material could be classified as

Table 3.5 Pore characteristics of cellulose based aerogels

Cellulose Concentration	BET Surface Area ($m^2 g^{-1}$)	Pore Diameter (nm)	Pore Volume ($cm^3 g^{-1}$)
NFC (7.5 wt%)	225	21.2	0.84
NFC (8 wt%)	230	22.6	0.87
NFC (8.5 wt%)	240	23.3	0.91
NFC (9 wt%)	245	25.2	0.93
NFC (9.5 wt%)	260	28.9	0.97
NFC (10 wt%)	270	30.8	0.96
NFC (10.5 wt%)	275	32.5	0.99
NFC (11 wt%)	280	34.1	1.03
NFC (11.5 wt%)	280	35.3	1.02
NFC (12 wt%)	290	36.4	1.06
NFC (12.5 wt%)	290	37.1	1.09

mesoporous [15]. SEM images also confirmed that the structure of the cellulose based aerogels majorly consisted of mesopores (Figure 3.4). The average mesopore diameter of the materials was observed in the range of 21 to 37 nm.

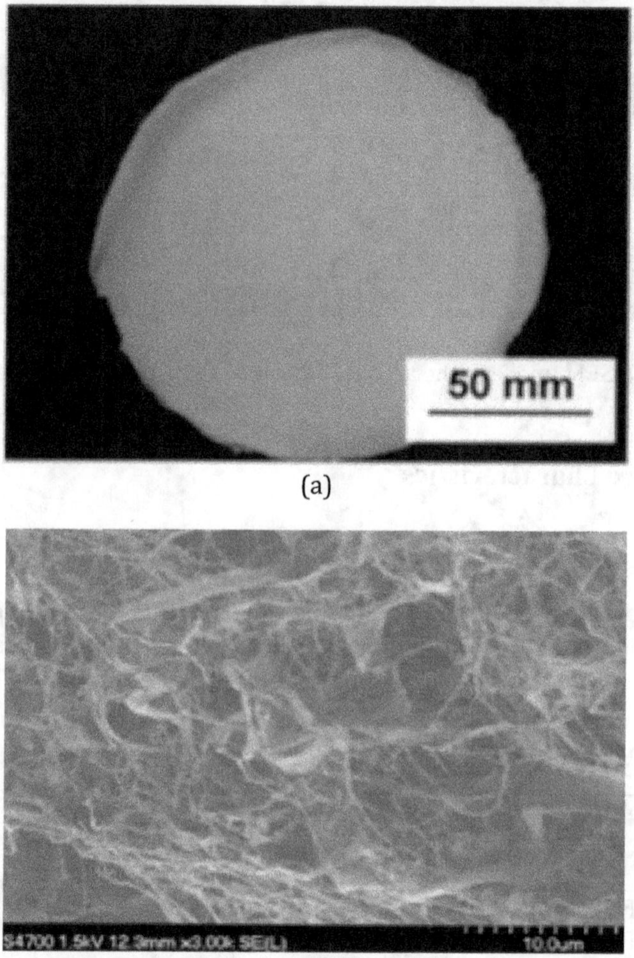

(a)

(b)

Figure 3.4 SEM image of the (a) aerogel and (b) pores in the aerogel.

3.4 Conclusion

Cellulose based aerogels were successfully synthesized using the vacuum drying process by employing different concentrations of nanofibrillated cellulose. The density of the aerogels ranged from 153 to 251 kg/m^3, whereas the pore diameter was between 21 to 37 nm. The average thermal conductivity of the samples was observed to be 0.05 $Wm^{-1}K^{-1}$, and the porosity percentage ranged from 75 to 85%. The observed results confirmed that the vacuum drying can be effectively used to prepare aerogels, thus, making the process much simpler and cost-effective as compared the conventional techniques requiring the use of supercritical dryers. Further, these aerogels are intended to be coated with functionalized graphene to improve the mechanical properties and make the material hydrophobic for oil spill clean-up applications.

References

1. Kistler, S. S. (1931) Coherent expanded aerogels and jellies. *Nature*, **127**, 741.
2. Chen, W., Yu, H., Li, Q., Liu, Y., and Li, J. (2011) Ultralight and highly flexible aerogels with long cellulose in nanofibers. *Soft Matter*, **7**(21), 10360-10368.
3. Chen, N., and Pan, Q. (2013) Versatile fabrication of ultralight magnetic foams and application for oil–water separation. *ACS Nano*, **7**(8), 6875-6883.
4. Pekala, R. (1989) Organic aerogels from the polycondensation of resorcinol with formaldehyde. *Journal of Materials Science*, **24**(9), 3221-3227.
5. Phalippou, J. Woignier, T. and Prassas, M. (1990) Glasses from aerogels. *Journal of Materials Science*, **25**(7), 3111-3117.
6. Biesmans, G., Mertens, A., Duffours, L., Woignier, T., and Phalippou, J. (1998) Polyurethane based organic aerogels and their transformation into carbon aerogels. *Journal of Non-Crystalline Solids*, **225**, 64- 68.
7. Reynolds, G., Fung, A., Wang, Z., Dresselhaus, M., and Pekala, R. (1995) The effects of external conditions on the internal structure of carbon aerogels. *Journal of Non-Crystalline Solids*, **188**(1), 27-33.
8. Husing, N., and Schubert, U. (2006) Aerogels. In: *Ullmann's Encyclopedia of Industrial Chemistry*, Wiley-VCH, Germany, doi:10.1002/14356007.c01_c01.
9. Bag, S., Arachchige, I. U., and Kanatzidis, M. G. (2008) Aerogels from metal chalcogenides and their emerging unique properties. *Journal of Materials Chemistry*, **18**(31), 3628-3632.
10. Farmer, J. C., Fix, D. V., Mack, G. V., Pekala, R. W., and Poco, J. F. (1996) Capacitive deionization of NaCl and $NaNO_3$ solutions with carbon aerogel electrodes. *Journal of the Electrochemical Society*,

143(1), 159-169.

11. Pierre, A. C., and Pajonk, G. M. (2002) Chemistry of aerogels and their applications. *Chemical Reviews*, **102**(11), 4243-4266.
12. Akimov, Y. K. (2003) Fields of application of aerogels (review). *Instruments and Experimental Techniques*, **46**(3), 287-299.
13. Emmerling, A., and Fricke, J. (1992) Small angle scattering and the structure of aerogels. *Journal of Non-Crystalline Solids*, **145**, 113-120.
14. Liebner, F., Haimer, E., Potthast, A., Loidl, D., Tschegg, S., Neouze, M.-A., Wendland, M., and Rosenau, T. (2009) Cellulosic aerogels as ultra-lightweight materials. Part 2: Synthesis and properties. *Holzforschung*, **63**(1), 3-11.
15. Rouquerol, F., Rouquerol, J., and Sing, K. (1999) *Adsorption by Powders and Porous Solids*, Academic Press, USA.

4

Polymer Nanocomposites for Wastewater Treatment

4.1 Introduction

Water is a fundamental component of life and a valuable natural resource. It performs a critical function in earth's ecosystem and is crucial for the survival of living beings on earth. Nevertheless, the quick industrialization is constantly degrading the water quality because of the inclusion of large quantities of pollutants into the water bodies [1,2]. With time, the pollutants present in water have emerged as threats to the complete biosphere, thus, the removal of these pollutants has become necessary. Almost seven hundred inorganic and organic micro-pollutants in water have been reported. A few of these micro-pollutants are profoundly toxic as well as carcinogenic, whereas certain pollutants have lengthy residence times in the surroundings and are not biotransformable or biodegradable. Some of the harmful organic pollutants include plasticizers, polybrominated diphenyl ethers (PBDEs), polynuclear aromatic hydrocarbons (PAHs), pesticides, drug residues, polychlorinated biphenyls (PCBs) and phenols, as well as highly toxic metals such as arsenic (As), chromium (Cr), lead (Pb), mercury (Hg), cadmium (Cd), etc.

In the recent past, different strategies for detecting/removing the heavy metals as well as dyes from water were broadly examined. The essential requirements for the purification materials are porosity, separation capacity, reusability and low cost [3,4]. Adsorption is considered as the most prevalent treatment procedure for separating the contaminants from an aqueous solution because of its high efficiency, simple recovery as well as reusability of the adsorbent [5,6]. The advancement of unique nanoparticles with improved adsorption ability, affinity and selectivity for heavy metals as well as alternate contaminants has resulted in immense research attention for the application of these materials [7]. Such nanoparticles with significantly high surface area can provide a great deal of help in the purification of water, however, their agglomeration limits the appli-

Haleema Saleem and Vikas Mittal, Khalifa University of Science and Technology, Abu Dhabi, UAE

cation [8]. The agglomeration tendency can generally be overcome by using these nanomaterials to generate the functional nanocomposite materials.

Nanocomposites are multi-phasic materials, where one of the phases possesses dimensions in the nanometer range (10-100 nm). Presently, the nanocomposite materials are being developed as appropriate substitutes for various engineering materials. These materials are stated to be the 21st century materials and can be categorized as non-polymer based and polymer based nanocomposites, as per their matrix. The nanocomposites possess superior stability, selectivity and adsorption ability, as compared to the nanoparticles. Accordingly, the nanocomposites find widespread applications in numerous areas. In this respect, the adsorption of different pollutants, for example dyes as well as heavy metal ions, from water with the assistance of nanocomposites, has generated remarkable attention. Out of the various nanocomposites, polymer nanocomposites have specifically attained distinguished status in the ongoing research and development efforts. Polymer nanocomposites possess several unique properties, such as activated functionalities, dimensional variability and film forming capability [9,10]. These composites are known for exceptionally tunable adsorption performance because of the presence of nanoparticles having high surface area. The advanced adsorption performance of the polymer nanocomposites enables the use of these materials for various applications such as water purification, chemical sensors, fuel cell technology and drug delivery.

Many techniques like melt mixing, solution casting, *in-situ* polymerization, phase separation, spinning, sol-gel, template synthesis and electro-chemical synthesis have been utilized for the generation of polymer nanocomposites. Nevertheless, the selection of appropriate preparation method is crucial for acquiring the functional properties. The prominent techniques utilized for the nanocomposite characterization include the microscopic techniques (transmission electron microscopy (TEM), atomic force microscopy (AFM) and scanning electron microscopy (SEM)), spectroscopic methods (Raman, Fourier-transform infrared spectroscopy (FT-IR)), thermal analysis (dynamic mechanical analysis (DMA), differential scanning calorimetry (DSC), differential thermal analysis (DTA), thermomechanical analysis (TMA) and thermo-gravimetric analysis (TGA)) and X-ray diffraction. The properties of the resultant nanocomposites rely upon the interactions between the matrix and dispersed phase, as well as content and dimensional features of the dispersed phase [11].

Numerous sulfides, metals and metal oxides have been employed as the catalysts for water as well as wastewater purification, both in the presence and absence of light [12]. Out of the different catalysts, zinc oxide (ZnO) and titanium dioxide (TiO_2) have attained considerable attention because of their stability, high photo-catalytic activity and low cost [13,14]. These materials have been utilized as photocatalysts for accelerating numerous surface redox reactions. On the other hand, out of the different polymer-based adsorbents, the ones based on conducting polymers such as polypyrrole (PPy), polyethelenamine (PEI), polyaniline (PANI) and their composites have acquired significant application in adsorbing different heavy metal ions, owing to simplicity in their preparation and operation, low-cost as well as mechanical and environmental stability. In this chapter, recent developments and prospects of polymer nanocomposites for wastewater treatment are presented. Four types of nanofillers, namely graphene, carbon nanotubes (CNTs), clay and metal oxides are analyzed for their effect on properties and applications of diverse polymers. Owing to the nanosize of these fillers, the generated polymer nanocomposites display enhanced characteristics, as compared to the pristine polymers or typical composites.

4.2 Different Polymer Nanocomposites and Application for Wastewater Treatment

Presently, the most widely examined nanocomposites for wastewater treatment primarily include the nanocomposites based on graphene, CNT, clay and metal oxides. The applications of these materials for wastewater treatment are reviewed here in detail. Overall, in consideration of the current speed of advancement and ever increasing applications, the polymer nanocomposites are very promising for water and wastewater treatment.

4.2.1 Graphene Based Polymer Nanocomposites

Graphene is a two-dimensional (2-D) single atom thick sheet material consisting of honeycomb rings generated by the sp2 bonded carbon atoms [15]. The extensive hypothetical specific surface area (SSA) of graphene (almost 2630 $m^2.g^{-1}$) and ample functional groups present in graphene oxide (GO) enable it as promising candidate for separating the organic pollutants as well as heavy metal ions [16]. Many research efforts have been undertaken for the modification of single graphene/GO sheets with different host molecules, for example polymers, metal nanoparticles and cylodextrin for generating hybrids which demonstrate superior properties and enhanced surface area, relative to the original graphite material. Beneficial characteristics are generated by the interdependent interactions between the graphene nanosheets and host molecules [17].

In this section, the utilization of graphene/polymer nanocomposites for the eradication of dyes (rhodamine B (RhB), methylene blue (MB), Congo red and Reactive Black 5), heavy metal ions (U(VI), Pb(II), Ni(II), Co(II) and Cr(VI)) and other pollutants like pesticides has been analyzed.

Removal of Dyes

The dyes are commonly utilized in paints and paper industries, tanneries, textile units and plastics processing. It is estimated that the quantity of dyes generated every year surpasses 700000 tons, with almost 10-15% being released into wastewater [18]. Nowadays, owing to massive industrialization, the organic dyes are observed to be regularly released to the wastewater without sufficient treatment. Indeed, even at lower levels, these dyes can generate various health hazards and environmental issues [18]. In the event that these dyes are released into rivers, a degradation in the quality of water takes place by discoloration of water, which subsequently impacts the photo-synthetic performance of the aquatic organisms by interrupting the solar light penetration. Thus, the development of robust separation strategies and materials to remove organic dyes from wastewater is critical to mitigate the harmful effects [19,20]. Specifically, large-scale utilization demands the prospective dye adsorbents to display a faster rate of dye removal and to be environmentally friendly. The adsorbents should also have the ability to be isolated from the aqueous media subsequent to the adsorption of waste dyes. GO sheets act as promising dye adsorbents due to their exclusive conjugated and 2-D framework, which shows exceptional adsorption ability with respect to different dye molecules through the π-π stacking interactions [21,22]. Furthermore, the negative charges in the GO sheets, resulting from different oxygen consisting functional groups (such as hydroxyl, carboxy and carbonyl groups), permit stronger electrostatic interactions with the cationic dye molecules [23,24]. The GO-based permeable materials have also been employed for the organic waste dye adsorption [25]. On the other hand, the GO-based hydrogels also represent an efficient solution for removing the dye-adsorbed materials from water, leading to its purification [26].

Guo *et al.* [27] reported the generation of GO based polyethylenimine (PEI) hydrogels as effective dye adsorbents. PEI was selected due to the presence of plentiful amine groups. The PEI/GO hydrogels were prepared by means of both hydrogen bonding as well as electrostatic interactions between GO sheets and amine-rich PEI. Both the XRD and SEM analyses demonstrated that the GO sheets were effectively crosslinked within the framework of PEI. For both RhB and MB dyes, the generated hydrogels exhibited removal within almost 4 hours following the pseudo-second-order model. The dye adsorption ability of hydrogels was attributed to the GO sheets, while PEI facilitated the gelation process of these sheets. Dye-adsorbed hydrogels could be advantageously isolated from an aqueous medium, confirming the promising application of the GO-based polymer hydrogels for the removal of organic dyes for wastewater treatment.

In polyaniline (PANI), π-electron conjugation framework can effectively exfoliate graphite as graphene sheets by the protonated aniline's intercalation polymerization [28]. In addition, PANI is beneficial for enhancing the adsorption characteristics of magnetic graphene-based adsorbents because of the inclusion of significant quantity of amine as well as imine groups [29]. Mu *et al.* [30] generated super-paramagnetic graphene/Fe_3O_4/PANI nanocomposites for dye removal, using an eco-friendly and simple technique. The graphite exfoliation as nanosheets and functionalization with Fe_3O_4 nanoparticles and PANI were concurrently accomplished by means of a one-pot reaction procedure consolidating aniline's intercalation polymerization and co-precipitation of the generated Fe^{2+} and the residual Fe^{3+}. The prepared graphene/PANI/Fe_3O_4 nanocomposites displayed superior adsorption ability for the Congo red dye, even in the presence of Brilliant green. The adsorption isotherms and adsorption kinetics were fitted with the Langmuir isotherm model and pseudo second-order kinetic model, respectively. Thus, the technique presented industrially feasible method to generate exceedingly proficient graphene based adsorbents on a large scale for the separation of dyes. Moreover, the developed technique could also be beneficial for exfoliating alternate 2-D materials, for example carbon nitride, boron nitride and MoS_2, for a variety of functional applications.

Chitosan is a multi-functional polymer which possesses primary as well as secondary hydroxyl groups, in addition to profoundly reactive amino groups. The polymer represents a useful support material for adsorption. Thus, various research studies have been focused on the chemical activation of chitosan to enhance its dye adsorption ability [31]. Indeed, magnetic chitosan has been developed as an advanced material for the removal of contaminants from the environment, in light of the fact that magnetic separation includes the straightforward use of an exterior magnetic field for extracting the adsorbents. Travlou *et al.* [32] prepared graphite oxide (GO)-chitosan nanocomposite for dye adsorption (Reactive Black 5). The nanocomposite possessed a significantly rougher surface, indicating that numerous minute magnetic chitosan particles segregated on the exterior of the GO layers. Additionally, the dye/adsorbent interactions were affirmed by FTIR spectroscopic analysis, by means of the peak shifts occurring after the adsorption of the dye molecules. The pH influenced the pollutant adsorption onto the material as well as its desorption from the material into an eluent, as shown in Figure 4.1. The adsorption assessment of the generated composite

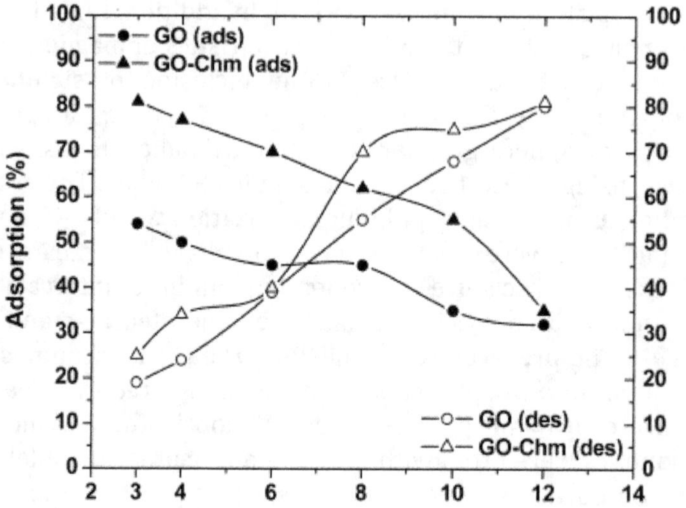

Figure 4.1 The impact of pH on adsorption and desorption. Reproduced from Reference 32 with permission from American Chemical Society.

illustrated that the basic conditions (pH 12) were optimal for the desorption process and acidic conditions were ideal for the adsorption procedure (pH 3).

Removal of Heavy Metal Ions

With the industrial advancement, wastewater consisting of heavy metals is often released into lakes, rivers, soils and other natural bodies. The heavy metals are exceedingly harmful and are not degradable in the natural surroundings. These might be aggregated by the living organisms and finally might penetrate into the food chain. There has been growing anxiety across the world regarding the release of heavy metals to the environment [33]. Out of the different techniques for heavy metal removal, adsorption is regarded to be eco-friendly as well as economical.

As mentioned earlier, PANI possesses numerous imine as well as amine functional groups, which could chelate the cationic metal ions, thus, indicating a remarkable affinity with the heavy metal ions. Hence, PANI could be utilized as a base material for removing the harmful heavy metals from the aqueous solutions. To achieve it effectively, *in-situ* aniline polymerization onto the exterior of different materials is regularly adopted, as bare PANI particles accumulate in the aqueous solutions, which brings about lesser adsorption ability as well as slower kinetics [34]. As a standout among the vital carbonaceous materials, the GO nanosheets comprise of numerous reactive oxygen containing functional groups and possess greater surface area which provide them a high possibility for supporting several functional nanomaterials [35,36]. Consequently, combining GO (good electron acceptor) with PANI (good electron donor) by means of electrostatic association can be an appropriate method to tackle the heavy metal contamination issue and has the potential to result in an optimal heavy metal adsorption ability for the adsorption from large aqueous volumes.

Liu *et al.* [29] prepared PANI/GO nanocomposites by dilute polymerization method. PANI molecules were observed to be effectively grafted on the surface of GO. GO based PANI nanocomposites displayed a greater surface area and a hierarchical framework. In addition, presence of PANI chains restricted the stacking of GO sheets and contributed additional functional groups. Accordingly, the GO based PANI nanocomposites were anticipated to possess additional adsorption sites for the heavy metal ion binding. The utilization of GO based PANI nanocomposites for the heavy metal adsorption from the aqueous solutions was examined under the ambient conditions. It was observed that the adsorption level of heavy metals enhanced with increment in the content of the adsorbent. With an enhancement in the adsorbent amount, more surface sites were accessible for binding the metal ions. Additionally, the solution pH performed a crucial role on the adsorption of heavy metal ions on the adsorbent. The maximum adsorption ability for various heavy metals on the GO/PANI nanocomposite was significantly greater than pure PANI. The maximum adsorption abilities for U(VI), Pb(II), Ni(II) and Co(II) ions on the GO/PANI composites obtained from the Langmuir models were 1552.31 $mg.g^{-1}$, 65.40 $mg.g^{-1}$, 25.67 $mg.g^{-1}$ and 22.28 $mg.g^{-1}$, respectively.

Cyclodextrins (CDs) are cyclic oligosaccharides consisting of eight, seven or six glucose units, which are conical shaped with a hydrophilic exterior and a hydrophobic internal cavity. CDs can precisely bind to different inorganic as well as organic pollutants for generating the inclusion complexes. Hence, incorporating polypyrrole and αCDs to GO for developing GO-αCD-PPy nanocomposites has been reported to be appropriate for separating heavy metals, specifically Cr(VI). GO-αCD-PPy nanocomposites have enhanced mechanical properties, high surface area, improved adsorption ability and good acid resistance. In a study carried out by Chauke *et al.* [17], the preparation of unique GO-αCD-PPy nanocomposites was stated and the material was employed for separating Cr(VI) from the aqueous solution. The developed material was efficient for separating Cr(VI) with fast kinetics and the Cr(VI) sorption rate was observed to adopt pseudo second order model. The highest adsorption abilities were 666.67 $mg.g^{-1}$, 625 $mg.g^{-1}$ and 606 $mg.g^{-1}$ at 45 °C, 35 °C and 25 °C, respectively, and the adsorption isotherms followed the Langmuir isotherm model. The presence of the co-existing ions exhibited no significant impact on the effectiveness of GO-αCD-PPy nanocomposites for the removal of Cr(VI). Additionally, the regeneration analysis confirmed that the adsorbent could be reused till 3 cycles.

Musico *et al.* [37] blended poly(N-vinylcarbazole) (PVK) with GO for generating GO-PVK nanocomposites possessing the removal capability of heavy metals from aqueous solutions. Uniform distribution of GO in the GO-PVK nanocomposites was demonstrated by the attenuated total reflectance – infrared spectroscopy (ATR-IR) and X-ray photoelectron microscopy (XPS) analyses. Results confirmed that the adsorption ability of GO-PVK nanocomposites for Pb^{2+} enhanced with an increment in the concentration of GO. It was attributed to the enhancement in the amount of oxygen-consisting functional groups accessible in the composite. Besides, the adsorption of Pb^{2+} onto the nanocomposite was affected by the variation in pH. It was noted that the higher pH resulted in a superior adsorption ability, due to the variation in the surface properties of GO-PVK nanocomposites. The highest adsorption ability of the GO based PVK nanocomposites for Pb^{2+} was observed to be 887.98 $mg.g^{-1}$ (Langmuir model), which was accomplished utilizing a 90:10 wt% ratio of GO:PVK with a contact time of 90 minutes at pH 7 ± 0.5.

Removal of Other Pollutants

Triazine pesticides are generally utilized for controlling weed in various crops. However, their residues have turned into a noteworthy source of environmental pollution. Conventional strategies for removing the triazine pesticide deposits in water are still commonly utilized. These methods have lesser efficiency and involve particle coagulation-flocculation, sedimentation as well as dual media filtration [38]. Developing a sorbent with an efficient separation will be a remarkable accomplishment for the pesticide removal process.

Zhang *et al.* [39] generated graphene based cellulose nanocomposites by combining the dissolved cellulose with GO and subsequently reducing it using hydrazine hydrate. Cellulose is a biodegradable, biocompatible and renewable polymer and is one of the most abundant natural polymer existing in nature. Utilizing cellulose is, thus, helpful for sustainable development of adsorbent materials [40]. The developed nanocomposites accomplished greater adsorption levels as compared to five alternate sorbents (graphene, graphite carbon black (GCB), graphite carbons, cellulose and primary secondary amine (PSA)) for six triazine pesticides. The adsorption required only 30 mg of cellulose-graphene composite for 10 mL of triazine pesticide solution. Equilibrium adsorption isotherm indicated that the adsorption process fitted with the Langmuir model. The thermodynamic parameters exhibited that the adsorption phenomenon was endothermic in nature, favorable as well as spontaneous. Besides, the cellulose-graphene composite was exceptionally stable and could be recycled effectively utilizing a simple organic solvent. In Figure 4.2, the results of six sequential adsorption and desorption cycles are presented. It was noted that the adsorption efficiency of the nanocomposite was still more than 85%, indicating it high potential of commercial use.

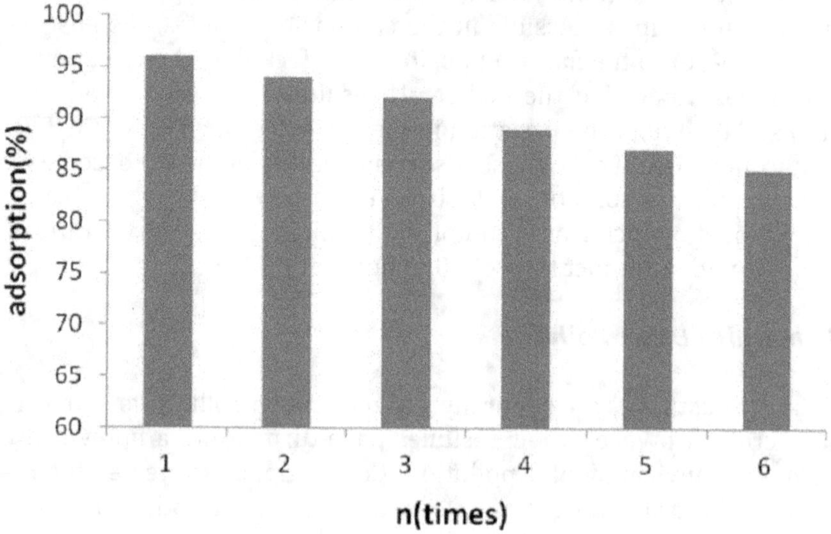

Figure 4.2 The regenerative ability of cellulose-graphene composite. Reproduced from Reference 39 with permission from American Chemical Society.

4.2.2 CNT Based Polymer Nanocomposites

CNTs exhibit extraordinary thermal, electrical and mechanical properties, which enable their use as extraordinary materials for an extensive range of specialized applications, including water purification [41,42]. Due to their superior structural features as well as high surface areas, CNTs have been utilized for the separation of heavy metal ions such as fluoride, Zn^{2+}, Pd^{2+}, Cu^{2+}, Cd^{2+} and Cr^{6+} as well as various organic matter in the environmental wastewater [43,44]. Compared to the conventional adsorbents such as activated carbon, several studies have confirmed that the CNTs demonstrated greater effectiveness for the separation of both dyes as well as heavy metal ions from wastewater [45,46]. In spite of the superior performance, the lower degree of dispersibility in aqueous solution poses an important restraint, thereby, resulting in lower adsorption ability. Hence, modification strategies to enhance their water dispersibility as well as adsorption ability are desirable [47]. In recent years, the modification of CNTs with polymers has attracted remarkable research interest [48,49]. Mostly, the generation of polymer modified CNTs and polymer/CNTs hybrids brings an enhancement in the adsorption as well as selectivity. In this section, the utilization of CNT/polymer nanocomposites has been summarized for the removal of dyes (methylene blue, methyl violet and malachite green) and heavy metal ions (nickel, copper, cadmium and zinc ions) from wastewater.

Removal of Dyes

Xie *et al.* [50] reported a unique procedure of consolidation of single-electron transfer living radical polymerization (SETLRP) and mussel motivated preparation for generating poly(sodium-p-styrene sulfonate) modified MWCNTs based effective adsorbents. The CNTs based polymer nanocomposites were analyzed for separating a cationic dye, methylene blue (MB), from water. The adsorption of the cationic dye MB was inversely proportional to temperature, whereas it was directly proportional to the contact time. The ultimate adsorption ability of CNT-polydopamine(PDA)-poly(sodium-p-styrenesulfonatehydrate)(PSPSH) composite for MB was noted to be 160 $mg.g^{-1}$. The technique depicted in the study for the generation of CNT based polymer nanocomposites was effective as well as simple, thus, confirming its potential for adoption as a universal approach for the generation of productive adsorbents for environmental applications.

It is recognized that the functional groups, for example carbonyl, amine, hydroxyl and thiol, have steady interactions with different synthetic dyes [51]. Thus, the water-soluble polymers consisting of appropriate functional groups are usually selected to modify the CNTs for enhancing the adsorption capacity as well as dispersion characteristics in aqueous solutions. Relative to the conventional linear polymers, the hyperbranched polymers (HPs) are more suitable for CNT modification, owing to their abundant surface functional groups and nearly spherical framework. Hu *et al.* [52] reported oxidized multi-walled carbon nanotubes (OMCNT) modified with water-soluble hyperbranched polyamine (WHPA) to generate an advanced nano-sorbent (OMCNT-WHPA) for effective removal of organic dyes. The prepared OMCNT-WHPA displayed superior water solubility as well as durable affinity towards the dyes due to the presence of plentiful amine, ether and hydroxyl groups. For example, OMCNT-WHPA exhibited equilibrium adsorption ability of 970.9 mg/g, 840.3 mg/g and 800.0 mg/g for methyl violet (MV), malachite green (MG) and methylene blue (MB) respectively, under optimal conditions. Langmuir model and pseudo-second order equation exhibited favorable correlation with the isotherm data and adsorption kinetics for the three pollutants. Thermodynamic results ($\Delta H < 0$, $\Delta S < 0$, $\Delta G < 0$) inferred that the adsorption process of MV, MG and MB was exothermic, feasible as well as instinctive in nature. An adsorption mechanism was also suggested, where the electrostatic attraction, hydrogen-bonding and π-π stacking interactions governed the organic dye adsorption. Likewise, the superior reproducibility confirmed the potential of OMCNT-WHPA for use in wastewater treatment.

Removal of Heavy Metals

One of the most commonly utilized polymers for CNTs modification is chitosan [53]. As mentioned earlier, chitosan is a polysaccharide biopolymer attained from chitin deacetylation and is noted to possess superior complexing capacity due to its amino group interactions with heavy metals present in wastewater [54,55]. Examples of some studies employing chitosan modified CNTs include Congo red adsorption [56] and the copper (II) determination by the anodic stripping voltammetry [57]. A nanocomposite of chitosan and MWCNTs (75:25 wt%) was also generated by Salam *et al.* [58]. Uniform distribution of MWCNTs in chitosan was observed. FTIR results demonstrated the presence of distinctive functional groups of both chitosan as well as MWCNTs inside the nanocomposite. MWCNT based chitosan nanocomposite was utilized for removing nickel, copper, cadmium and zinc ions from the aqueous solution. The results demonstrated that the MWCNT based chitosan nanocomposite could separate majority of the metal ions from the aqueous solution effectively. The results also confirmed the competition among the metal particles under examination. The removal of metal ions from the aqueous solution followed the order Cu(II) > Zn(II) ≈ Cd(II) > Ni(II).

4.2.3 Clay Based Polymer Nanocomposites

Widespread attention has been received by clays as well as their modified forms for utilization as metal ion adsorbents from aqueous medium due to the fact that these are abundantly available and substantially less expensive than membranes, zeolites and activated carbon. Clay/polymer nanocomposites treat the water by adsorption as well as flocculation of both organic and inorganic micro-pollutants from aqueous solutions. Additionally, when modified with biocides, certain clay/polymer nanocomposites effectively separate the microorganisms, for example, *Escherichia coli, Pseudomonas aeruginosa, Staphylococcus aureus* and *Candida albicans* from water. The clay based polymer nanocomposites are also much more effortlessly recovered from the aqueous media, relative to pristine clay. These nanocomposites additionally display effective treatment durations as compared to clay or polymer adsorbents. In addition, these composites possess better life cycles and greater adsorption ability relative to clay. Hence, these nanocomposites demonstrate high potential as effective water as well as wastewater treatment agents. In this section, the use of clay based polymer nanocomposites has been explored for the removal of inorganic pollutants (Pb(II), Zn(II), Cd(II) and Cu(II)), organic pollutants (trinitrophenol, trichlorophenol and phenol) and dyes from wastewater.

Removal of Inorganic Pollutants

Probably, the most intensely examined application of clay/polymer nanocomposites is the separation of inorganic micro-pollutants from aqueous solutions. In a study by Solener *et al.* [59], clay based polymethoxyethylacrylamide (PMEA) nanocomposite was generated to separate Pb(II) from the aqueous solution. The nanocomposite exhibited an excellent adsorption ability of 81.02 $mg.g^{-1}$. The Pb(II) adsorption on the clay/polymer nanocomposite proceeded by means of physisorption and was observed to be endothermic. The nanocomposite demonstrated no variation in the adsorption ability within 4.0-6.0 pH range. Indeed, even at lesser pH of around 3.0, the adsorbent displayed an optimal adsorption ability for Pb(II). In another study, Anirudhan and Ramachandran [60] generated organo-bentonite ammoxidated polyacrylonitrile (PAN) nanocomposite possessing about 6.3 pH_{PZC} and an adsorption ability of 52.61 $mg.g^{-1}$, 65.4 $mg.g^{-1}$ and 77.43 $mg.g^{-1}$ for the separation of Cd(II), Zn(II) and Cu(II), respectively, from aqueous medium. Nevertheless, it was observed that the presence of ions, such as Na(I), decreased the productivity of the polymer nanocomposite. In the treatment of crude wastewater effluents from electroplating as well as fertilizer industries, the organobentonite ammoxidated PAN nanocomposite accomplished almost 99% expulsion of the metal ions Cd(II), Zn(II) and Cu(II) in solution. Figure 4.3 demonstrates the quantity of

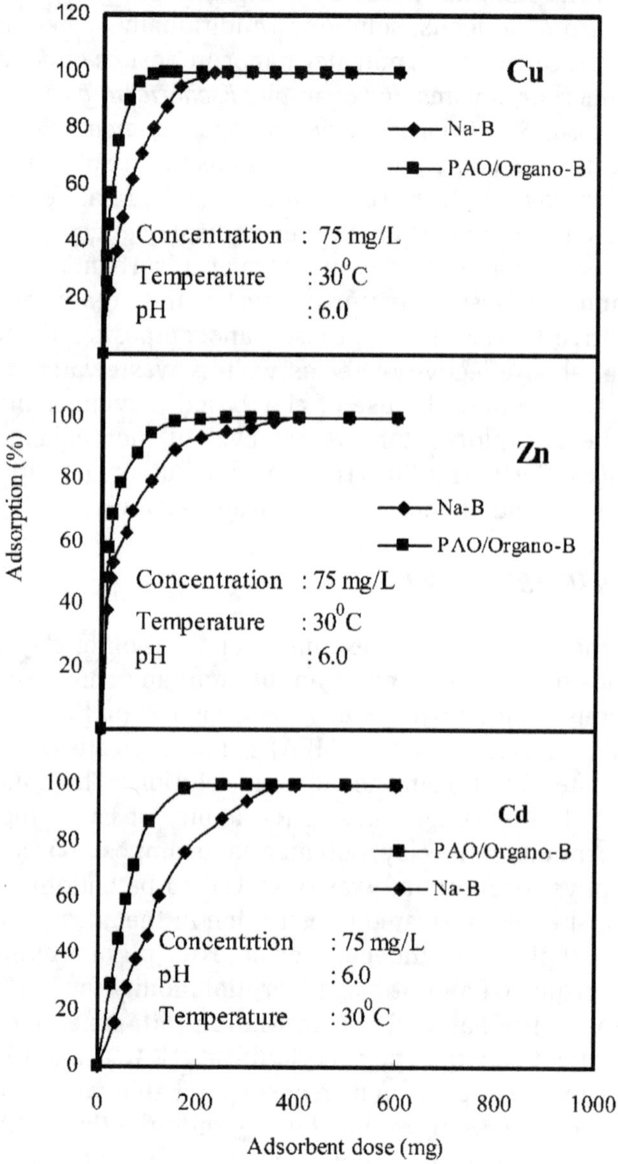

Figure 4.3 Impact of the adsorbent dose on the adsorption of Cu(II), Zn(II), and Cd(II). Reproduced from Reference 60 with permission from American Chemical Society.

Cd(II), Zn(II), and Cu(II) metal ions adsorbed as a function of adsorbent concentration, at pH of about 6.0. Primary concentration of metal ions was almost 75 mg.L^{-1}, and the adsorbent dosage changed from 25 mg to 600 mg and was equilibrated for approximately 4 hours. The measure of Na-saturated bentonite (Na-B) needed for the entire separation of Zn(II), Cd(II), and Cu(II) ions from their solutions was about 500, 400 and 225 mg, respectively. Nevertheless, the quantity of nanocomposite needed for the total separation of Zn(II), Cd(II) and Cu(II) ions from their solutions was 250, 175 and 125 mg, respectively. Hence, the adsorption ability of the nanocomposite was 1.8-2.3 times greater, relative to Na-B for the separation of Zn(II), Cd(II) and Cu(II) ions. Remarkable increment in the metal ion adsorption capacity was ascribed to the presence of amidoxime groups in the composite structure. It was suggested that an ion exchange followed by metal chelation with the amidoxime was the most advantageous separation mechanism for the metal ions using the developed composites.

The natural polymers have likewise been utilized for generating clay based polymer nanocomposites yielding effective adsorbents for separating heavy metal ion from aqueous solutions. Shawky [61] prepared montmorillonite (Mt) based alginate nanocomposite which provided an adsorption ability of 238.1 mg.g^{-1} for Pb(II) separation from aqueous solution. It was noted that the adsorption ability of the clay/polymer nanocomposites diminished marginally with enhanced solution temperature. The fixed bed analysis proposed that 100 mg.L^{-1} of Pb(II) was the utmost concentration the clay/polymer nanocomposites could effectively treat in aqueous solution. The raw wastewater treatment with the developed clay/polymer nanocomposite demonstrated that almost 100% of Zn(II), Al(III), Pb(II), Cr(III), Fe(III), Ni(II) and Mn(II) were removed. In another study, Bleiman and Mishael [62] generated montmorillonite reinforced chitosan (CS) nanocomposites possessing an adsorption ability of 18.4 mg.g^{-1} for removing selenium from well water. It was noted that the adsorption ability was greater relative to Fe-oxide as well as Al-oxide adsorbents. The variation in pH exhibited no critical impact on the adsorption ability of the Mt-CS nanocomposite for the removal of selenium. The clay/polymer nanocomposite was observed to remove selenium to levels lesser than the level specified by the WHO standard. In spite of the benefits of the Mt based CS nanocomposite, the separation of selenium from the aqueous solution in fixed bed frameworks was less effective due to the material's lesser hydraulic conductivity. Hence, the authors proposed the granulation of the clay/polymer nanocomposite in order to achieve effective performance.

Liu *et al.* [63] consolidated the benefits of functional polymers and clay nanomaterials to generate an inexpensive chemical adsorbent possessing superior mechanical characteristics for industrial treatment of water. The technique included preparing attapulgite based poly(acrylic acid) (ATP/PAA) nanocomposite hydrogels by means of "one-pot" inverse suspension polymerization using multi-functionalized attapulgite as crosslinker (Figure 4.4). Almost all acrylic acid monomer was observed to graft on the surface of attapulgite to generate a three-dimensional crosslinking hydrogel network. The attapulgite/poly(acrylic acid) nanocomposite hydrogels exhibited selective Pb(II) ion adsorption with a maximal adsorption ability of 42 $mg.g^{-1}$. The adsorbed Pb(II) ions could also be eluted totally. Additionally, the optimized attapulgite/poly(acrylic acid) nanocomposite hydrogel adsorbent exhibited a remarkable reusability as the desorption rate and adsorption ability stayed unaltered for 10 cycles of adsorption-desorption. The observed performance confirmed the potential of the developed material as a promising adsorbent for the heavy metal removal from polluted water.

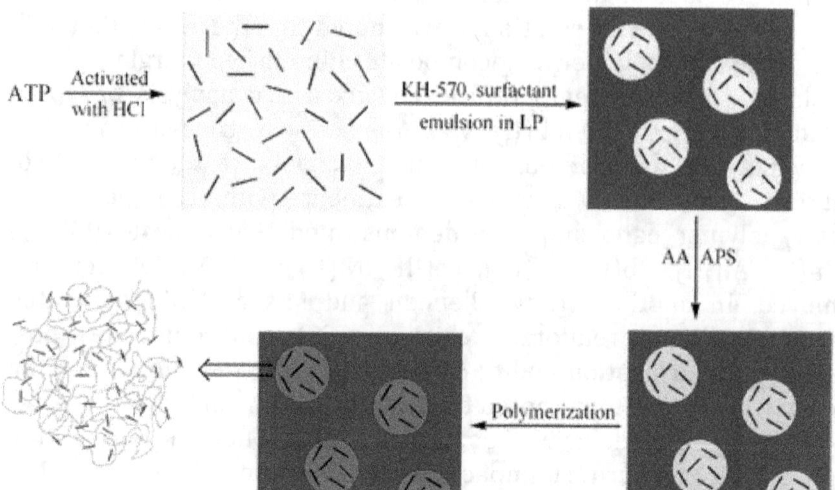

Figure 4.4 Schematic of the nanocomposite synthesis. Reproduced from Reference 63 with permission from American Chemical Society.

Removal of Organic Pollutants

In addition to the more classical separation of the inorganic pollutants, the clay/polymer nanocomposites have also been utilized for the organic micro-pollutant removal from water. In the adsorbent, the presence of an organic component (polymer chains) enhances the adsorbent capacity to separate the organic micro-pollutants from water due to improved hydrophobicity of the adsorbent. Ganigar *et al.* [64] utilized Mt-polydiallydiamethylammonium and Mt-poly-4-vinylpyridine-co-styrene nanocomposites for the separation of trinitrophenol (TNP) and trichlorophenol (TCP) from water. The Mt–poly-4-vinylpyridine-co-styrene nanocomposite adsorbent provided superior adsorption effectiveness for the contaminants. For this composite, the TNP separation effectiveness was almost 99.5%, though the TCP separation effectiveness was between 40% to 60% of the quantity existing in the solution. From the adsorption isotherms of the contaminants, it was proposed that the hydrophobic synergy was the fundamental means for the adsorption of the micro-pollutants onto Mt–poly-4-vinylpyridine-co-styrene nanocomposite, whereas for the TNP, the electrostatic interactions also took place. Also, it was found that the hydration properties of TNP enhanced its binding capacity with wet composites. On the other hand, hydrophobic TCP was adsorbed by the dehydrated nanocomposites.

In another study, Wang *et al.* [65] demonstrated that the bentonite modification using cetyltrimethyl ammonium bromide along with intercalation by polyacrylamide generated a clay based polymer nanocomposite, whose effectiveness for nitrobenzene as well as phenol adsorption from aqueous solution was increased by 28% and 26%, respectively. Further, Zhu *et al.* [66] generated a clay/polymer nanocomposite utilizing cetyl-trimethylammonium and polydiallyldimethylammonium modified montmorillonite. The clay/polymer nanohybrid exhibited an enhanced proficiency for removing phenol from water, relative to the Mt modified using surfactant alone. Besides, the authors demonstrated that utilizing a cationic polymer for managing the cationic surfactant arrangement in the interlayer spacing of Mt could be advantageous to enhance the adsorption affinity of the organo-Mt. The authors opined that the polydiallyldimethylammonium intercalation might have generated a variation in the cetylammonium micelle arrangement that enhanced the affinity of the clay/polymer nanohybrid for phenol, thus, exhibiting optimal performance.

Removal of Dyes

As mentioned before, cellulose is an inexpensive, biodegradable and renewable natural polymer, broadly utilized to develop hydrogels [67]. By the consolidation of the beneficial features of cellulose and clay, it is possible to develop superabsorbent cellulose/clay hydrogels with high specific area frameworks plugged by a considerable measure of water. Peng *et al.* [68] reported clay/cellulose hydrogels having excellent mechanical performance, superabsorbent characteristics and superior dye removal effectiveness. The principle methodology for the generation of superabsorbent hydrogels was the chemical crosslinking of carboxy-methyl cellulose (CMC), cellulose and intercalated clay in aqueous solution of NaOH/urea. The generated hydrogels exhibited high absorption ability with respect to methylene blue solution, by means of impulsive physisorption that fitted well with the Langmuir isotherm and pseudo second order models. Ultimate removal effectiveness of hydrogel specimens for methylene blue solution with 100 mg/L and 10 mg/L concentration was 98% and 96.6%, respectively. For understanding the interaction mechanism between the dye and hydrogels, the FT-IR spectra of specimens are presented in Figure 4.5. The spectra demonstrated

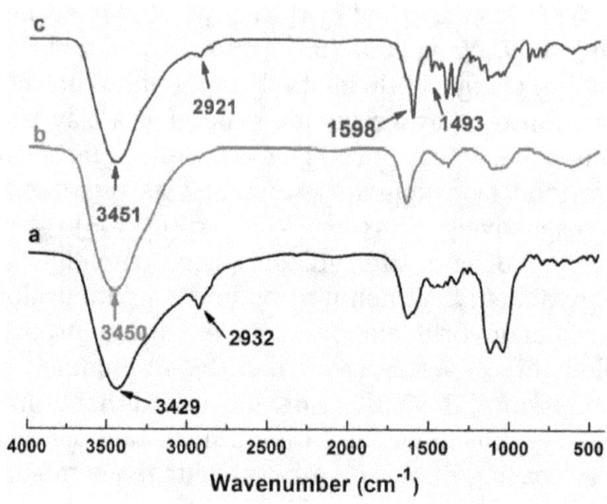

Figure 4.5 FT-IR spectra of clay based cellulose nanocomposite hydrogels before (a) and after methylene blue dye absorption (b), and methylene blue dye (c). Reproduced from Reference 68 with permission from American Chemical Society.

that the presence of a strong hydrophobic interaction between methylene blue and hydrogel network restricted the motility of methyl groups of methylene blue and methylene groups of hydrogel.

For Coagulation and Flocculation

Agricultural effluents (e.g., from wineries, coffee bean or soy industry) consist of a significant amount of dispersed and organic matter. It is generally difficult to effectively remove these dispersed solids from water by conventional means due to the fact that these are primarily present in the colloidal state, which makes the filtration process challenging due to clogging of the filters [69]. Certain pretreatment procedures have been effectively utilized for the reduction or complete removal of suspended solids in water, which require three stages: coagulation, flocculation and separation by means of filtration or decanting.

Utilizing sepiolite-polydiallyldimethylammoniumchloride nanocomposite, Rytwo [70] noted that a winery effluent having total suspended solids of about 1610 $mg.L^{-1}$ was totally treated in less than 2 minutes, as compared to approximately 24 hours for alum and clay treatments. In another study by Rytwo *et al.* [71], sepiolite-polydiallyldimethylammoniumchloride based nanocomposite, sepiolite-chitosan and volclay-polydiallyldimethylammoniumchloride nanocomposite adsorbents were used for effectively treating the wastewater from winery between 15 and 60 min. The developed materials, thus, signify the potential to completely avoid the vast sedimentation tanks usually required in the typical flocculation operations.

4.2.4 Metal Oxide Based Polymer Nanocomposites

Inorganic metal oxide based materials have received extensive research attention owing to beneficial and tunable properties. On transforming the size of metal oxides to nanometer scale, these materials demonstrate improved hybrid characteristics, compared to the pristine materials. Such transformation to the nanoscale size enhances the surface area, however, it also renders the metal oxides sensitive to agglomeration due to the presence of van der Waals interactions [72]. As a result, the metal oxides may experience reduction in their capacity, mechanical strength as well as selectivity [72,73]. The blending of inorganic metal oxides and polymers presents a useful alternative to generate materials with remarkable magnetic, electrical, mechanical and thermal properties, as compared to the inorganic metal oxides and bulk polymers [74]. In these nanocomposites, the polymer contributes efficient luminescence, photoconductivity, advantageous processing, adjustable electronic properties and structural flexibility [75], whereas the inorganic metal oxides provide high mechanical and thermal stability, carrier mobility, band gap tunability as well as magnetic and dielectric properties. Among various metal oxides, titanium dioxide (TiO_2) is one of the most broadly analyzed semiconductors due to its simple preparation, nontoxic nature, relatively low cost, photo-degradation ability and good stability [76]. In this section, the application of metal oxides based polymer nanocomposites has been reviewed for the removal of dyes (Acid Red G and methylene blue) and heavy metal ions from wastewater.

Removal of Dyes

Wang *et al.* [77] reported a nanocomposite adsorbent based on TiO_2/polyaniline (TiO_2/PANI). The nanocomposite exhibited a remarkable adsorption effectiveness of greater than 90% for 500 mg/L Acid Red G solution. Though the impact of ion concentration on the adsorption was insignificant, the effect of pH on the adsorption of Acid Red G by TiO_2/PANI nanocomposite was important. The adsorption stability tests demonstrated that the *in-situ* regeneration of TiO_2/PANI was effectively accomplished using HCl as activator and NaOH as the stripping agent. The adsorption equilibrium of TiO_2 based PANI nanocomposite for Acid Red G could be accomplished within 5 minutes. Adsorption kinetics of TiO_2/PANI for Acid Red G was explained by the pseudo-second-order model. As indicated by the Langmuir adsorption isotherm model, a monolayer adsorption took place and the utmost adsorption ability of Acid Red G on TiO_2/PANI was noted to be 454.55 $mg.g^{-1}$. Despite regeneration for 8 times, the TiO_2/PANI retained superior adsorption ability with respect to both real Acid Red G effluent and pristine Acid Red G solution.

Feng *et al.* [78] generated TiO_2/polypyrrole (TiO_2/PPy) nanocomposites in acetic acid (AC), citric acid (CI), tartaric acid (TA) and succinic acid (SU), and utilized these nanocomposites for adsorbing cationic dye methylene blue and anionic dye Acid Red G. The surface physico-chemical characteristics of the generated nanocomposites were observed to be different from each other because of the utilization of differed carboxylic acids with/without hydroxyl groups, and the dissimilarities additionally affected the adsorption ability of the generated nanocomposites. The negative surface charge was favorable for methylene blue adsorption onto TiO_2/PPy nanocomposites, while the positive surface charge was noted to be ideal for Acid Red G adsorption. Further, it was noted that all the composites could separate the chosen dyes from aqueous solutions quickly with adsorption equilibrium time close to 30 minutes. The CI-TiO_2/PPy and TA-TiO_2/PPy generated in carboxylic acids having the hydroxyl group displayed greater adsorption ability, as compared to TiO_2 based PPy nanocomposites generated in the carboxylic acids with no hydroxyl groups. The adsorption ability of TiO_2/PPy composites demonstrated no loss subsequent to 4 adsorption-desorption cycles. The adsorption of the chosen dyes on the composites was noted to be endothermic in nature. The major adsorption mechanism was hydrogen bonding and electrostatic interaction.

Removal of Heavy Metals

The harmful heavy metals of specific concern in the industrial wastewater treatment are Cr, Zn, Pb, Ni, Cd, Hg and Cu [79-81]. Neeta *et al.* [82] reported nanosized zinc oxide (ZnO) particles generated by heating zinc sulfate in nitrate eutectic melt at 450 °C. The generated ZnO nanoparticles were encapsulated in urea-formaldehyde (UF) resin. The ZnO based UF nanocomposite was utilized for cost effective and efficient separation of Cu^{2+} ions from the solution. The nanocomposites exhibited a removal ability of 80% within 15 minutes. The Cu(II) ion adsorption by UF/ZnO nanocomposite was observed to be significant in the 6.6 to 7.0 pH range and followed the Langmuir-type adsorption isotherm.

PANI has been extensively employed for removing Cr(VI) and other organic pollutants because of the presence of plentiful imine and amine groups [83]. PANI possesses superior acid as well as alkali resistance, which makes it appropriate for separating both inorganic and organic contaminants as these bind with imine and amine groups [84]. Recent research studies have recommended the use of PANI composites (with fillers including metal oxides) as adsorbent for the removal of Cr(VI) and phenolic compounds [85,86].

As mentioned earlier, ZnO is an efficient adsorbent and photocatalyst for water purification, attributable to its activity and eco-friendly features [87]. Hence, ZnO incorporation in polymers is observed to upgrade the electrical, mechanical, thermal as well as optical properties, along with adsorption characteristics. Ahmad and Hasan [88] prepared xanthan gum (XG) grafted PANI-ZnO nanocomposite by an oxidative free-radical graft copolymerization process (Figure 4.6). Batch experimental design was utilized for the assessment of optimum conditions of adsorption of 2,4-dinitrophenol and Cr(VI) over XG grafted PANI- ZnO nanocomposite. Adsorption needed a contact time of about 120 min to achieve equilibrium, utilizing an initial concentration of adsorbate of about 120 mg/L with 0.04 g adsorbent. The optimum pH for the separation of Cr(VI) and 2,4-dinitrophenol was observed to be 2.0 and 4.0 respectively. The process associated with the separation of Cr(VI) and 2,4-dinitrophenol ions was adsorption-coupled reduction. The nanocomposite could be reused for five or more successive sequences. The adsorptive effectiveness for Cr(VI) remained at 51.67% and at 45.97% for 2,4-dinitrophenol after five cycles, proposing that the nanocomposite possessed superior regeneration ability.

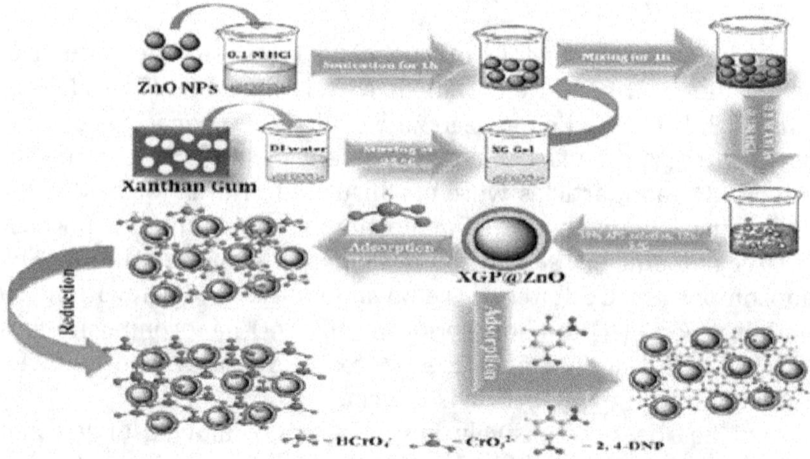

Figure 4.6 Schematic of the nanocomposite synthesis. Reproduced from Reference 88 with permission from American Chemical Society.

4.3 Conclusions

The chapter reviews the developments in the application of polymer nanocomposites for the removal of various pollutants from wastewater. More specifically, the wastewater treatment application of polymer nanocomposites with nanofillers such as graphene, CNTs, clay and metal oxides has been reviewed. The different preparation techniques, properties and characterization of the polymer nanocomposites have also been assessed. Adsorption as well as thermodynamic and kinetic models for the removal of contaminants have additionally been summarized.

Similar to other hybrid materials generated for different fields, polymer nanocomposites for wastewater treatment consolidate the beneficial properties of their constituents, for example high surface area, chemical and mechanical stability, surface charge, etc. The improved thermal and mechanical performance relative to the polymer enables the polymer nanocomposites to withstand high pressures as well as temperatures encountered in a fixed bed framework. The graphene based polymer nanocomposites have been observed as superior adsorbents for removing dyes (rhodamine B (RhB), methylene blue (MB), Congo red and Reactive Black 5), heavy metal ions (U(VI), Pb(II), Ni(II), Co(II) and Cr(VI)) and other pollutants like pesticides from wastewater. Also, the CNT based polymer nanocomposites have proved to be effective adsorbents for the separation of dyes (methylene blue, methyl violet and malachite green) and heavy metal ions (nickel, copper, cadmium and zinc ions) from wastewater. The clay/polymer nanocomposites treat the water by adsorption as well as flocculation of both organic and inorganic micro-pollutants from aqueous solutions. The usage of clay/polymer nanocomposites for removing inorganic pollutants (Pb(II), Zn(II), Cd(II) and Cu(II)), organic pollutants (trinitrophenol, trichlorophenol and phenol) and dyes from wastewater has proved to be effective. In addition, the metal oxides based polymer nanocomposites have also been confirmed to be effective for separating Acid Red G and methylene blue dyes as well as heavy metal ions from wastewater.

References

1. Shannon, M. A., Bohn, P. W., Elimelech, M., Georgiadis, J. G., Marinas, B. J., and Mayes, A. M. (2008) Science and technology for water purification in the coming decades. *Nature*, **452**(7185), 301-310.
2. Qu, X., Alvarez, P. J., and Li, Q. (2013) Applications of nanotechnology in water and wastewater treatment. *Water Research*, **47**(12), 3931-3946.

3. Malaviya, P., and Singh, A. (2011) Physicochemical technologies for remediation of chromium-containing waters and wastewaters. *Critical Reviews in Environmental Science and Technology*, **41**(12), 1111-1172.
4. Bhatnagar, A., and Sillanpaa, M. (2009) Applications of chitin-and chitosan-derivatives for the detoxification of water and wastewater - a short review. *Advances in Colloid and Interface Science*, **152**(1), 26-38.
5. Garg, V. K., Gupta, R., Yadav, A. B., and Kumar, R. (2003) Dye removal from aqueous solution by adsorption on treated sawdust. *Bioresource Technology*, **89**(2), 121-124.
6. Aksu, Z. (2005) Application of biosorption for the removal of organic pollutants: A review. *Process Biochemistry*, **40**(3), 997-1026.
7. Daer, S., Kharraz, J., Giwa, A., and Hasan, S. W. (2015) Recent applications of nanomaterials in water desalination: A critical review and future opportunities. *Desalination*, **367**, 37-48.
8. Ray, P. Z., and Shipley, H. J. (2015) Inorganic nano-adsorbents for the removal of heavy metals and arsenic: A review. *RSC Advances*, **5**(38), 29885-29907.
9. Jordan, J., Jacob, K. I., Tannenbaum, R., Sharaf, M. A., and Jasiuk, I. (2005) Experimental trends in polymer nanocomposites - a review. *Materials Science and Engineering A*, **393**(1), 1-11.
10. Lofrano, G., Carotenuto, M., Libralato, G., Domingos, R. F., Markus, A., Dini, L., Gautam, R. K., Baldantoni, D., Rossi, M., Sharma, S. K., and Chattopadhyaya, M. C. (2016) Polymer functionalized nanocomposites for metals removal from water and wastewater: an overview. *Water Research*, **92**, 22-37.
11. Bahmanyar, M., Sedaghat, S., Ramazani A. S. A., and Baniasadi, H. (2015) Preparation of ethylene vinyl acetate copolymer/graphene oxide nanocomposite films via solution casting method and determination of the mechanical properties. *Polymer-Plastics Technology and Engineering*, **54**(2), 218-222.
12. Petronella, F., Truppi, A., Ingrosso, C., Placido, T., Striccoli, M., Curri, M. L., Agostiano, A., and Comparelli, R. (2017) Nanocomposite materials for photocatalytic degradation of pollutants. *Catalysis Today*, **281**, 85-100.
13. Baruah, S., Pal, S. K., and Dutta, J. (2012) Nanostructured zinc oxide for water treatment. *Nanoscience & Nanotechnology-Asia*, **2**(2), 90-102.
14. Lazar, M. A., Varghese, S., and Nair, S. S. (2012) Photocatalytic water treatment by titanium dioxide: recent updates. *Catalysts*, **2**(4), 572-601.
15. Xu, M., Liang, T., Shi, M., and Chen, H. (2013) Graphene-like two-dimensional materials. *Chemical Reviews*, **113**(5), 3766-3798.
16. Duru, I., Ege, D., and Kamali, A. R. (2016) Graphene oxides for rem-

oval of heavy and precious metals from wastewater. *Journal of Materials Science*, **51**(13), 6097-6116.

17. Chauke, V. P., Maity, A., and Chetty, A. (2015) High-performance towards removal of toxic hexavalent chromium from aqueous solution using graphene oxide-alpha cyclodextrin-polypyrrole nanocomposites. *Journal of Molecular Liquids*, **211**, 71-77.
18. Hai, F. I., Yamamoto, K., and Fukushi, K. (2007) Hybrid treatment systems for dye wastewater. *Critical Reviews in Environmental Science and Technology*, **37**(4), 315-377.
19. Georgakilas, V., Otyepka, M., Bourlinos, A. B., Chandra, V., Kim, N., Kemp, K. C., Hobza, P., Zboril, R., and Kim, K. S. (2012) Functionalization of graphene: covalent and non-covalent approaches, derivatives and applications. *Chemical Reviews*, **112**(11), 6156-6214.
20. Kemp, K. C., Seema, H., Saleh, M., Le, N. H., Mahesh, K., Chandra, V., and Kim, K. S. (2013) Environmental applications using graphene composites: water remediation and gas adsorption. *Nanoscale*, **5**(8), 3149-3171.
21. Geim, A. K., and Novoselov, K. S. (2007) The rise of graphene. *Nature Materials*, **6**(3), 183-191.
22. Huang, X., Qi, X., Boey, F., and Zhang, H. (2012) Graphene-based composites. *Chemical Society Reviews*, **41**(2), 666-686.
23. Sharma, P., and Das, M. R. (2012) Removal of a cationic dye from aqueous solution using graphene oxide nanosheets: investigation of adsorption parameters. *Journal of Chemical & Engineering Data*, **58**(1), 151-158.
24. Huang, X., Yin, Z., Wu, S., Qi, X., He, Q., Zhang, Q., Yan, Q., Boey, F., and Zhang, H. (2011) Graphene-based materials: synthesis, characterization, properties, and applications. *Small*, **7**(14), 1876-1902.
25. Liu, F., Chung, S., Oh, G., and Seo, T. S. (2012) Three-dimensional graphene oxide nanostructure for fast and efficient water-soluble dye removal. *ACS Applied Materials & Interfaces*, **4**(2), 922-927.
26. Huang, Y., Zeng, M., Ren, J., Wang, J., Fan, L., and Xu, Q. (2012) Preparation and swelling properties of graphene oxide/poly (acrylic acid-co-acrylamide) super-absorbent hydrogel nanocomposites. *Colloids and Surfaces A: Physicochemical and Engineering Aspects*, **401**, 97-106.
27. Guo, H., Jiao, T., Zhang, Q., Guo, W., Peng, Q., and Yan, X. (2015) Preparation of graphene oxide-based hydrogels as efficient dye adsorbents for wastewater treatment. *Nanoscale Research Letters*, **10**(1), 272.
28. Li, Y., Zhao, X., Yu, P., and Zhang, Q. (2012) Oriented arrays of polyaniline nanorods grown on graphite nanosheets for an electrochemical supercapacitor. *Langmuir*, **29**(1), 493-500.
29. Liu, Z., Yang, J., Li, C., Li, J., Jiang, Y., Dong, Y., and Li, Y. (2014) Adsorption of Co (II), Ni (II), Pb (II) and U (VI) from aqueous solu-

tions using polyaniline/graphene oxide composites. *Korean Chemical Engineering Research*, **52**(6), 781-788.

30. Mu, B., Tang, J., Zhang, L., and Wang, A. (2017) Facile fabrication of superparamagnetic graphene/polyaniline/Fe_3O_4 nanocomposites for fast magnetic separation and efficient removal of dye. *Scientific Reports*, **7**(1), 5347.
31. Crini, G. (2005) Recent developments in polysaccharide-based materials used as adsorbents in wastewater treatment. *Progress in Polymer Science*, **30**(1), 38-70.
32. Travlou, N. A., Kyzas, G. Z., Lazaridis, N. K., and Deliyanni, E. A. (2013) Functionalization of graphite oxide with magnetic chitosan for the preparation of a nanocomposite dye adsorbent. *Langmuir*, **29**(5), 1657-1668.
33. Wang, Y., Feng, X., Villalobos, M., Tan, W., and Liu, F. (2012) Sorption behavior of heavy metals on birnessite: relationship with its Mn average oxidation state and implications for types of sorption sites. *Chemical Geology*, **292**, 25-34.
34. Zhang, Y., Zhu, C., and Kan, J. (2008) Synthesis and characterization of ferromagnetic polyaniline with conductivity in an applied magnetic field. *Journal of Applied Polymer Science*, **109**(5), 3024-3029.
35. Liu, M., Chen, C., Hu, J., Wu, X., and Wang, X. (2011) Synthesis of magnetite/graphene oxide composite and application for cobalt (II) removal. *The Journal of Physical Chemistry C*, **115**(51), 25234-25240.
36. Sun, Y., Wang, Q., Chen, C., Tan, X., and Wang, X. (2012) Interaction between Eu (III) and graphene oxide nanosheets investigated by batch and extended X-ray absorption fine structure spectroscopy and by modeling techniques. *Environmental Science & Technology*, **46**(11), 6020-6027.
37. Musico, Y. L. F., Santos, C. M., Dalida, M. L. P., and Rodrigues, D. F. (2013) Improved removal of lead (II) from water using a polymer-based graphene oxide nanocomposite. *Journal of Materials Chemistry A*, **1**(11), 3789-3796.
38. Lykins, Jr., B. W., Koffskey, W. E., and Miller, R. G. (1986) Chemical products and toxicologic effects of disinfection. *Journal-American Water Works Association*, **78**(11), 66-75.
39. Zhang, C., Zhang, R. Z., Ma, Y. Q., Guan, W. B., Wu, X. L., Liu, X., Li, H., Du, Y. L. and Pan, C. P. (2015) Preparation of cellulose/graphene composite and its applications for triazine pesticides adsorption from water. *ACS Sustainable Chemistry & Engineering*, **3**(3), 396-405.
40. Siro, I., and Plackett, D. (2010) Microfibrillated cellulose and new nanocomposite materials: a review. *Cellulose*, **17**(3), 459-494.
41. Sui, Z., Meng, Q., Zhang, X., Ma, R., and Cao, B. (2012) Green synthesis of carbon nanotube–graphene hybrid aerogels and their use as

versatile agents for water purification. *Journal of Materials Chemistry*, **22**(18), 8767-8771.

42. Mittal, G., Dhand, V., Rhee, K. Y., Park, S. J., and Lee, W. R. (2015) A review on carbon nanotubes and graphene as fillers in reinforced polymer nanocomposites. *Journal of Industrial and Engineering Chemistry*, **21**, 11-25.
43. Zhang, X., Ji, J., Zhang, X., Yang, B., Liu, M., Liu, W., Tao, L., Chen, Y., and Wei, Y. (2013) Mussel inspired modification of carbon nanotubes using RAFT derived stimuli-responsive polymers. *RSC Advances*, **3**(44), 21817-21823.
44. Zhang, X., Liu, M., Zhang, Y., Yang, B., Ji, Y., Feng, L., Tao, L., Li, S., and Wei, Y. (2012) Combining mussel-inspired chemistry and the Michael addition reaction to disperse carbon nanotubes. *RSC Advances*, **2**(32), 12153-12155.
45. Bulut, Y., and Karaer, H. (2015) Adsorption of methylene blue from aqueous solution by crosslinked chitosan/bentonite composite. *Journal of Dispersion Science and Technology*, **36**(1), 61-67.
46. Fu, J., Chen, Z., Wang, M., Liu, S., Zhang, J., Zhang, J., Han, R., and Xu, Q. (2015) Adsorption of methylene blue by a high-efficiency adsorbent (polydopamine microspheres): kinetics, isotherm, thermodynamics and mechanism analysis. *Chemical Engineering Journal*, **259**, 53-61.
47. Sun, J. Z., Qin, A., and Tang, B. Z. (2013) Functional polyacetylenes: hybrids with carbon nanotubes. *Polymer Chemistry*, **4**(2), 211-223.
48. Abdel, S. M., and Burk, R. (2009) Solid phase extraction of polyhalogenated pollutants from freshwater using chemically modified multi-walled carbon nanotubes and their determination by gas chromatography. *Journal of Separation Science*, **32**(7), 1060-1068.
49. Salam, M. A., and Burk, R. (2008) Novel application of modified multiwalled carbon nanotubes as a solid phase extraction adsorbent for the determination of polyhalogenated organic pollutants in aqueous solution. *Analytical and Bioanalytical Chemistry*, **390**(8), 2159-2170.
50. Xie, Y., He, C., Liu, L., Mao, L., Wang, K., Huang, Q., Liu, M., Wan, Q., Deng, F., Huang, H., and Zhang, X. (2015) Carbon nanotube based polymer nanocomposites: biomimic preparation and organic dye adsorption applications. *RSC Advances*, **5**(100), 82503-82512.
51. Ghorai, S., Sarkar, A., Raoufi, M., Panda, A. B., Schonherr, H., and Pal, S. (2014) Enhanced removal of methylene blue and methyl violet dyes from aqueous solution using a nanocomposite of hydrolyzed polyacrylamide grafted xanthan gum and incorporated nanosilica. *ACS Applied Materials & Interfaces*, **6**(7), 4766-4777.
52. Hu, L., Yang, Z., Wang, Y., Li, Y., Fan, D., Wu, D., Wei, Q., and Du, B. (2017) Facile preparation of water-soluble hyperbranched polyamine functionalized multiwalled carbon nanotubes for high-

efficiency organic dye removal from aqueous solution. *Scientific Reports,* **7**(1), 3611.

53. Ke, G., Guan, W. C., Tang, C. Y., Hu, Z., Guan, W. J., Zeng, D. L., and Deng, F. (2007) Covalent modification of multiwalled carbon nanotubes with a low molecular weight chitosan. *Chinese Chemical Letters,* **18**(3), 361-364.
54. Rhazi, M., Desbrieres, J., Tolaimate, A., Rinaudo, M., Vottero, P., and Alagui, A. (2002) Contribution to the study of the complexation of copper by chitosan and oligomers. *Polymer,* **43**(4), 1267-1276.
55. Rhazi, M., Desbrieres, J., Tolaimate, A., Rinaudo, M., Vottero, P., Alagui, A., and El Meray, M. (2002) Influence of the nature of the metal ions on the complexation with chitosan.: Application to the treatment of liquid waste. *European Polymer Journal,* **38**(8), 1523-1530.
56. Chatterjee, S., Lee, M. W., and Woo, S. H. (2010) Adsorption of congo red by chitosan hydrogel beads impregnated with carbon nanotubes. *Bioresource Technology,* **101**(6), 1800-1806.
57. Janegitz, B. C., Marcolino-Junior, L. H., Campana-Filho, S. P., Faria, R. C., and Fatibello-Filho, O. (2009) Anodic stripping voltammetric determination of copper (II) using a functionalized carbon nanotubes paste electrode modified with crosslinked chitosan. *Sensors and Actuators B: Chemical,* **142**(1), 260-266.
58. Salam, M. A., Makki, M. S., and Abdelaal, M. Y. (2011) Preparation and characterization of multi-walled carbon nanotubes/chitosan nanocomposite and its application for the removal of heavy metals from aqueous solution. *Journal of Alloys and Compounds,* **509**(5), 2582-2587.
59. Solener, M., Tunali, S., Ozcan, A. S., Ozcan, A., and Gedikbey, T. (2008) Adsorption characteristics of lead (II) ions onto the clay/poly (methoxyethyl) acrylamide (PMEA) composite from aqueous solutions. *Desalination,* **223**(1-3), 308-322.
60. Anirudhan, T. S., and Ramachandran, M. (2008) Synthesis and characterization of amidoximated polyacrylonitrile/organobentonite composite for Cu (II), Zn (II), and Cd (II) adsorption from aqueous solutions and industry wastewaters. *Industrial & Engineering Chemistry Research,* **47**(16), 6175-6184.
61. Shawky, H. A. (2011) Improvement of water quality using alginate/montmorillonite composite beads. *Journal of Applied Polymer Science,* **119**(4), 2371-2378.
62. Bleiman, N., and Mishael, Y. G. (2010) Selenium removal from drinking water by adsorption to chitosan–clay composites and oxides: batch and columns tests. *Journal of Hazardous Materials,* **183**(1), 590-595.
63. Liu, P., Jiang, L., Zhu, L., and Wang, A. (2013) Attapulgite/poly (acrylic acid) nanocomposite (ATP/PAA) hydrogels with multifunctionalized attapulgite (org-ATP) nanorods as unique cross-

linker: preparation optimization and selective adsorption of Pb (II) ion. *ACS Sustainable Chemistry & Engineering*, **2**(4), 643-651

64. Ganigar, R., Rytwo, G., Gonen, Y., Radian, A., and Mishael, Y. G. (2010) Polymer–clay nanocomposites for the removal of trichlorophenol and trinitrophenol from water. *Applied Clay Science*, **49**(3), 311-316.
65. Wang, T., Zhu, R. L., Ge, F., Zhu, J. X., He, H. P., and Chen, W. X. (2010) Sorption of phenol and nitrobenzene in water by CTMAB/CPAM organobentonites. *Huan Jing Ke Xue*, **31**(2), 385-389.
66. Zhu, J., Wang, T., Zhu, R., Ge, F., Wei, J., Yuan, P., and He, H. (2011) Novel polymer/surfactant modified montmorillonite hybrids and the implications for the treatment of hydrophobic organic compounds in wastewaters. *Applied Clay Science*, **51**(3), 317-322.
67. Chang, C., and Zhang, L. (2011) Cellulose-based hydrogels: present status and application prospects. *Carbohydrate Polymers*, **84**(1), 40-53.
68. Peng, N., Hu, D., Zeng, J., Li, Y., Liang, L., and Chang, C. (2016) Superabsorbent cellulose–clay nanocomposite hydrogels for highly efficient removal of dye in water. *ACS Sustainable Chemistry & Engineering*, **4**(12), 7217-7224.
69. Rytwo, G., Rettig, A., and Gonen, Y. (2011) Organo-sepiolite particles for efficient pretreatment of organic wastewater: application to winery effluents. *Applied Clay Science*, **51**(3), 390-394.
70. Rytwo, G. (2012) The use of clay-polymer nanocomposites in wastewater pretreatment. *The Scientific World Journal*, 2012, doi:10.1100/2012/498503.
71. Rytwo, G., Lavi, R., Rytwo, Y., Monchase, H., Dultz, S., and Konig, T. N. (2013) Clarification of olive mill and winery wastewater by means of clay–polymer nanocomposites. *Science of the Total Environment*, **442**, 134-142.
72. Pan, B., Pan, B., Zhang, W., Lv, L., Zhang, Q., and Zheng, S. (2009) Development of polymeric and polymer-based hybrid adsorbents for pollutants removal from waters. *Chemical Engineering Journal*, **151**(1), 19-29.
73. Hua, M., Zhang, S., Pan, B., Zhang, W., Lv, L., and Zhang, Q. (2012) Heavy metal removal from water/wastewater by nanosized metal oxides: a review. *Journal of Hazardous Materials*, **211**, 317-331.
74. Xu, T. (2005) Ion exchange membranes: state of their development and perspective. *Journal of Membrane Science*, **263**(1), 1-29.
75. Fischer, H. (2003) Polymer nanocomposites: from fundamental research to specific applications. *Materials Science and Engineering C*, **23**(6), 763-772.
76. Tan, X., Fang, M., Li, J., Lu, Y., and Wang, X. (2009) Adsorption of Eu (III) onto TiO_2: effect of pH, concentration, ionic strength and soil

fulvic acid. *Journal of Hazardous Mmaterials*, **168**(1), 458-465.

77. Wang, N., Li, J., Lv, W., Feng, J., and Yan, W. (2015) Synthesis of polyaniline/TiO_2 composite with excellent adsorption performance on acid red G. *RSC Advances*, **5**(27), 21132-21141.
78. Feng, J., Chen, J., Wang, N., Li, J., Shi, J., and Yan, W. (2016) Enhanced adsorption capacity of polypyrrole/TiO_2 composite modified by carboxylic acid with hydroxyl group. *RSC Advances*, **6**(48), 42572-42580.
79. Ali, I. (2012) New generation adsorbents for water treatment. *Chemical Reviews*, **112**(10), 5073-5091.
80. Gupta, V. K., Al Khayat, M., Singh, A. K., and Pal, M. K. (2009) Nano level detection of Cd (II) using poly (vinyl chloride) based membranes of Schiff bases. *Analytica Chimica Acta*, **634**(1), 36-43.
81. Tesh, S. J., and Scott, T. B. (2014) Nano-composites for water remediation: A review. *Advanced Materials*, **26**(35), 6056-6068.
82. Neeta, P., Shukla, S. K., and Singh, N. B. (2015) Zinc oxide-urea formaldehyde nanocomposite film as low-cost adsorbent for removal of Cu(II) from aqueous solution. *Advanced Materials Letters*, **6**(2), 172-178.
83. Karthik, R., and Meenakshi, S. (2014) Removal of hexavalent chromium ions using polyaniline/silica gel composite. *Journal of Water Process Engineering*, **1**, 37-45.
84. Saravanan, R., Sacari, E., Gracia, F., Khan, M. M., Mosquera, E., and Gupta, V. K. (2016) Conducting PANI stimulated ZnO system for visible light photocatalytic degradation of coloured dyes. *Journal of Molecular Liquids*, **221**, 1029-1033.
85. Chavez-Guajardo, A. E., Medina-Llamas, J. C., Maqueira, L., Andrade, C. A., Alves, K. G., and de Melo, C. P. (2015) Efficient removal of Cr (VI) and Cu (II) ions from aqueous media by use of polypyrrole/maghemite and polyaniline/maghemite magnetic nanocomposites. *Chemical Engineering Journal*, **281**, 826-836.
86. Guo, Y., Wang, H., He, C., Qiu, L., and Cao, X. (2009) Uniform carbon-coated ZnO nanorods: microwave-assisted preparation, cytotoxicity, and photocatalytic activity. *Langmuir*, **25**(8), 4678-4684.
87. Khan, T. A., Nazir, M., Ali, I., and Kumar, A. (2017) Removal of Chromium (VI) from aqueous solution using guar gum–nano zinc oxide biocomposite adsorbent. *Arabian Journal of Chemistry*, **10**, S2388-S2398.
88. Ahmad, R., and Hasan, I. (2017) Efficient Remediation of an Aquatic Environment Contaminated by Cr(VI) and 2,4-Dinitrophenol by XG-g-Polyaniline@ ZnO Nanocomposite. *Journal of Chemical & Engineering Data*, **62**(5), 1594-1607.

5

Polymeric Materials for Oil Spill Clean Up and Phenol Removal

5.1 Introduction

Today, large amount of oil-wastewater is generated by the oil and gas industry [1], which is hazardous to the environment [2]. In addition, refined petroleum products and crude oil are accidently spilled into water bodies from the sources such as offshore platforms, oil cargo ships and tankers, which causes serious damage to marine environment [3]. There are many instances of oil spill accidents like Torrey Canyon oil spill in 1967, Mexico oil spill in 2010 (210 million gallons of oil spilled into sea) and Bohai Bay oil spill in 2011. In order to protect the marine ecosystem, oil industries and environment protection authorities are in continuous need of effective methods for successful separation of the water-oil mixture. Oil spills lead to different forms with water, such as stable water-in-oil emulsion, unstable water-in-oil emulsion and free-floating oil [4]. Separation of unstable oil-in-water emulsion or free-floating oil is applicable by the conventional techniques (flotation, gravity separation, coagulation, ultrasonic separation, skimming and centrifugation), however, stable oil-in-water emulsions cannot be separated by applying these techniques [5-7]. High operational cost and low efficiency are the other disadvantages of the conventional techniques. To overcome these challenges, some promising membrane separation based methods have also been developed, such as reverse osmosis [8], ultrafiltration [9,10], microfiltration [11], dehydration of oil emulsion by pervaporation [12], membrane distillation [13] and flocculation followed by microfiltration [14]. Current day techniques employed for the cleaning of oil spills are of three types such as extraction of oil from the surface of water using absorbents [15-19], applying dispersant agents for oil degradation [20,21], and controllable burning of oil spill. Among these techniques, oil removal by absorbents is considered a promising method for allowing effective removal of oil from water. Oil absorbents are basically categorized into three classes, i.e.,

Naman Arora and Vikas Mittal, Khalifa University of Science and Technology, Abu Dhabi, UAE

inorganic mineral product (zeolites [22,23], perlite [24] and silica [25,26]), synthetic polymers [27, 28] and natural organics (rice straw [29], corn cob [30], wool fiber [31], kapok fiber [32] and milkweed fiber [18]). For oil absorption process, absorbents are facilitated with distinct advantages such as simple operation, cost effectiveness, environment friendliness and ability to clean oil spills without any secondary pollution [5,33,34]. An ideal oil absorbent material should also possess low density, high selectivity, high oil absorption ability and excellent reusability [15]. Research studies have also underlined the advantages of special wettability-controlled oil-water mixture separation method over traditional separation methods. Superhydrophobic or superoleophobic surfaces possess a water or oil contact angle (CA) higher than 150° and sliding angle less than 10° [35-38]. These materials have generally been fabricated by either designing with superoleophobic-superhydrophilic surface or developing with superhydrophobic - superoleophilic surface. There are two strategies to fabricate superhydrophobic - superoleophilic materials, either by constructing a rough structure on surface or modifying a surface with low surface energy chemicals [39]. Generally, superoleophobic surface exhibits superhydrophobicity [40-42] and superhydrophilic surface shows superoleophilicity [43,44]. However, it has been observed that hydrophilic and oleophobic properties can be generated simultaneously due to some stimuli-responsive surfaces, based on unfavourable interaction with non-polar liquids (hexadecane) and a favourable interaction with polar liquids (water) [45-47]. Materials with superhydrophobic - superoleophilic or superoleophobic-superhydrophilic surfaces have a significant role in oil-water separation application.

Phenols and its derivatives are one of the major organic pollutants which lead to significant water pollution. Presence of phenolic compounds in water is very concerning due to their high oxygen demand, low biodegradability and high toxicity. For the treatment of industrial wastewater containing phenolic derivatives, many techniques have been employed such as adsorption, catalytic oxidation and separation via membrane [48-52]. Phenolic compounds are introduced into water bodies by effluents of industries like petroleum, plastic, pharmaceutical, coke manufacturing, rubber, textiles and pesticides manufacturing [53-55]. Long-term intake of toxic phenol and its derivatives can cause severe chronic insomnia, rapid fatigue and headache [56]. Overall, adsorption technology is widely employed for organic and inorganic micro-pollutants removal from industrial wastewater.

In this review, various polymer based sponges/foams (Section 5.2), aerogels (Section 5.3), fibers (Section 5.4), membranes (Section 5.5) and other polymer absorbents (Section 5.6) have been reviewed for the purpose of removal of oil from water. In Section 5.7, polymeric adsorbents for the adsorption of phenol and its derivatives from aqueous and other solutions have also been reviewed in detail.

5.2 Polymer Based Sponges and Foams for Oil Spill Clean Up

A wide variety of sponges and foams have been employed as oil absorbents for the efficacious absorption of oil from water bodies. Wang *et al.* [57] reported the preparation of carbon nanotube (CNT)/polydimethylsiloxane (PDMS) coated polyurethane (PU) sponge by using dip coating method to absorb variety of organic impurities or oils from water. PU sponges as 3D porous materials are commercially available with the capability to absorb water as well as organic solvents. By anchoring CNT/PDMS coatings onto the frame of hydrophilic PU sponge, the wettability of sponge was changed from hydrophilic to superhydrophobic. It was observed that the water contact angle of the PDMS-coated PU sponge was 140±3° which could not be classified as superhydrophobic material. However, the combination of micro-scale surface roughness of PU sponge and hydrophobic CNT/PDMS nanocomposites provided the hybrid with a water contact angle of 162±2°. The resulting superhydrophobic sponge exhibited an absorption capacity 15 - 25 times its own weight to remove the organic solvents or oils, depending on their viscosities and densities. The surface of sponge revealed its crater like nanostructures which featured Cassie-Baxter surfaces. This kind of structure increased the roughness of surface and further led to a composite interface where the air got trapped within the grooves beneath the liquid and as a result, induced the superhydrophobicity. In order to achieve continuous removal of 20 L of oil (gasoline, n-hexane, or n-hexadecane) from water surface, a small piece (0.42 g) of CNT/PDMS-coated PU sponge (2.0 x 2.0 x 3.0 cm^3) was connected as a filter in a tube in conjunction with a vacuum system (Figure 5.1). The resulting assembly was capable to separate large amount of oil up to 35000 times its own weight from water surface. Jiang *et al.* [58] reported the fabrication of amine functionalized silica nanoparticles coated polyurethane/polytetrafluoroethylene (PTFE) sponge. In this study, commercially available PU sponge was deposited with PTFE. Subsequently, PU/PTFE sponge was immersed into A-SiO_2 (amine treated) nanoparticles suspension followed by the immersion in SiO_2 nanoparticles suspension. The resulting sponge was the treated with silane (octyltrichlorosilane) by using vapor phase deposition technique, providing a large surface roughness. It resulted in increasing the water contact angle (WCA) and roll-off angle which confirmed it

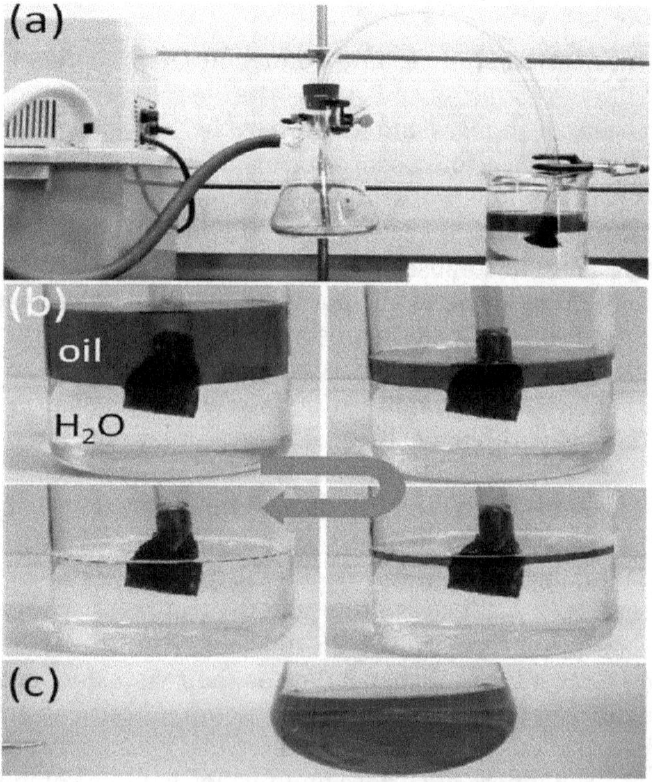

Figure 5.1 Images of (a) continuous oil/water separation system, (b) progress of the continuous absorption and removal of an organic solvent from the water surface, and (c) oil collected in part b. Reproduced from Reference 57 with permission from American Chemical Society.

as superhydrophobic material to remove oil or organic solvent from water. Zhang *et al.* [59] reported the synthesis of nanocellulose sponges which are flexible, ultra-light in weight and hydrophobic by using a novel and efficient silylation process in water. The materials had very high porosity (≥99%), which was engineered by freeze drying of the nanofibrillated cellulose (NFC) water suspension in the presence of methylmethoxysilane (MTMS) sol. The silylated sponges were able to remove dodecane spills from water (Figure 5.2). The authors also investigated the absorption capacity of sponge for various oils and other organic solvents which were found to be 49-102 (g/g) depending on the nature of the liquids. Wang *et al.* [60] reported the microfibrillated cellulose fibers (MCF sponges) modified with MTMS (methylmethoxysilane) via a chemical vapour deposition process to make it a superhydrophobic material. The sponges exhibited excellent recyclability (more than 30 times), elasticity, flexibility and porosity (up to 99.84%). The absorption capacities of sponge for pump

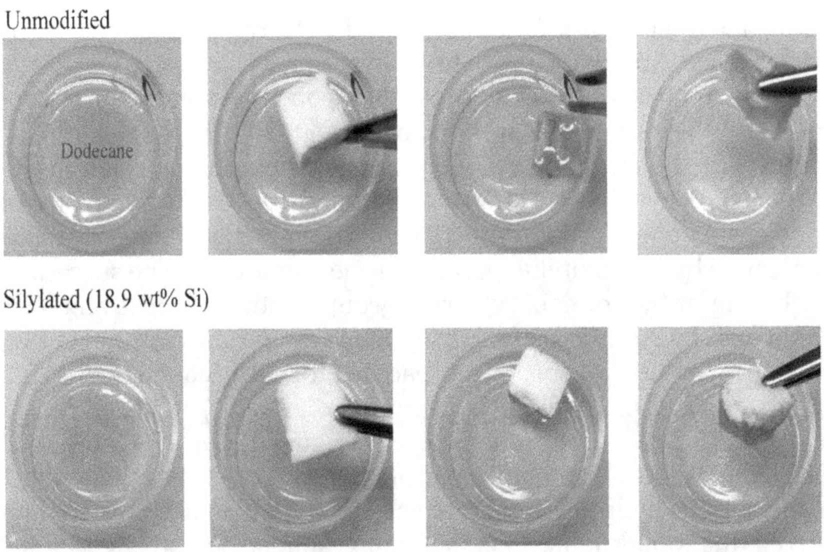

Figure 5.2 Removal of a red colored dodecane spill (0.02 g) from water with the silylated NFC sponge (0.02 g). In comparison, the unmodified material was not selective and lost its original shape. Reproduced from Reference 59 with permission from American Chemical Society.

oil, motor oil, white oil and silicone oil were determined as 197, 198, 178 and 228 g/g, respectively. The sponge also exhibited high absorption capacities for a variety of organic solvents, ranging from 88 g/g to 163 g/g depending on the viscosity and density of the liquid. Wu *et al.* [61] reported the polyurethane/silica nanoparticles (PU/SiO_2 NPs) sponge subsequently modified with gasoline as a hydrophobic material. The absorption capacity of the sponge for diesel, peanut oil and motor oil were calculated as 95±3, 108±4, and 103±3 g/g, respectively. The resultant sponge was subjected to 15 absorption-squeeze cycles, which gradually reduced its oil sorption capacity from 103 to 76 g/g. Liu *et al.* [62] reported reduced graphene oxide (rGO) coated polyurethane sponge fabricated by using a facile method. Thermo-gravimetric analysis (TGA) revealed that the amount of rGO coated onto the PU sponge was too less to change the decomposition behaviour of the PU sponge. The sponge exhibited good strength, elasticity, low density (0.0088 g/cm^3), porosity (99.29%), excellent recyclability (50 times without deterioration) and high hydrophobicity, the properties which confirmed its suitability as absorbent for oil spill clean up. It was subjected to variety of organic solvents and oils, the absorption capacities were observed to be higher than 80 g/g for all the liquids and the highest value was recorded as 160 g/g for chloroform (Figure 5.3).

Pan *et al.* [63] reported hydrophobic polyvinyl-alcohol formaldehyde (PVF) sponges. The authors modified the PVF sponge by treating it with the mixture of acetonitrile and stearoyl chloride in the presence of pyridine at different temperatures. Hydrophobicity of sponge was enhanced with increasing the reaction temperature, as confirmed by the solubility parameter measurements. The successful anchoring of hydrophobic stearoyl groups onto the PVF frame was confirmed by ATR-IR spectroscopy. The resultant hydrophobic PVF sponges exhibited absorption capacities for oil products from 13.7 g/g to 56.6 g/g, maintaining high porosity (≥94.8%) and good reusability for 35 absorption- squeeze cycles. Nguyen *et al.* [64] fabricated graphene based melamine sponge via dip coating and facile method. Anchoring of hydrophobic graphene onto the sponge skeleton occurred due to mechanical flexibility of graphene and strong van der Waals interactions between the graphene and sponge skeleton, which led the water contact angle to change from 0° to 162°. The authors observed that the sponge with graphene loading ≤ 5.1% did not restrict the absorption of water, however, as the loading of graphene was increased to ≥7.3%, water droplets attained the water contact angle of about 160° with quasi-spherical shapes on the surface of the sponge, representing superhydrophobicity. The sponge exhibited superhydrophobic and superoleophilic surfaces because of the combined behaviour of the graphene micro/nano-framework, hydrophobic property of graphene and the micro-porous structure of melamine sponge. The absorption capacities of resultant sponge were up to 165 times its own weight for different liquids, along with high selectivity and excellent recyclability. However, during recycling of such coated sponges, the linkage between graphene and the sponge skeleton could be affected which can further reduce the hydrophobic

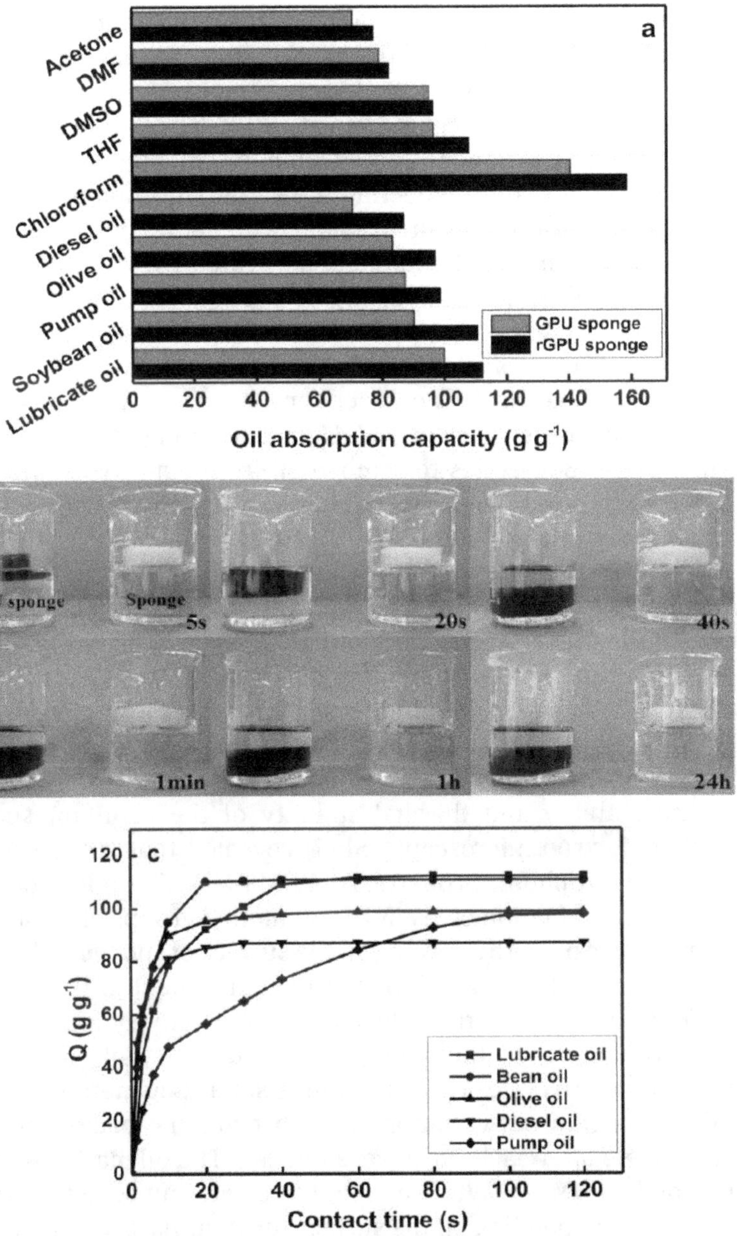

Figure 5.3 (a) Absorption capacity of the GPU and rGPU sponges for various organic liquids, (b) images of the absorption processes for the PU and rGPU sponges, (c) absorption capacities of the rGPU sponge versus contact time with various oils. Reproduced from Reference 62 with permission from American Chemical Society.

property of the coated sponge. For ensuring the adhesion between graphene nano-sheets and sponge skeleton, the sponge was immersed in dilute solution of PDMS in xylene (0.25 mg/ml), followed by overnight drying in vacuum oven at 120 °C. The formation of PDMS thin film onto the graphene-coated sponge cemented graphene with sponge skeleton without affecting the superhydrophobicity and superoleophilicity of the resultant sponge. Zhu *et al.* [65] reported robust superhydrophobic PU sponge, fabricated by one-step solution immersion method. The polyurethane sponge was poured into hexane solution containing methyltrichlorosilane (0.5 - 3%, v/v) for 30 min followed by air drying with a relative humidity of 30 - 70% to evaporate hexane and further dried for 2 h at 60 °C to obtain a water contact angle in the range of 142-157°. The sponge could be reused for more than 300 cycles to remove oil from aqueous solution and exhibited absorption capacities about 12 - 25 times its own weight for oil and organic solvents.

Turco *et al.* [66] reported embedded magnetic multiwalled carbon nanotubes (MWNTs) into a PDMS porous sponge prepared by polymerizing the PDMS prepolymer in the presence of magnetic MWNTs covered hard template. The authors generated 3% MWNTs - PDMS and 1% MWNTs - PDMS sponges, with water contact angles of 153.4±6.9° and 146.6±4.6°, respectively. The presence of magnetic MWNTs improved the oil absorption capacity, mechanical strength, chemical stability and thermal stability of the resultant sponges. Scanning electron microscopy (SEM) revealed that superoleophilic and superhydrophobic properties of PDMS-MWNTs sponge occurred due to the combination of MWNTs and methyl groups of PDMS. Basically, presence of methyl groups on the surface of sponge reduced the surface energy of sponge, along with enhancement of this effect by MWNTs also enhanced this effect, thereby effectively increasing the surface roughness of the sponge. The reusability of sponges was checked up to 10 cycles by using simple squeezing method, suggesting that no significant variation of oil absorption capacity and water contact angle after each cycle were noticed. The oil mass absorption ability of the PDMS - MWNT sponges were determined in the range of 8.5 to 20 g/g depending on the surface tension, viscosity and density of oils and organic solvents.

Zhang *et al.* [67] reported swellable porous PDMS oil absorbent, which had three dimensional interconnected porous structure, good selectivity for oil/water separation, excellent reusability for 20 absorbing-recovering cycles without losing absorption capacity. High absorption capacities of 4 - 34 g/g were determined for various oils and organic solvents depending on their viscosity and density. For crude oil, diesel and gasoline, the absorption capacities were determined as 9, 12, and 22 g/g, respectively. Swellable porous PDMS was prepared by a modified sugar-template method. The method involved the direct curing of a PDMS prepolymer in the solution of *p*-xylene in the presence of commercially available sugar particles. The authors also observed that on increasing the amount of *p*-xylene from 0 to 2 times that of PDMS, the absorption capacity of PDMS oil absorbent for petroleum ether also increased from 13.8 to 20.8 g/g. More importantly, a gel state of PDMS occurred after curing due to the presence of *p*-xylene, which resulted in a swellable structure of PDMS. The porosity of swellable porous PDMS was observed in the range of 50 - 81% depending on the sugar-particle template size. The water contact angle of the sponge was determined as 144°.

Nagappan *et al.* [68] reported bio-inspired hybrid- PDMS based melamine sponge with high selectivity for oil/water separation, excellent recyclability, water contact angle > 175° and high absorption capacities ranging from 2000 to 3600 w/w%. The sponge was prepared by using polymethylhydoxysiloxane (PMHOS), lotus leaf powder and phenyl substituted silica ormosil (PSiOr). The authors observed that the superhydrophobic property of hybrid occurred due to the presence of PMHOS and PSiOr additionally enhanced the surface roughness, surface area, adhesiveness and water contact angle of hybrid on the substrate. Bio-inspired micro/nano-composites hybrid could, thus, be confirmed as an effective and environmental friendly absorbent for oil spill clean up application. In another study reporting the use of sponges for oil spill clean up, Zhu *et al.* [69] studied the fabrication of copper-film coated superhydrophobic and superoleophilic PU sponges by using solution-immersion method. It was observed that the as-prepared sponges did not sink beneath the water surface, while the original sponge submerged into the water bath. Silver-mirror like surface appeared when the as-prepared sponges were partially or totally immersed in water by an external force, which exhibited Cassie-Baxter non-wetting behaviour owing to a continuous air layer between the water and superhydrophobic surface (Figure 5.4).

Li *et al.* [70] reported three-dimensional graphene/polypyrrole (PPy) foams for the clean up of oil spills. Tanobe *et al.* [71] reported the modification of post-consumed polyurethane foams by grafting with polystyrene, thereby generating an oil absorbent for cleaning of oil spills. Calcagnile *et al.* [72] reported the functionalization of commercially available polyurethane foam with oleic acid-capped colloidal iron oxide nanoparticles, followed by the treatment with PTFE particles to obtain superhydrophobic and superoleophilic material for the absorption of oil from polluted regions.

Figure 5.4 Images of a piece of the (a) as-prepared sponge, (b) original sponge, (c) as-prepared sponge placed on a water bath and (d) as-prepared sponge immersed in the water bath by an external force. Reproduced from reference 69 with permission from American Chemical Society.

The details of various polymeric sponges or foams are also listed in Table 5.1.

Table 5.1 Polymeric sponges or foams for oil spill clean up

Authors	Sponge/Foams	Method	Oils/Organic solvents	Absorption capacity (g/g) or (times) or (w/w %)	WCA	OCA	Ref.
Wang *et al.*	CNT-PDMS coated PU sponge	Dip-coating	n-Hexadecane n-Hexane Gasoline Soybean oil Used motor oil Diesel	[15-25 times its own weight]	162±2°	0° 0° 0° - - -	[57]
Jiang *et al.*	PU/PTFE/A-SiO_2/SiO_2 coated PU sponge	Vapour phase deposition	[n-Hexane, Pentane, Heptane, Benzene, Toluene, Silicone oil]	-	~165°	-	[58]
Zhang *et al.*	Silylated nano-fibrillated cellulose sponge	Freeze drying method	[Dodecane, Acetone, Silicone oil, Chloroform, Toluene, Ethanol, Mineral oil, Dichloromethane]	[49-102 g/g]	136°	-	[59]
Wang *et al.*	MTMS modified micro-fibrillated cellulose sponge	Freeze drying method	White oil Silicone oil Pump oil Motor oil [Cyclohexane, DMF, Acetone, DMSO, Hexane, Ethanol, Toluene, Dichloromethane, Isopropyl alcohol]	178 g/g 228 g/g 197 g/g 198 g/g [88-163 g/g]	154°	-	[60]

Table 5.1 (continued)

Authors	Sponge/Foams	Method	Oils/ Organic solvents	Absorption capacity (g/g) or (times) or (w/w %)	WCA	OCA	Ref.
Wu *et al.*	SiO_2 NPs/polyurethane sponge modified with gasoline	Dip coating method	Motor oil Peanut oil Diesel	103±3 g/g 108±4 g/g 95±3 g/g	126°	-	[61]
Liu *et al.*	Reduced graphene oxide coated PU sponge	Dip coating method	[Acetone, DMF, DMSO,THF, Chloroform, Diesel, Olive oil, Pump oil, Soybean oil, Lubricate oil]	[80-160 g/g]	127°	0°	[62]
Pan *et al.*	Hydrophobic PVF sponge	Wet chemical synthesis	[Toluene, n-Hexane, Kerosene, soybean oil, Hydraulic oil, Crude oil]	[13.7-56.6 g/g]	138±1.5°	-	[63]
Nguyen *et al.*	Graphene based Melamine sponge	Dip coating method	[Chloroform, DMF, Hexane, Acetone, Ethanol, Methanol, Used pump oil, Pump oil, Soybean oil, Motor oil]	[54-165 times its own weight]	162°	-	[64]
Zhu *et al.*	Methyltrichlorosilane treated polyurethane sponge	one-step solution immersion method	[Decane, Dodecane, Octane, Crude oil, Gasoline, Bean oil, Lubricating oil]	[15-25 times its own weight]	157°	0°	[65]

Table 5.1 (continued)

Authors	Sponge/Foams	Method	Oils/ Organic solvents	Absorption capacity (g/g) or (times) or (w/w %)	WCA	OCA	Ref.
Turco *et al.*	MWNT-PDMS sponge	Template assisted	[Dichloromethane, Petroleum ether, Hexane, Chloroform, THF, Toluene, Gasoline]	[8.5 to 20 g/g]	153.4±6.9°	0°	[66]
Zhang *et al.*	Swellable porous PDMS absorbent	modified sugar-template method	Crude oil Diesel Gasoline [Chloroform, Cyclohexane, dichloromethane, Petroleum ether, n-Hexane, Toluene, Acetone]	9 g/g 12 g/g 22 g/g [4-34 g/g]	144°	0°	[67]
Nagappan *et al.*	Bio-inspired hybrid-PDMS/Melamine sponge	Dip coating method	[Chloroform, Toluene, Benzene, Decane, Diesel oil, Canola oil, Corn oil, Soybean oil]	[2000 to 3600 w/w%]	> 175°	<5°	[68]
Zhu *et al.*	Copper film-coated PU sponge	Solution-immersion process	[Lubricating oil, Octane, Decane and Dodecane]	[13 times its own weight]	> 170°	-	[69]
Li *et al.*	3D-graphene/polypyrrole foam	Vacuum freeze drying method	Kerosene Diesel	101.8 g/g 108.76 g/g	-	-	[70]
Calcagnile *et al.*	PU/NPs/PTFE based foam	Deposition technique	Mixed oil with blue dye	-	>160°	-	[72]

5.3 Polymer Based Aerogels for Oil Spill Clean Up

Aerogels have been employed as effective oil absorbents due to their low density, large porosity and high specific surface area. These lowest-density solids are composed of about 99.98% air by volume, but are still extremely rigid. Aerogels can easily bear many times more weight than their own weight and have high ability for oil spill clean up [73]. Chin *et al.* [74] reported the preparation of cellulose aerogel with magnetic and hydrophobic properties via *in-situ* incorporation of magnetic Fe_3O_4 nanoparticles (NPs) in it and further followed by TiO_2 thin layer coating on the surface of the aerogels. After the CO_2 supercritical point drying of cellulose gel, the magnetic cellulose aerogel was obtained. TEM revealed that the mean particle size of Fe_3O_4 NPs ranged between 7 and 10 nm and were encapsulated between the cellulose fibers. SEM analysis also exhibited that the magnetic cellulose aerogel was comprised of nanofiber networks and the coating of TiO_2 led to nanofibers' surfaces rougher and thicker. Hydrophobicity and oleophilicity of the aerogel was attributed to the thin layer of TiO_2 which was coated onto the cellulose aerogel. The oil absorption capacity of the magnetic cellulose aerogel for paraffin oil was up to 28 times its own weight and it could be easily removed by the usage of external magnet during the absorption process. Korhonen *et al.* [17] reported the nanocellulose aerogels possessing high porosity, selectivity for oil-water separation and good recyclability for up to 10 cycles without deterioration of absorption capacity and weight. The authors functionalized the cellulose nanofibrils of aerogel with uniform nanoscopic coating of TiO_2 NPs (7 nm), which was responsible for the hydrophobic and oleophilic properties of nanocellulose aerogel. Absorption capacities of the aerogel varied from 20-40 g/g, depending on the density of the different liquids. Different oils and organic solvents were examined for absorption capacity of the nanocellulose aerogel such as hexadecane, paraffin oil, petroleum benzene, mineral oil, octane, etc.

Nguyen *et al.* [73] reported highly porous aerogel prepared from the cellulose fibers of paper waste. The authors functionalized the cellulose aerogel with methyltrimethoxysilane (MTMS) to develop its hydrophobic and oleophilic properties. MTMS-coated cellulose aerogel exhibited water contact angles of 143° and 145°, as shown in Figure 5.5. In this study, Ruby (RB), Te Giac Trang (TGT) and Rang Dong (RD) crude oils were used to examine the oil absorption capacity of the aerogel at 25 °C. The absorption capacity of RB, TGT and RD were determined as 18.4, 18.5 and 20.5 g/g, respectively. The absorption capacity of the absorbent is controlled by the hydrophobic interaction between the oils and absorbents, oil viscosity, van der Waals forces, pore morphology and capillary effect [16, 75-79]. This is probably a reason that RB and TGT had almost similar oil absorption capacity because of their comparable viscosity values (0.0090 and 0.0088, respectively) and the absorption capacity of RD was observed

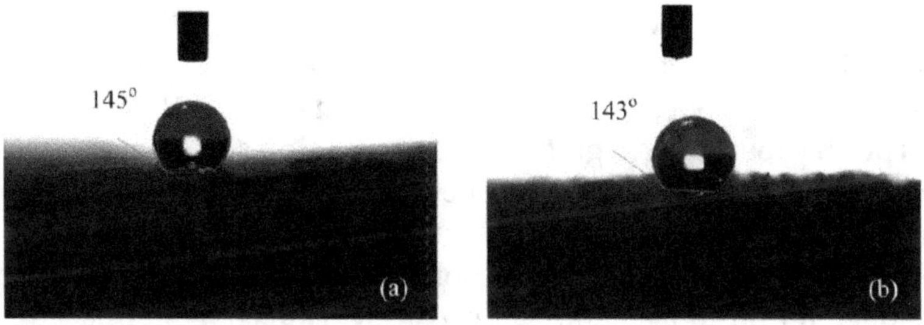

Figure 5.5 (a) Water contact angle on the external surface of the coated aerogel and (b) on the cut surface of the coated aerogel. Reproduced from Reference 73 with permission from American Chemical Society.

to be higher because of lower viscosity value (0.0062) of RD crude oil. The authors also examined the effect of temperature on oil absorption capacity of MTMS-coated cellulose aerogel. As the temperature was increased from 10 to 25 °C, the crude oil (RB) absorption capacity also increased from 13.9 to 18.4 g/g, achieving the highest oil sorption value (24.4 g/g) at 40 °C and subsequently decreased to 19.9 g/g at 60 °C. Sai *et al.* [80] also reported surface modified bacterial cellulose aerogel, which exhibited high porosity (~99.6%), high surface area (≥169.1 m^2/g) and low density (≤6.77 mg/cm^3). The water contact angle and absorption capacities for various oils of the aerogel were observed to be 146.5° and ~185 g/g, respectively. Si *et al.* [81] reported superhydrophobic-superoleophilic polyacrylonitrile (PAN)/SiO_2 nanofiber based aerogels for the separation of the water-in-oil emulsions. SEM revealed that the fiber aerogels had open-cell geometry with a major cellular pore size of 10-50 μm. Due to the numerous minor cellular pores (1-5 μm sizes), the major cells were observed to be highly interconnected. With increasing content of SiO_2 (from 0.01 to 2 wt%), the average pore size of the aerogels got reduced from 18.9 to 4.2 μm and in-contrast, the specific surface area increased from 2.66 to 76.54 m^2/g. After introducing the SiO_2 nanoparticles into the aerogels, the surface roughness, wettability and

WCA (162°) were increased. Fibrous aerogels also exhibited anti-fouling property for long usage, excellent reusability and super elasticity. Jiang *et al.* [82] also reported the ultra-porous (99.5 to 99.9%) and ultra-light (1.7 to 8.1 mg/cm^3) triethoxyl(octyl)silane)-coated aerogels, prepared by using cellulose nanofibrils (defibrillated from rice straw cellulose). Zheng *et al.* [83] prepared the polyvinyl alcohol (PVA)-cellulose nanofibril (CNF) based aerogels which were treated with methyltrichlorosilane by using thermal chemical vapour deposition, rendering the water contact angle of (150.3±1.2°). The authors observed that methyltrichlorosilane-coated PVA/CNF aerogels were highly porous (>98%) and had ultra-low densities (< 15 Kg/m^3). The absorption capacities ranging from 44 to 96 times its own weight for various oils and organic solvents such as diesel oil, gasoline, crude oil, corn oil, pump oil, hexane, chloroform and toluene were observed.

In summary, large number of polymer based aerogels have been developed for oil spill cleaning purposes. Various polymer based aerogels for cleaning of oil spills are also listed in Table 5.2.

5.4 Polymer Based Fibers for Oil Spill Clean Up

Synthetic fibers have potential application in the removal process of oil spills from water. Zhu *et al.* [19] employed electrospinning process to fabricate cost-effective polyvinyl chloride (PVC)/polystyrene (PS) fibers with low density (2-3 mg/cm^3) and high porosity of ~99.7%. SEM analysis revealed that the sorbent was made up of fibers having diameter in the range of 1.5 to 3 μm which enhanced the surface area and oil adsorption on the surface of the fibers. Additionally, large number of interconnected voids (size ranging between 20-70 μm) were observed among fibers (Figure 5.6). These voids had the tendency to store large amounts of oil, thus, avoiding the escape of oil from PVC/PS sorbent at the time of draining. The authors calculated the absorption capacities of PVC/PS sorbent for ethylene glycol, diesel, peanut oil and motor oil as 81 g/g, 38 g/g, 119 g/g and 146 g/g, respectively. The voids among the fibers were considered as crucial for high absorption capacity of the sorbent. In another study reporting the use of fibers as an absorbent for oil spill clean-up applications, Lin *et al.* [78] prepared nanoporous polystyrene fibers with high specific surface area by employing electrospinning method. High sorption capacities of hydrophobic-oleophilic fibers were achieved for sunflower seed oil, bean oil and motor oil as 96.89, 111.80 and 113.87 g/g, respectively. In another study, hydrophobic - oleophilic fibrous

Table 5.2 Various polymer based aerogels for cleaning of oil spills

Authors	Aerogels	Methods	Oils/ Organic solvents	Absorption capacity (g/g) or (times) or (w/w %)	WCA	OCA	Ref.
Chin *et al.*	Magnetic cellulose/TiO_2 aerogel	CO_2 supercritical point drying and Sol-gel process	Paraffin oil	28 times its own weight	-	-	[74]
Korhonen *et al.*	Nanocellulose/TiO_2 aerogel	Vacuum freeze drying	[Hexadecane, Octane, Mineral oil, Chloroform, Toluene, Paraffin oil, Octanol, Dodecane, Petroleum benzene, Hexane]	[20-40 g/g]	>90°	-	[17]
Nguyen *et al.*	MTMS-coated recycled cellulose aerogel	Simple alkaline/urea method and freeze drying	RB crude oil TGT crude oil RD crude oil	18.4 g/g 18.5 g/g 20.5 g/g	>140°	-	[73]
Jiang *et al.*	Triethoxyl(octyl) silane-coated cellulose nanofibril aerogel	Chemical vapor deposition and freeze drying	[Chloroform, DMSO, Ethylene glycol, DMF, Soybean oil, Toluene, Pump oil, Acetone, Cyclohexane, Decane, Octane, Hexane]	[139-356 g/g]	-	-	[82]

Table 5.2 (continued)

Authors	Aerogels	Methods	Oils/ Organic solvents	Absorption capacity (g/g) or (times) or (w/w %)	WCA	OCA	Ref.
Zheng *et al.*	Silane treated PVA-cellulose nanofibril hybrid aerogel	Simple thermal chemical vapor deposition pocess and freeze drying	[Diesel, Gasoline, Crude oil, Corn oil, Pump oil, Hexane, Chloroform, Toluene]	[44-96 times its own weight]	150.3±1.2°	-	[83]
Sai *et al.*	Hydrophobic and oleophilic bacterial cellulose aerogel	Liquid phase reaction followed by freeze drying	[n-Hexane, Acetone, Paraffin liquid, Plant oil, Gasoline, Diesel, Toluene, Chlorobenzene, Dichloromethane, Chloroform]	[Up to 185 g/g]	146.5°	-	[80]
Si *et al.*	Superhydrophobic-superoleophilic PAN/SiO_2 nanofibers based aerogel	Freeze drying method	Petroleum ether	-	162°	0°	[81]

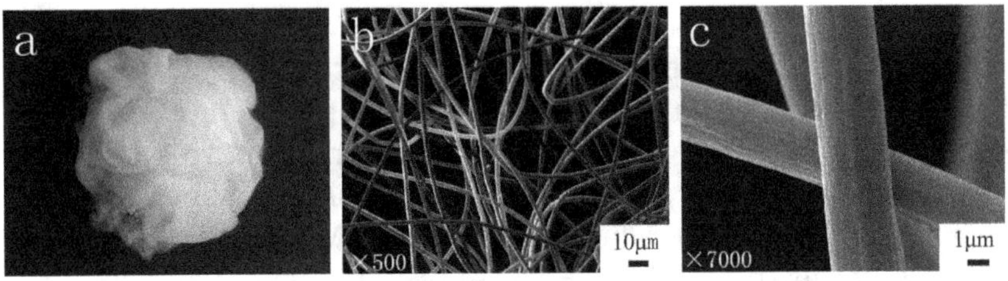

Figure 5.6 (a) Photo, (b) low and (c) high magnification SEM images of PVC/PS sorbent. Reproduced from reference 19 with permission from American Chemical Society.

mats were developed by Lin *et al.* [84]. The work demonstrated the fabrication of PS-PU fibrous mats by using co-axial electrospinning method, where PS and PU solutions were used as the shell and core solutions, respectively. By manipulating the factors such as solvent compositions in the shell solution, concentration of core solution and spinning voltages, the structure of fibrous mats could be controlled. The porous structure of the PS-PU fibers ranged between 20 and 80 nm, when investigated by synchrotron radiation small-angle X-ray scattering (SAXS) and Brunauer-Emmett-Teller (BET). The sorption capacities of fibrous mats for sunflower seed oil and motor oil were 47.78 and 64.40 g/g, respectively and were strongly affected by the inter-fiber voids among the fibers. It was noticed that the sorption capacity of the sorbent was maintained even after five cycles of absorption-desorption. Lin *et al.* [85] also fabricated fibrous mats consisting of PS fibers with micro-nanostructures by using electrospinning technique, mainly focusing on the porous structures of the fibers. The authors investigated the properties of PS fibers by tuning the composition of solvent, concentration of solution and molecular weight of polystyrene. The absorption ability of fibrous mats for sunflower seed oil and motor oil were determined as 79.62 and 84.41 g/g, respectively. Thus, different studies confirmed that polystyrene fibers had better oil absorption capacities than commercial polypropylene non-woven fabric [78,84,85].

In another study, Wu *et al.* [79] reported the fabrication of PS fibrous films with pores formation on the fibers via electrospinning method. The diameter and surface morphology of fibers impacted the oil adsorption capacity as well as oil/water selectivity. With decreasing the diameter of the fibers, the uptake rate and adsorption capacity of the oil increased. The oil adsorption capacities of PS fibrous film for motor oil, peanut oil, silicon oil and diesel oil were around 131.63,

112.30, 81.40 and 7.13 g/g, respectively. The WCA and OCA (motor oil) were measured to be 151.3±1.6° and 0°, respectively, indicating that the PS fibrous film exhibited superhydrophobic and superoleophilic properties. Zhao *et al.* [86] employed melt blown method for the preparation of polypropylene (PP) and poly (butylmethacrylate-co- hydroxyethylmethacrylate) (PBMA-co-HEMA) non-woven oil absorbents. In this study, PP and PBMA-co-HEMA blends were prepared with different composition ratios by weight. It was observed that with increasing the content of PBMA-co-HEMA in the resultant non-woven material, water contact angle and absorption capacity of the material increased. PBMA-co-HEMA (20 wt%) exhibited the highest WCA (136.8±1.19°) and absorption capacities for crude oil, diesel and toluene as 9.02, 8.88, and 7.74 g/g, respectively. After eight cycles of absorption-desorption, only 1.5% reduction in the absorption capacity of the non-woven material was noticed. For the continuous separation of oil-water mixture, Xue *et al.* [87] proposed superhydrophobic and superoleophilic poly(ethylene terephthalate) (PET) textiles prepared via sol-gel coating method. Firstly, PET textiles were treated with NaOH to make the fiber surface rough and also to improve compatibility between the fibers and silica sols. Subsequently, by coating the PET fibers with $-Si(CH_3)_3$ functionalized SiO_2 nanoparticles, superhydrophobic-superoleophilic PET textiles were obtained. The water contact angle of the PET textiles was observed to be more than 150° and gasoline drops spread immediately upon contacting the textile. PET textiles indicated excellent superhydrophobic property even after 700 cycles. Li *et al.* [88] reported polyester textile with superhydrophobic and superoleophilic properties and employed as an effective membrane for oil/water separation and selective removal of oil from water (Figure 5.7). Under hydrothermal conditions, fabrication of the textile was performed via direct *in-situ* growth of layered double hydroxides (LDH) micro-crystals on the surface of polyester textile and followed by further modification of the textile with sodium laurate to reduce the surface energy. TEM analysis revealed that LDH micro-crystals were perpendicularly aligned to the fiber surface of the textile, creating a nest-like microstructure that also led to the micro-nano sized spaces among LDH micro-crystals, contributing to the fiber surface roughness of the textile. Superhydrophobic and superoleophilic properties of the polyester textile were achieved with the combination of low surface energy molecules and hierarchical micro-nanostructure. The water contact angles were measured to be 144±1.2° and 154±1.6° for LDH coated textile and modified LDH coated textile with sodium laurate, respectively and in contrast, oils were spread easily on the resultant textile. The authors also subjected the as-prepared textile to 20 adsorption-squeezing cycles. No change in respect to the water contact angle was observed even after 20 cycles. In another study, Zhang *et al.* [89] employed chemical vapor deposition technique to conduct one-step growth of silicone nano-filaments onto the surface of polyester textile to produce superhydrophobic-superoleophilic polyester material. Wu *et al.* [90] also reported superhydrophobic-superoleophilic polyester material with high selectivity and recyclability for oil-water separation. The authors coated a commercial polyester material by dipping it into the solution of Fe_3O_4@hexadecyltriethoxysilane (HD)-silica/HD-polymer to produce superhydrophobic-superoleophilic material. The oil absorption capacities of coated polyester material for crude oil, diesel and petrol were determined as 396%, 406% and 355%, respectively. By using a magnet, the as-prepared material could be magnetically driven to clean the oil spills. Polymeric fibers for removal of oil or organic solvents from water are also listed in Table 5.3.

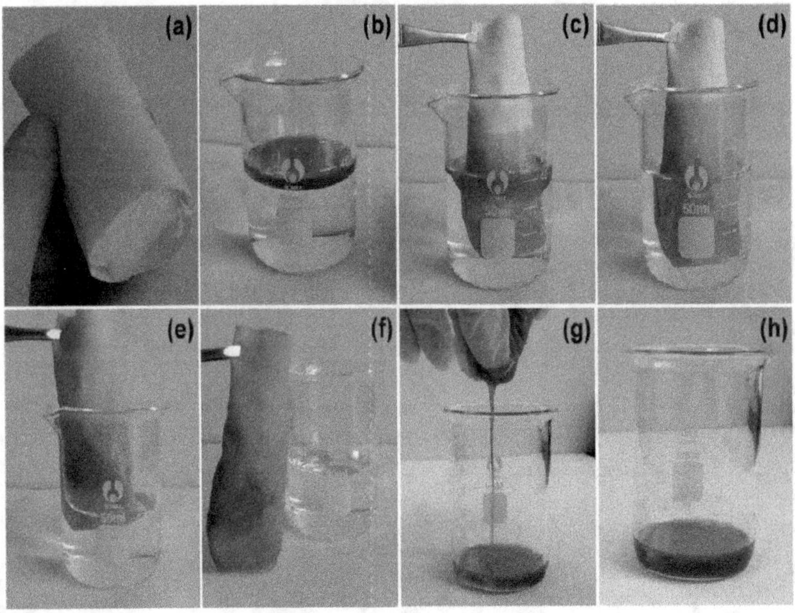

Figure 5.7 Photographs of (a) the textile bag, and (b-g) the oil adsorption and recycling process by using the superhydrophobic textile bag. Reproduced from Reference 88 with permission from American Chemical Society.

Table 5.3 Polymer based fibers for oil spill clean up

Authors	Fibers/Mats/Textiles	Methods	Oils/ Organic solvents	Absorption capacity (g/g) or (times) or (w/w %)	WCA	OCA	Ref.
Zhu *et al.*	Polyvinyl chloride (PVC)/polystyrene (PS) fibers	Electrospinning process	Motor oil Peanut oil Diesel Ethylene glycol	146 g/g 119 g/g 38 g/g 81 g/g	-	-	[19]
Wu *et al.*	PS porous fibers	Electrospinning process	Motor oil Peanut oil Silicon oil Diesel oil	131.63 g/g 112.30 g/g 81.40 g/g 7.13 g/g	151.3±1.6°	0°	[79]
Lin *et al.*	PS-PU fibrous mats	Co-axial electrospinning	Motor oil Sunflower oil	64.40 g/g 47.48 g/g	Obtuse angle	-	[84]
Lin *et al.*	PS fibers	One step electrospinning	Motor oil Bean oil Sunflower oil	113.87 g/g 111.80 g/g 96.89 g/g	-	-	[78]
Lin *et al.*	PS fibers with micro - nanostructures	Electrospinning process	Motor oil Sunflower oil	84.41 g/g 79.62 g/g	Obtuse angle	-	[85]
Zhao *et al.*	PP/PBMA-co-HEMA	Melt blown method	Crude oil Diesel Toluene	9.02 g/g 8.88 g/g 7.74 g/g	136.8±1.19°	0°	[86]
Xue *et al.*	PET textiles	Sol-gel coating	Crude oil, Gasoline	-	>150°	-	[87]
Liu *et al.*	LDH coated polyester textile bag	Hydrothermal process	[n-Hexane, Chloroform, Diesel, Gasoline]	-	154±1.6°	-	[88]
Wu *et al.*	Fe_3O_4@HD-silica/HD polymer coated polyester	Dip coating method	Petrol Diesel Crude oil	355 % 406 % 396 %	-	-	[90]

5.5 Polymeric Membranes for Oil-Water Mixture Separation

Membrane technology has been proved to a promising method to separate oil-in-water emulsions, which otherwise cannot be separated by applying the conventional techniques (flotation, gravity separation, coagulation, ultrasonic separation, skimming and centrifugation). Membrane based separation process is highly applicable for the separation of oil-in-water emulsions with promising methods such as reverse osmosis [8], ultrafiltration [9,10], microfiltration [11], dehydration of oil emulsion by pervaporation [12] and membrane distillation [13]. Li *et al.* [91] reported the aminated polyacrylonitrile (APAN) electrospun nano-fibrous membranes exhibiting superhydrophobic, superoleophilic and excellent recyclable properties for oil-water separation. Firstly, the authors aminated the electrospun PAN nanofibers with diethylenetriamine (DETA) solution, followed by anchoring of silver (Ag) nano-clusters onto the APAN nano-fibrous membranes via electroless plating technique. Subsequently, the modification of the membrane surface with long alkyl chain thiols was performed. The surface roughness of APAN nano-fibrous membranes could be controlled by adjusting the electroless plating time. The permeate fluxes of APAN-Ag-SR-50 nano-fibrous membrane with 100 μm thickness were found to be higher for low viscosity oils, and comparatively lower for the high viscosity oils such as vegetable oil (954.9±38.5 L/m^{2}h) and vacuum pump oil (447.5±100.3 L/m^{2}h). The water and oil contact angles of as-prepared membrane were determined as 162±1.9° and 0°, respectively. Tai *et al.* [92] employed electrospinning technique for the preparation of flexible electrospun SiO_2-carbon nanocompositesbased nano-fibrous membranes, where PAN was used as a carrying polymer for SiO_2 precursor and a precursor for carbon nanofiber (CNF). At 125 °C, ultra-hydrophobic (WCA = 144.2±1.2°) and superoleophilic (OCA = 0°) properties were attained after coating of silicone oil onto the surface of the membrane via vapor deposition method. The chemical and thermal stability tests suggested that the membrane was capable to maintain its ultra-hydrophobicity in a pH range 2-14. The wettability of the as-prepared membrane was resistant up to 350 °C. The authors also determined the average value of permeate flux of membrane (with thickness 213.3±21.8 μm) for hexane, iso-octane and petroleum spirit as 2648.8±89.7, 1719.1±36.2, and 3032.4±234.6 L/m^{2}h, respectively. These permeate flux values are equivalent or even higher than the flux achieved in advanced membranes such as the superhydrophobic polyvinylidene fluoride membrane (PVDF) and mineral coated polypropylene micro-filtration membrane having permeate flux values of (700-3500 L/m^{2}h) [93] and (>2000 L/m^{2}h) [94], respectively. Chakrabarty *et al.* [95] also reported the modified polysulfone (PSF) membranes for the removal of oil from stable oil-in-water emulsion. In another study, Zhang *et al.* [96] reported the polymer of intrinsic microporosity (PIM-1) and polyhedral oligomeric silsesquioxane (POSS) based electrospun microfibrous membranes which could be used to separate immiscible oil-water mixtures and water-in-oil emulsions with efficiencies higher than 99.95% and 99.97%, respectively. Higher separation efficiencies were gained due to two reasons, i. e. superhydrophobic-superoleophilic surface of the membrane and pore sizes (1-5 μm) of membrane which were smaller than the emulsion droplet sizes (5-20 μm). It was observed that PIM-1 fibrous membrane exhibited the WCA of 132° but with an increase of POSS concentration to 40 wt%, hierarchical structures (surface roughness) were gained on the fiber surface to achieve the superhydrophobic membrane with WCA 155°. On immersing PIM-1/POSS fibrous membrane in water, the superhydrophobicity of membrane was confirmed by the presence of a silvery plastron layer on the surface of fibrous membrane which was caused by the formation of air cushion between the membrane surface and water [97]. The authors also employed the PIM-1/POSS fibrous membrane to separate the n-hexane-water mixture and determined the permeate flux of about 2890±100 L/m^{2}h. The as-prepared membranes were also capable to adsorb oil red O and blue 35 with corresponding adsorption capacities of 7.49 and 8.33 mg/g, respectively. In another study, superhydrophobic-superoleophilic nanofibrous membranes were reported by Huang *et al.* [98] for the effective separation of water-in-oil emulsions. The membrane was prepared by the combination of *in-situ* polymerized fluorinated benzoxazine (F-PBZ), silica nanofibrous membranes (SNF) and alumina (Al_2O_3) nanoparticles. The membranes (F-SNF/ Al_2O_3 NPs) exhibited excellent antifouling properties, durability, reusability and the high flux of 892±50 L/m^{2}h.

Tang *et al.* [99] designed novel superhydrophobic-superoleophilic nanofibrous membranes for gravity driven oil-water separation. Firstly, multi-walled carbon nanotubes (MWCNTs) reinforced electrospun poly (m-phenylene isophthalamide) (PMIA) nano-fibrous membranes were prepared via electrospinning technique, followed by the fabrication of *in-situ* polymerized fluorinated-polybenzoxazine functional layer incorporated with SiO_2 nanoparticles (F-

PBZ/SiO_2 NPs layer) on the surface of PMIA/MCNTs nano-fibrous membrane. Modification of hydrophilic PMIA nano-fibrous membrane with F-PBZ/SiO_2 NPs layer resulted into a superhydrophobic and superoleophilic membrane. The as-prepared membrane exhibited the water contact angle and oil contact angle of 161° and 0°, respectively. The permeate flux of membrane was determined as 3311 L/m^2h, when employed to separate the mixture of dichloromethane and water. The membranes exhibited high separation efficiency for oil-water mixtures along with excellent stability over the pH range of 2-12. In a similar study, Shang *et al.* [100] reported similar F-PBZ/SiO_2 NPs modified electrospun cellulose acetate nano-fibrous membrane with superhydrophobic (WCA = 161°) and superoleophilic (OCA = 3°) properties for the separation of dichloromethane from water. Chen *et al.* [101] reported hydrophilic cellulose acetate (CA)-graft-PAN membranes to achieve high permeation flux as well as excellent fouling resistance for oil-water mixture separation. The hydrophilic CA-graft-PAN membranes were fabricated by employing phase inversion method. Membrane material was synthesized by grafting PAN onto CA powder by using free-radical polymerization. Thus, by embedding the PAN segments into the matrix of membrane during phase inversion method, low surface energy of the membrane was achieved to enhance the hydrophilic property of the CA-graft-PAN membrane for the successful oil-water separation. Modified PSF electrospun nano-fiber mats as effective oil-water separation membranes were reported by Obaid *et al.* [102] The authors incorporated NaOH nanoparticles in the PSF nano-fibers, followed by the formation of thin polyamide layer via interfacial polymerization of m-phenylenediamine (MPD) and 1,3,5-benzenetricarbonyl chloride (TMC) on the surface of the electrospun mats. The water contact angle of PSF nanofiber membrane decreased from ~130° to 13° due to the combined effect of NaOH nanoparticles and thin polyamide layer. NaOH nanoparticles were observed to absorb water due to the formation of hydrogen bonds, which led to a decrease in the water contact angle. Similarly, the presence of amide content also reduced the water contact angle [103,104]. The as-prepared membrane was subjected to three successive cycles for separation of soybean oil-water mixture and water flux of the membrane reduced from 8 to 5.5 m^3/m^2 day due to membrane fouling. Zhu *et al.* [105] employed surface-initiated atom transfer radical polymerization (SI-ATRP) technique for the fabrication of a novel zwitterionic polyelectrolyte grafted PVDF membrane with ultrahigh efficiency to separate oil from water. Xiang *et al.*

[106] also prepared superoleophobic PVDF membrane coated with polydopamine via facile solution-immersion method to separate both surfactant-stabilized and surfactant-free oil-water emulsions. After the hydrophilic coating of polydopamine onto the PVDF membrane surface, the water contact angle of membrane reduced from 118±1.5° to 53±2.3°. The oil contact angle was found to be 152±0.3°, when a 3 μl chloroform droplet was dropped on the membrane. The authors also studied the flux values of toluene-in-water (T/W), chloroform-in-water (C/W) and dichloroethane-in-water (DCE/W) emulsions (both surfactant stabilized and surfactant-free emulsions). Dry-wet spinning technique was employed by Li *et al.* [107] to prepare a hollow fiber ultrafiltration membrane for the separation of oil-water mixture. Another superhydrophilic and oleophobic nano-fibrous membranes for ultrafast oil-water separation was reported by Raza *et al.* [108]. The fabrication of the membrane was achieved by the deposition of *in-situ* crosslinked polyethylene glycol diacrylate (PEGDA) nano-fibers onto the polyacrylonitrile/polyethylene glycol (PG NF) membrane. The as-prepared PEGDA@PG NF membranes exhibited high flux values of 10975, 11565, 11976, and 12656 L/m^2h for soybean oil-water (S-W), crude oil-water (C-W), petrol-water (P-W) and hexane-water (H-W) mixtures, respectively. For separation of oil-water emulsion, Chen *et al.* [109] reported polyethersulfone membranes modified with amphiphilic copolymer Pluronic F127. Zhang *et al.* [110] employed the salt-induced phase-inversion approach for the preparation of the superhydrophilic-underwater superoleophilic polyacrylic acid-grafted-polyvinylidene fluoride (PAA-g-PVDF) membranes for effective separation of oil-in-water emulsions. Under applied pressure of 0.1 bar, PAA-g-PVDF membranes exhibited high flux values for surfactant-free and surfactant-stabilized (sodium dodecyl sulfate (SDS) and Tween 80) oil-in-water emulsions.

Various superhydrophobic-superoleophilic or superhydrophilic-superoleophobic polymeric membranes used for oil spill clean up are listed in Table 5.4.

5.6 Other Polymer Absorbents for Oil Spill Clean Up

Li *et al.* [111] reported a smart fiber membrane endowed with switchable wetting states, i.e., superhydrophilicity-underwater superoleophobicity and superhydrophobicity-superoleophilicity. With the combination of the underwater oleophilicity/hydrophilicity of poly(methyl methacrylate) (PMMA) and the pH responsive poly (4-

Table 5.4 Polymeric membranes for oil-water mixture separation

Authors	Membranes	Methods	Oils/ Organic solvents	Permeate Flux (L/m^2h) or (m^3/m^2 day) or (L/ m^2h bar)	WCA	OCA	Ref.
Li *et al.*	Superhydrophobic-superoleophilic APAN-Ag-SR nanofibrous membrane	Electrospinning and electroless plating technique	Vacuum pump oil Vegetable oil [Petroleum ether, n-hexane, chloroform]	447.5±100.3 L/m^2h 954.9±38.5 L/m^2h 2250-3375 L/m^2h	162.4 ± 1.9°	0°	[91]
Tai *et al.*	Hydrophobic carbon-silica nanofibrous membrane	Electrospinning method	Petroleum spirit Iso-octane Hexane	3032.4±234.6 L/m^2h 1719.1±36.2 L/m^2h 2648.8±89.7 L/m^2h	144.2 ± 1.2°	0°	[92]
Chen *et al.*	Superhydrophilic mineral coated PAA-grafted polypropylene membrane	Coating deposition	[Petroleum ether, Hexadecane, Gasoline, Diesel, Dichloroethane]	>2000 L/m^2h	-	>150°	[94]
Zhang *et al.*	Superhydrophobic-superoleophilic PIM-1/POSS fibrous membrane	Electrospinning method	n-Hexane Cyclohexane n-Heptane Petroleum ether Soybean oil Motor oil	2890±100 L/m^2h - - - - -	155°	0° - - - - -	[96]
Tang *et al.*	Superhydrophobic-superoleophilic F-PBZ/SiO_2 NPs modified PMIA nanofibrous membrane	Electrospinning method and *in-situ* polymerization	Dichloromethane	3311 L/m^2h	161°	0°	[99]
Shang *et al.*			Dichloromethane	-	161°	0°	[100]
Chen *et al.*	Hydrophilic cellulose acetate (CA)-graft-polyacrylonitrile (PAN) membrane	Phase inversion method	Vacuum pump oil (GS-1)	227.8-334.5 L/m^2h (Pure water flux)	(55.4 ± 1.7) - (56.9 ± 0.9)	-	[101]

Table 5.4 (continued)

Authors	Membranes	Methods	Oils/ Organic solvents	Permeate Flux (L/m²h) or (m³/m² day) or (L/ m²h bar)	WCA	OCA	Ref.
Obaid *et al.*	Hydrophilic modified polysulfone nanofibrous membrane	Electrospinning method	Soybean oil	5.5-8 (m³/m²day) (water flux)	13°	-	[102]
Huang *et al.*	Superhydrophobic-superoleophilic (F-SNF/ Al_2O_3 NPs) nanofibrous membrane	*In-situ* polymerization approach	Petroleum ether	892±50 L/m²h	161°	0°	[98]
Zhu *et al.*	Superhydrophilic-superoleophobic PMAPS-grafted-PVDF membrane	Surface-initiated atom transfer radical polymerization (SI-ATRP) technique	Diesel oil	11000-13500 (L/ m²h bar) water flux	-	>150°	[105]
			Petroleum ether	-		>150°	
			Soybean oil	-		>150°	
			Hexane	-		>150°	
			Isooctane	-		>150°	
Xiang *et al.*	Superoleophobic PVDF membrane coated with polydopamine	Facile solution-immersion method	Surfactant stabilized (C/W) emulsion	3969 L/m²h	53 ± 2.3°	152 ± 0.3°	[106]
			Surfactant-free (C/W) emulsion	14037 L/m²h		152 ± 0.3°	
			Surfactant stabilized (T/W) emulsion	1991 L/m²h		-	
			Surfactant-free (T/W) emulsion	15882 L/m²h		-	

Table 5.4 (continued)

Authors	Membranes	Methods	Oils/ Organic solvents	Permeate Flux (L/m^2h) or (m^3/m^2 day) or (L/m^2h bar)	WCA	OCA	Ref.
			Surfactant stabilized (DCE/W) emulsion	2351 L/m^2h		-	
			Surfactant-free (DCE/W) emulsion	10183 L/m^2h		-	
Raza *et al.*	Superhydrophilic and oleophobic PEGDA@PG NF membrane	Electrospinning method	Hexane	12656 L/m^2h	-	-	[108]
			Petrol	11976 L/m^2h		-	
			Crude oil	11565 L/m^2h		-	
			Soybean oil	10975 L/m^2h		-	
Zhang *et al.*	Superhydrophilic and underwater superoleophobic PAA-g-PVDF membrane	Salt induced phase-inversion method	1,2-Dichlorophenol	-	0°	160°	[110]
			Hexadecane	2320 L/m^2h		-	
			Toluene	1680 L/m^2h		-	
			Diesel	1550 L/m^2h		-	
			SDS-Hexadecane	1140 L/m^2h		-	
			Tween 80-Toulene	1020 L/m^2h		-	
			SDS-Diesel	720 L/m^2h		-	

vinylpyridine) (P4VP), the as-prepared membrane attained switchable surface wettability towards oil and water. The smart fiber membrane responded through switching the pH medium for on-demand oil/water separation. Due to the protonation and deprotonation of pyridyl groups, the wettability of the as-prepared membrane could be switchable at different pH [112-115]. At pH 3, P4VP (weak polybase) became protonated and due to extended conformation of P4VP segments caused by the electrostatic repulsion between pyridyl groups, the protonated P4VP segments stretched out. This promoted the water uptake into the membrane texture, leading to the supehydrophilic surface of the membrane and subsequently, water selectively passed through the membrane. At pH 7, the separation process got reversed after the deprotonation of P4VP (Figure 5.8). Zhang *et al.* [116] employed the solvothermal method to prepare superhydrophobic-superoleophilic nano-porous polydivinylbenzene (PDVB) adsorbents for the adsorption of various organic contaminants and volatile organic compounds such as n-heptane, kerosene, benzene and nitrobenzene. Large surface area, adjustable pore sizes and nano-porus structures were exhibited by the PDVB material. Atta *et al.* [117] reported the cross-linked poly (octadecene-alt-maleic anhydride) copolymers, which were used as crude oil sorbers. Similarly, Yuan *et al.* [28] reported cross-linked polyolefin terpolymer as an oil superabsorbent, having oil uptake capacity about 45 times its own weight. The tea-bag nanocomposite system as an oil absorbent material have been reported by Avila *et al.* [118]. This system is based on PS superhydrophobic membranes, surrounding the exfoliated graphite. The absorption rates of tea-bag system for vacuum pump oil, new and used motor oil and water-oil emulsions ranged between 2.5 and 40 g/g. Gu *et al.* [119] reported highly hydrophobic and superoleophilic magnetic polymer nanocomposites prepared via emulsion polymerization. Styrene (St), divinylbenzene (DVB) and methylmethacrylate (MMA) were used for the preparation of core-shell magnetic polymer nanocomposites. Fe_3O_4@poly(St/DVB) magnetic nanocomposites and poly(MMA/St/DVB) layer were employed as core and shell, respectively for preparing highly selective absorbent for the successful separation of oil from water. The absorption capacities of as-prepared magnetic polymer nanocomposites for different oils were found to be 3.63 times their own weight. Another highly hydrophobic and superoleophilic core-shell Fe_3O_4/polystyrene magnetic nanocomposites with absorption capacities of 3 times their own weight for lubricating oils were reported by Chen *et al.* [120].

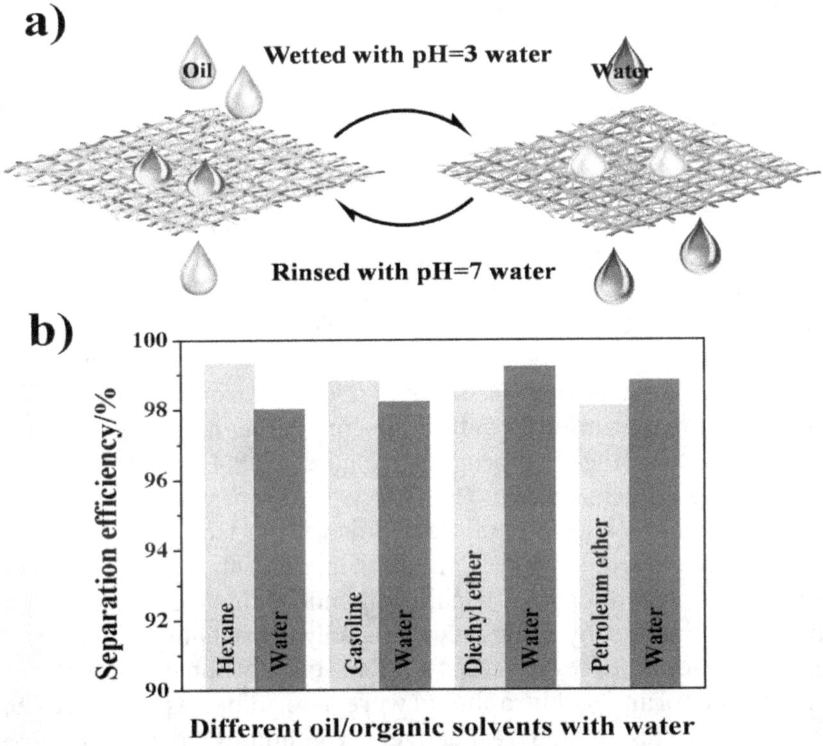

Figure 5.8 (a) Schematic diagrams of pH-switchable oil/water separation and (b) Oil/organic solvent separation efficiency (dry state) and water separation efficiency (after wetting with pH 3 water) of four different mixtures. Reproduced from Reference 111 with permission from American Chemical Society.

β-Cyclodextrin based polymeric microspheres (POAMs) as oil absorbent were reported by Song *et al.* [121]. Butyl acrylate and octadecyl acrylate were employed as co-monomers for the preparation of POAMs via suspension polymerization. The absorption capacities of POAMs for diesel, kerosene, gasoline, toluene, xylenes, $CHCl_3$, CCl_4 were determined as 18.2, 27.1, 30, 42.8, 48.7, 75.1, and 83.4 g/g, respectively. Mao *et al.* [122] reported the preparation of another porous magnetic poly (styrene-divinylbenzene)/Fe_3O_4 microspheres via seed swelling polymerization process. The absorption capacities of resultant microspheres for various oils were up to 8.42 times their own weight. The microspheres also exhibited good reusability for 10 cycles, though a slight reduction in oil absorption capacity was noticed after the 10th cycle.

5.7 Polymeric Adsorbents for Phenol Removal

A wide range of polymer based adsorbents have been employed for the adsorption of phenol from industrial wastewater. Huang *et al.* [123] reported the HJ-1 resin (hyper-crosslinked polymeric adsorbent) for adsorption of phenol and *p*-cresol from industrial wastewater. Adsorption kinetics was obsered to follow pseudo-second-order kinetics and langmuir model could be fitted with the adsorption isotherm data. Caetano *et al.* [124] reported the removal of phenol using three polymeric resins, i.e., MN 200 (crosslinked polystyrene), AuRIX 100 (styrene divinylbenzene with guanidine functional group) and Dowex XZ (styrene divinylbenzene with t-butylamine functional group). To define the controlling mechanism of adsorption process, the shell progressive model (SPM) and homogeneous particle diffusion model (HPDM) were used, which fitted well with the kinetic experimental data. Adsorption isotherm data was successfully described by the langmuir model. Senel *et al.* [125] reported the removal of 2,4,6-trichlorophenol, *o*-chlorophenol, phenol and *p*-chlorophenol by using recyclable polyamide hollow fibers. The maximum adsorption capacities of 2,4,6-trichlorophenol, *o*-chlorophenol, phenol and *p*-chlorophenol were determined as 179.2, 202.8, 145.9, and 194.5 μmol/g, respectively. The hollow fibers were also observed to be recyclable for more than 10 cycles without any observable decrease in the adsorption capacity. The kinetic studies indicated that the adsorption of chlorophenols onto hollow fibers fitted the pseudo-second-order model. An *et al.* [126] reported polyethyleneimine-grafted-silica gel particles (PEI/SiO_2) for the adsorption of phenol from aqueous solution. Due to the coupling effect of γ-chloropropyl trimethoxysilane, the surface of silica gel particle was grafted with the macromolecule polyethyleneimine. The adsorption ability of PEI/SiO_2 for phenol was determined to be 160 mg/g. The equilibrium adsorption data was well described by Freundlich isotherm model. In another study, An *et al.* [127] also employed molecular imprinting technique to prepare the imprinted polymer MIP-PEI/SiO_2 with excellent affinity and selectivity for phenol. Shao *et al.* [128] employed plasma-induced grafting technique to prepare the polyaniline (PANI)-grafted-multiwalled carbon nanotubes (MWCNTs) for the removal of phenol from contaminated water. It was also observed that due to the grafting of PANI onto MWCNTs, there was a strong conjugate effect between PANI and phenol, which jointly enhanced the adsorption capacity to adsorb phenol. Zeng *et al.*

[129] reported polymeric adsorbents, PDM-1 (macroporous styrene-co-divinylbenzene copolymer with ester groups) and PDM-2 (PDM-1 copolymer was post-crosslinked by Friedel-Crafts reaction of pendant vinyl groups), which were employed for the adsorption of phenol from aqueous solution. Adsoprtion kinetics were observed to follow pseudo-second-order model and Freundlich isotherm model was well fitted with the equilibrium adsorption data. Another water-compatible hypercrosslinked NJ-8 (oxygen modified polystyrene) polymeric adsorbent was reported by Li *et al.* [130] for the adsorption of phenolic derivatives such as *p*-nitrophenol, *p*-cresol, *p*-chlorophenol and phenol. The equilibrium adsorption data was well described by using the Freundlich isotherm. Ming *et al.* [131] reported the comparison of NDA 103 (aminated polystyrene), amberlite XAD4 (polystyrene) and Amberlite IRA96C (weak base polystyrene) for the adsorption of phenol from water solution at temperatures 293-313 K. Langmuir and Freundlich isotherms for three different polymeric adsorbents were well fitted to the equilibrium adsorption data. It was observed that phenol adsorption capacity of NDA 103 (driven by hydrogen bonding and van der Waals, simultaneously) was higher than Amberlite XAD4 and Amberlite IRA96C (driven by van der Waals interaction only and hydrogen bonding only, respectively). Zhang *et al.* [132] also investigated the phenol adsorption performance of aminated polystyrene resins NDA-101 with surface area of 845 m^2/g and NDA-103 with surface area of 611 m^2/g. It was observed that due to the combination of π-π stacking interactions with phenol and multiple hydrogen bonding, aminated resins achieved better adsorption capacity. Both Freundlich and Langmuir equations fitted well to the adsorption isotherms. Pan *et al.* [133] prepared NDA-701 (styrene-divinylbenzene) as a hyper-crosslinked polymer adsorbent with a unique bimodal pore size distribution and high surface area (824 m^2/g) for adsorption of 4-nitrophenol from aqueous solution. 4-nitrophenol adsorption isotherm was well described by Freundlich model. Kinetic studies demonstrated the adsorption of 4-nitrophenol onto NDA-701 to follow the pseudo-second-order model. In another study, Pan *et al.* [134] also reported macroporous polystyrene, its chemically modified animated derivative and a weak anion exchanger (CHA-111, MCH-111 and ND-900, respectively) for the adsorption of phenolic compouds such as phenol, *p*-cresol, *p*-chlorophenol, *p*-nitrophenol, 2,4-dichlorophenol and 2,4-dinitrophenol. Li *et al.* [135] compared the adsorption capacity of Amberlite XAD-4 and MX-4 polymeric adsorbents for different phenolic compounds from the

aqueous solution. To obtain MX-4, which could be used directly without any wetting process, Amberlite XAD-4 was chemically modified with acetyl group. The equilibrium adsorption capacities for phenolic compounds were observed to increase by 20 % after the modification of Amberlite XAD-4. Isotherm data for phenol, *p*-cresol, *p*-chlorophenol and *p*-nitrophenol onto Amberlite XAD-4 and MX-4 well fitted to the Freundlich model. Effective removal of phenol from aqueous solution by employing Amberlite XAD-7 (porous acrylic ester polymer) was also reported by Pan *et al.* [136]. The kinetic analysis fitted with pseudo-second-order kinetic model and adsorption isotherms were well described by both Langmuir and Freundlich models. Thermodynamic parameters also revealed that phenol adsorption process was exothermic and spontaneous in nature. Juang *et al.* [137] reported non-ionic macroreticular resins for adsorption isotherms of phenol and 4-chlorophenol from aqueous solution. Zeng *et al.* [138] prepared a series of post-crosslinked methyl methacrylate/divinylbenzene (MMA-co-DVB) and ethylene glycol dimethacrylate/ divinylbenzene (EGDMA-co-DVB) copolymer adsorbents by modifying them with different amounts of MMA and EGDMA for the enhanced adsorption of phenol from aqueous solution. Phenol adsoprtion isotherms onto post-crosslinked copolymers were well desribed by both Frendlich and Langmuir models. Wagner *et al.* [139] used polystyrene and polymethacrylate resins which were crosslinked with divinylbenzene for the adsorption of different phenolic derivatives from polluted water. The reins were used for phenol and chlorophenols adsorption at temperatures ranging from 294.15 to 318.15 K. Adsorption isotherms for different phenols were well fitted to Redlich-Peterson model.

The adsorption of phenol from cyclohexane by using polymethylmethacrylate-co-divinylbenzene, polymethylmethacrylate-co-divinylbenzene/ polystyrene-co-divinylbenzene and poly (N-p-vinylbenzyl acetylamide) was reported by Li *et al.* [140]. Dursun *et al.* [141] reported chitin, a natural occuring polymer, for the adsorption of phenol from aqueous solution [141]. Freundlich model fitted well to the adsorption data and pseudo-second-order model correlated to the experimental data. Thermodynamics parameters revealed that phenol adsorption process onto chitin was endothermic and spontaneous in nature. Kawabata *et al.* [142] reported the adsorption of phenol from aqueous solution by using vinylpyridine-divinylbenzene copolymer. Fowkes *et al.* [143] reported PMMA films for the adsorption of phenol from methylene iodide solution. Burleigh *et al.* [144]

reported the adsorption of 4-methylphenol, 4-nitrophenol and 4-chlorophenol by using porous polysilsesquioxanes. Huang *et al.* [145] reported HJ-Z01 resin (a hypercrosslinked resin modified with N-methylacetamide) for the adsorption of phenol from aqueous solution. Isotherms for adsorption of phenol onto HJ-Z01 could be well correlated by Langmuir and Freundlich models. Experimental data fitted with pseudo-second-order model. Gupta *et al.* [146] reported the pervaporation performances of hydroxyterminated polybutadiene based polyurethaneurea membranes for the separation of phenol from water-phenol mixture. All the polymeric adsorbents for the adsorption of phenol and its derivatives from aqueous solution are also listed in Table 5.5.

5.8 Conclusion

Oil spills and phenol-contaminated water are serious concern for the ecosystem, which need to be addressed with responsible actions. In recent years, diverse polymeric materials have been developed to effectively deal with these issues.

In this chapter, different advanced polymer systems exhibiting functionalities for adsorption, absorption and separation of these contaminants from aqueous bodies have been reviewed. Firstly, diverse polymeric materials such as sponges, aerogels, fibers, membranes and other polymeric materials (polymer nanocomposites, textile bag, microspheres, tea bag, smart materials, etc.) have been focused for the effective removal of oil spill clean up. Such materials have been largely fabricated by either designing with superoleophobic-superhydrophilic or superhydrophobic-superoleophilic surfaces to separate the water-oil mixture. Secondly, various synthetic polymeric resins as adsorbents have been reviewed for the removal of phenol and its derivatives from industrial wastewater.

Table 5.5 Various polymeric adsorbents for the adsorption of phenol and its derivative

Authors	Adsorbent	Structure	Pollutant	Model	Kinetics order	Ref.
Huang *et al.*	HJ-1 resin	Formaldehyde carbonyl groups modified polystyrene	Phenol, p-Cresol	Langmuir	Pseudo-second order	[123]
Caetano *et al.*	MN 200 Dowex XZ AuRIX 100	Crosslinked polystyrene Styrene divinylbenzene with t-butylamine functional group Styrene divinylbenzene with guanidine	Phenol	Langmuir	-	[124]
Senel *et al.*	Polyamide hollow fibers carrying Reactive Green HE 4BD	Polyamide hollow fibers	2,4,6-Trichlorophenol, *o*-Chlorophenol, Phenol and *p*-Chlorophenol	-	Pseudo-second order	[125]
An *et al.*	PEI/SiO_2	Macromolecule polyethyleneimine-grafted-silica gel particles	Phenol	Freundlich	-	[126]
An *et al.*	MIP-PEI/SiO_2	Molecular imprinted PEI/SiO_2polymer	Phenol and *p*-Nitrophenol	Freundlich	-	[127]
Shao *et al.*	PANI/MWCNTs nanocomposites	Polyaniline-grafted-MWCNT nanocomposites	Phenol	-	-	[128]
Zeng *et al.*	PDM-1 PDM-2	macroporous styrene-co-DVB with ester groups PDM-1 copolymer post-cross-linked by Friedel –Crafts reaction of pendant vinyl groups	Phenol	Freundlich	Pseudo-second order	[129]
Li *et al.*	NJ-8	Oxygen modified polystyrene	[Phenol, p-Cresol, p-Chlorophenol and p-Nitrophenol]	Freundlich	-	[130]

Table 5.5 (continued)

Authors	Adsorbent	Structure	Pollutant	Model	Kinetics order	Ref.
Ming *et al.*	NDA-103 XAD4 Amberlite IRA96C	Aminated polystyrene Polystyrene Weak base polystyrene	Phenol	Langmuir, Freundlich	-	[131]
Zhang *et al.*	NDA-101 NDA-103	Aminated polystyrene with surface area of 845 m^2/g Aminated polystyrene with surface area of 611 m^2/g	Phenol	Langmuir, Freundlich	-	[132]
Pan *et al.*	NDA-701	Styrene-Divinylbenzene.	4-Nitrophenol	Freundlich	Pseudo-second order	[133]
Pan *et al.*	CHA-111 MCH-111 ND-900	Macroporous polystyrene Animated derivative of CHA-111 Weakly anion exchanger (polystyrene)	[phenol, *p*-Cresol, *p*-Chlorophenol, *p*-Nitrophenol, 2,4-Dichlorophenol, 2,4-Dinitrophenol]	Freundlich	-	[134]
Li *et al.*	XAD-4 MX-4	Polystyrene Polystyrene modified with acetyl group	[phenol, *p*-Cresol, *p*-Chlorophenol, *p*-Nitrophenol]	Freundlich	-	[135]
Pan *et al.*	Amberlite XAD-7	Porous acrylic ester polymer	Phenol	Langmuir, Freundlich	Pseudo-second order	[136]
Zeng *et al.*	MMA-co-DVB EGDMA-co-DVB	Methylmethacrylate-divinylbenzene copolymer Ethylene glycol dimethacrylate- divinylbenzene copolymer	Phenol	Langmuir, Freundlich	-	[138]

Table 5.5 (continued)

Authors	Adsorbent	Structure	Pollutant	Model	Kinetics order	Ref.
Wagner *et al.*	Amberlite XAD-4 Serdolite PAD I, II and III TRP-100 Lewatit VP-OC 1163	Polystyrene and polymethacrylate resins cosslinked with divinylbenzene	[Phenol, 4-Chlorophenol, 2- Chlorophenol, 2,4,6-trichlorophenol]	Redlich-Peterson	-	[139]
Li *et al.*	Various copolymers	P(MMA-co-DVB) Interpenetrating polymer based on P(MMA-co-DVB) and P(S-co-DVB) Poly (N-p-vinylbenzyl acetylamide)	Phenol	Freundlich	-	[140]
Dursun *et al.*	Chitin	Natural occurring polymer	Phenol	Freundlich	Pseudo-second order	[141]
Kawabata *et al.*	Vinylpyridine-Divinylbenzene	Vinylpyridine-Divinylbenzene copolymer	Phenol	-	-	[142]
Fowkes *et al.*	PMMA films	PMMA films	Phenol	-	-	[143]
Burleigh *et al.*	Polysilsesquioxanes	Arylene- and ethylene-bridged Polysilsesquioxanes	[4-Methylphenol, 4-Nitrophenol, 4-Chlorophenol]	-	-	[144]
Huang *et al.*	HJ-Z01	Hypercrosslinked resin modified with N-methylacetamide	Phenol	Langmuir and Freundlich	Pseudo-second order	[145]
Gupta *et al.*	PUUSD membrane	Hydroxyterminated polybutadiene based polyurethaneurea membranes	Phenol	-	-	[146]

References

1. Neff, J. M. (2002) *Bioaccumulation in Marine Organisms: Effect of Contaminants from Oil Well Produced Water*, Elsevier, USA.
2. Fakhru'l-Razi, A., Pendashteh, A., Abdullah, L. C., Biak, D. R. A., Madaeni, S. S., and Abidin, Z. Z. (2009) Review of technologies for oil and gas produced water treatment. *Journal of Hazardous Materials*, **170**, 530-551.
3. Fingas, M. (2010) *Oil Spill Science and Technology*, 2nd edition, Gulf Professional Publishing, USA.
4. Um, M.-J., Yoon, S.-H., Lee, C.-H., Chung, K.-Y., and Kim, J.-J. (2001) Flux enhancement with gas injection in crossflow ultrafiltration of oily wastewater. *Water Research*, **35**, 4095-4101.
5. Gaaseidnes, K., and Turbeville, J. (1999) Separation of oil and water in oil spill recovery operations. *Pure and Applied Chemistry*, **71**, 95-101.
6. Cheryan, M., and Rajagopalan, N. (1998) Membrane processing of oily streams. Wastewater treatment and waste reduction. *Journal of Membrane Science*, **151**, 13-28.
7. Xue, Z., Cao, Y., Liu, N., Feng, L., and Jiang, L. (2014) Special wettable materials for oil/water separation. *Journal of Materials Chemistry A*, **2**, 2445-2460.
8. Mohammadi, T., Kazemimoghadam, M., and Saadabadi, M. (2003) Modeling of membrane fouling and flux decline in reverse osmosis during separation of oil in water emulsions. *Desalination*, **157**, 369-375.
9. Karakulski, K., Kozlowski, A., and Morawski, A. (1995) Purification of oily wastewater by ultrafiltration. *Separations Technology*, **5**, 197-205.
10. Bodzek, M., and Konieczny, K. (1992) The use of ultrafiltration membranes made of various polymers in the treatment of oil-emulsion wastewaters. *Waste Management*, **12**, 75-84.
11. Ohya, H., Kim, J., Chinen, A., Aihara, M., Semenova, S., Negishi, Y., Mori, O., and Yasuda, M. (1998) Effects of pore size on separation mechanisms of microfiltration of oily water, using porous glass tubular membrane. *Journal of Membrane Science*, **145**, 1-14.
12. Deng, S., Sourirajan, S., Chan, K., Farnand, B., Okada, T., and Matsuura, T. (1991) Dehydration of oil-water emulsion by pervaporation using porous hydrophilic membranes. *Journal of Colloid and Interface Science*, **141**, 218-225.
13. Gryta, M., and Karakulski, K. (1999) The application of membrane distillation for the concentration of oil-water emulsions. *Desalination*, **121**, 23-29.
14. Zhong, J., Sun, X., and Wang, C. (2003) Treatment of oily wastewater produced from refinery processes using flocculation and ceramic

membrane filtration. *Separation and Purification Technology*, **32**, 93-98.

15. Chu, Y., and Pan, Q. (2012) Three-dimensionally macroporous Fe/C nanocomposites as highly selective oil-absorption materials. *ACS Applied Materials & Interfaces*, **4**, 2420-2425.
16. Gui, X., Zeng, Z., Lin, Z., Gan, Q., Xiang, R., Zhu, Y., Cao, A., and Tang, Z. (2013) Magnetic and highly recyclable macroporous carbon nanotubes for spilled oil sorption and separation. *ACS Applied Materials & Interfaces*, **5**, 5845-5850.
17. Korhonen, J. T., Kettunen, M., Ras, R. H., and Ikkala, O. (2011) Hydrophobic nanocellulose aerogels as floating, sustainable, reusable, and recyclable oil absorbents. *ACS Applied Materials & Interfaces*, **3**, 1813-1816.
18. Choi, H. M., and Cloud, R. M. (1992) Natural sorbents in oil spill cleanup. *Environmental Science & Technology*, **26**, 772-776.
19. Zhu, H., Qiu, S., Jiang, W., Wu, D., and Zhang, C. (2011) Evaluation of electrospun polyvinyl chloride/polystyrene fibers as sorbent materials for oil spill cleanup. *Environmental Science & Technology*, **45**, 4527-4531.
20. Lessard, R. R., and DeMarco, G. (2000) The significance of oil spill dispersants. *Spill Science & Technology Bulletin*, **6**, 59-68.
21. Kujawinski, E. B., Kido Soule, M. C., Valentine, D. L., Boysen, A. K., Longnecker, K., and Redmond, M. C. (2011) Fate of dispersants associated with the Deepwater Horizon oil spill. *Environmental Science & Technology*, **45**, 1298-1306.
22. Yang, C., Kaipa, U., Mather, Q. Z., Wang, X., Nesterov, V., Venero, A. F., and Omary, M. A. (2011) Fluorous metal–organic frameworks with superior adsorption and hydrophobic properties toward oil spill cleanup and hydrocarbon storage. *Journal of the American Chemical Society*, **133**, 18094-18097.
23. Sakthivel, T., Reid, D. L., Goldstein, I., Hench, L., and Seal, S. (2013) Hydrophobic high surface area zeolites derived from fly ash for oil spill remediation. *Environmental Science & Technology*, **47**, 5843-5850.
24. Bastani, D., Safekordi, A., Alihosseini, A., and Taghikhani, V. (2006) Study of oil sorption by expanded perlite at 298.15 K. *Separation and Purification Technology*, **52**, 295-300.
25. Syed, S., Alhazzaa, M., and Asif, M. (2011) Treatment of oily water using hydrophobic nano-silica. *Chemical Engineering Journal*, **167**, 99-103.
26. Majano, G., and Mintova, S. (2010) Mineral oil regeneration using selective molecular sieves as sorbents. *Chemosphere*, **78**, 591-598.
27. Zhou, X. M., and Chuai, C. Z. (2010) Synthesis and characterization of a novel high-oil-absorbing resin. *Journal of Applied Polymer Science*, **115**, 3321-3325.

28. Yuan, X., and Chung, T. M. (2012) Novel solution to oil spill recovery: using thermodegradable polyolefin oil superabsorbent polymer (oil–SAP). *Energy & Fuels*, **26**, 4896-4902.
29. Sun, X.-F., Sun, R., and Sun, J.-X. (2002) Acetylation of rice straw with or without catalysts and its characterization as a natural sorbent in oil spill cleanup. *Journal of Agricultural and Food Chemistry*, **50**, 6428-6433.
30. Husseien, M., Amer, A., El-Maghraby, A., and Hamedallah, N. (2009) A comprehensive characterization of corn stalk and study of carbonized corn stalk in dye and gas oil sorption. *Journal of Analytical and Applied Pyrolysis*, **86**, 360-363.
31. Radetic, M. M., Jocic, D. M., Jovancic, P. M., Petrovic, Z. L., and Thomas, H. F. (2003) Recycled wool-based nonwoven material as an oil sorbent. *Environmental Science & Technology*, **37**, 1008-1012.
32. Ali, N., El-Harbawi, M., Jabal, A.A., and Yin, C.-Y. (2012) Characteristics and oil sorption effectiveness of kapok fibre, sugarcane bagasse and rice husks: oil removal suitability matrix. *Environmental Technology*, **33**, 481-486.
33. Nordvik, A.B. (1995) The technology windows-of-opportunity for marine oil spill response as related to oil weathering and operations. *Spill Science & Technology Bulletin*, **2**, 17-46.
34. Wayment, E., and Wagstaff, B. (1999) Appropriate technology for oil spill management in developing nations. *Pure and Applied Chemistry*, **71**, 203-208.
35. Burton, Z., and Bhushan, B. (2006) Surface characterization and adhesion and friction properties of hydrophobic leaf surfaces. *Ultramicroscopy*, **106**, 709-719.
36. Poynor, A., Hong, L., Robinson, I.K., Granick, S., Zhang, Z., and Fenter, P.A. (2006) How water meets a hydrophobic surface. *Physical Review Letters*, **97**, 266101.
37. Onda, T., Shibuichi, S., Satoh, N., and Tsujii, K. (1996) Super-water-repellent fractal surfaces. *Langmuir*, **12**, 2125-2127.
38. Bico, J., Marzolin, C., and Quéré, D. (1999) Pearl drops. *Europhysics Letters*, **47**, 220.
39. Bhushan, B., and Jung, Y. C. (2011) Natural and biomimetic artificial surfaces for superhydrophobicity, self-cleaning, low adhesion, and drag reduction. *Progress in Materials Science*, **56**, 1-108.
40. Darmanin, T., and Guittard, F. (2009) Super oil-repellent surfaces from conductive polymers. *Journal of Materials Chemistry*, **19**, 7130-7136.
41. Steele, A., Bayer, I., and Loth, E. (2008) Inherently superoleophobic nanocomposite coatings by spray atomization. *Nano Letters*, **9**, 501-505.
42. Zeng, J., Wang, B., Zhang, Y., Zhu, H., and Guo, Z. (2014) Strong amphiphobic porous films with oily-self-cleaning property beyond na-

ture. *Chemistry Letters*, **43**, 1566-1568.

43. Drelich, J., and Chibowski, E. (2010) Superhydrophilic and superwetting surfaces: definition and mechanisms of control. *Langmuir*, **26**, 18621-18623.
44. Song, S., Jing, L., Li, S., Fu, H., and Luan, Y. (2008) Superhydrophilic anatase TiO_2 film with the micro-and nanometer-scale hierarchical surface structure. *Materials Letters*, **62**, 3503-3505.
45. Hutton, S., Crowther, J., and Badyal, J. (2000) Complexation of fluorosurfactants to functionalized solid surfaces: smart behavior. *Chemistry of Materials*, **12**, 2282-2286.
46. Howarter, J. A., and Youngblood, J. P. (2007) Self-Cleaning and Antifog surfaces via stimuli-responsive polymer brushes. *Advanced Materials*, **19**, 3838-3843.
47. Howarter, J. A., Genson, K. L., and Youngblood, J. P. (2011) Wetting behavior of oleophobic polymer coatings synthesized from fluorosurfactant-macromers. *ACS Applied Materials & Interfaces*, **3**, 2022-2030.
48. Melian, E. P., Diaz, O. G., Arana, J., Rodriguez, J. D., Rendon, E. T., and Melian, J. H. (2007) Kinetics and adsorption comparative study on the photocatalytic degradation of o-, m-and p-cresol. *Catalysis Today*, **129**, 256-262.
49. Babic, K., Driessen, G., Van der Ham, A., and De Haan, A. (2007) Chiral separation of amino-alcohols using extractant impregnated resins. *Journal of Chromatography A*, **1142**, 84-92.
50. Arana, J., Melian, E. P., Lopez, V. R., Alonso, A. P., Rodriguez, J. D., Diaz, O. G., and Pena, J. P. (2007) Photocatalytic degradation of phenol and phenolic compounds: Part I. Adsorption and FTIR study. *Journal of Hazardous Materials*, **146**, 520-528.
51. Ahmaruzzaman, M., and Sharma, D. (2005) Adsorption of phenols from wastewater. *Journal of Colloid and Interface Science*, **287**, 14-24.
52. Bulut, Y., and Baysal, Z. (2006) Removal of Pb (II) from wastewater using wheat bran. *Journal of Environmental Management*, **78**, 107-113.
53. Kujawski, W., Warszawski, A., Ratajczak, W., Porebski, T., Capała, W., and Ostrowska, I. (2004) Removal of phenol from wastewater by different separation techniques. *Desalination*, **163**, 287-296.
54. Otero, M., Zabkova, M., and Rodrigues, A.E. (2005) Adsorptive purification of phenol wastewaters: Experimental basis and operation of a parametric pumping unit. *Chemical Engineering Journal*, **110**, 101-111.
55. Gupta, V. K., Mohan, D., Suhas, A., and Singh, K. P. (2006) Removal of 2-aminophenol using novel adsorbents. *Industrial & Engineering Chemistry Research*, **45**, 1113-1122.
56. Slein, M., and Sansone, E. (1981) Degradation of chemical carcino-

gens. *Journal of Occupational and Environmental Medicine*, **23**, 162.
57. Wang, C.-F., and Lin, S.-J. (2013) Robust superhydrophobic/superoleophilic sponge for effective continuous absorption and expulsion of oil pollutants from water. *ACS Applied Materials & Interfaces*, **5**, 8861-8864.
58. Jiang, G., Hu, R., Xi, X., Wang, X., and Wang, R. (2013) Facile preparation of superhydrophobic and superoleophilic sponge for fast removal of oils from water surface. *Journal of Materials Research*, **28**, 651-656.
59. Zhang, Z., Sebe, G., Rentsch, D., Zimmermann, T., and Tingaut, P. (2014) Ultralightweight and flexible silylated nanocellulose sponges for the selective removal of oil from water. *Chemistry of Materials*, **26**, 2659-2668.
60. Wang, S., Peng, X., Zhong, L., Tan, J., Jing, S., Cao, X., Chen, W., Liu, C., and Sun, R. (2015) An ultralight, elastic, cost-effective, and highly recyclable superabsorbent from microfibrillated cellulose fibers for oil spillage cleanup. *Journal of Materials Chemistry A*, **3**, 8772-8781.
61. Wu, D., Fang, L., Qin, Y., Wu, W., Mao, C., and Zhu, H. (2014) Oil sorbents with high sorption capacity, oil/water selectivity and reusability for oil spill cleanup. *Marine Pollution Bulletin*, **84**, 263-267.
62. Liu, Y., Ma, J., Wu, T., Wang, X., Huang, G., Liu, Y., Qiu, H., Li, Y., Wang, W., and Gao, J. (2013) Cost-effective reduced graphene oxide-coated polyurethane sponge as a highly efficient and reusable oil-absorbent. *ACS Applied Materials & Interfaces*, **5**, 10018-10026.
63. Pan, Y., Shi, K., Peng, C., Wang, W., Liu, Z., and Ji, X. (2014) Evaluation of hydrophobic polyvinyl-alcohol formaldehyde sponges as absorbents for oil spill. *ACS Applied Materials & Interfaces*, **6**, 8651-8659.
64. Nguyen, D. D., Tai, N.-H., Lee, S.-B., and Kuo, W.-S. (2012) Superhydrophobic and superoleophilic properties of graphene-based sponges fabricated using a facile dip coating method. *Energy & Environmental Science*, **5**, 7908-7912.
65. Zhu, Q., Chu, Y., Wang, Z., Chen, N., Lin, L., Liu, F., and Pan, Q. (2013) Robust superhydrophobic polyurethane sponge as a highly reusable oil-absorption material. *Journal of Materials Chemistry A*, **1**, 5386-5393.
66. Turco, A., Malitesta, C., Barillaro, G., Greco, A., Maffezzoli, A., and Mazzotta, E. (2015) A magnetic and highly reusable macroporous superhydrophobic/superoleophilic PDMS/MWNT nanocomposite for oil sorption from water. *Journal of Materials Chemistry A*, **3**, 17685-17696.
67. Zhang, A., Chen, M., Du, C., Guo, H., Bai, H., and Li, L. (2013) Poly (dimethylsiloxane) oil absorbent with a three-dimensionally interconnected porous structure and swellable skeleton. *ACS Applied Materials & Interfaces*, **5**, 10201-10206.
68. Nagappan, S., Park, J. J., Park, S. S., Lee, W.-K., and Ha, C.-S. (2013)

Bio-inspired, multi-purpose and instant superhydrophobic–superoleophilic lotus leaf powder hybrid micro–nanocomposites for selective oil spill capture. *Journal of Materials Chemistry A*, **1**, 6761-6769.

69. Zhu, Q., Pan, Q., and Liu, F. (2011) Facile removal and collection of oils from water surfaces through superhydrophobic and superoleophilic sponges. *The Journal of Physical Chemistry C*, **115**, 17464-17470.
70. Li, H., Liu, L., and Yang, F. (2013) Covalent assembly of 3D graphene/polypyrrole foams for oil spill cleanup. *Journal of Materials Chemistry A*, **1**, 3446-3453.
71. Tanobe, V., Sydenstricker, T., Amico, S., Vargas, J., and Zawadzki, S. (2009) Evaluation of flexible postconsumed polyurethane foams modified by polystyrene grafting as sorbent material for oil spills. *Journal of Applied Polymer Science*, **111**, 1842-1849.
72. Calcagnile, P., Fragouli, D., Bayer, I. S., Anyfantis, G. C., Martiradonna, L., Cozzoli, P. D., Cingolani, R., and Athanassiou, A. (2012) Magnetically driven floating foams for the removal of oil contaminants from water. *ACS Nano*, **6**, 5413-5419.
73. Nguyen, S. T., Feng, J., Le, N. T., Le, A. T., Hoang, N., Tan, V. B., and Duong, H. M. (2013) Cellulose aerogel from paper waste for crude oil spill cleaning. *Industrial & Engineering Chemistry Research*, **52**, 18386-18391.
74. Chin, S. F., Romainor, A. N. B., and Pang, S. C. (2014) Fabrication of hydrophobic and magnetic cellulose aerogel with high oil absorption capacity. *Materials Letters*, **115**, 241-243.
75. Choi, H.-M., Kwon, H.-J., and Moreau, J. P. (1993) Cotton nonwovens as oil spill cleanup sorbents. *Textile Research Journal*, **63**, 211-218.
76. Deschamps, G., Caruel, H., Borredon, M.-E., Bonnin, C., and Vignoles, C. (2003) Oil removal from water by selective sorption on hydrophobic cotton fibers. 1. Study of sorption properties and comparison with other cotton fiber-based sorbents. *Environmental Science & Technology*, **37**, 1013-1015.
77. Wang, J., Zheng, Y., and Wang, A. (2013) Coated kapok fiber for removal of spilled oil. *Marine Pollution Bulletin*, **69**, 91-96.
78. Lin, J., Shang, Y., Ding, B., Yang, J., Yu, J., and Al-Deyab, S. S. (2012) Nanoporous polystyrene fibers for oil spill cleanup. *Marine Pollution Bulletin*, **64**, 347-352.
79. Wu, J., Wang, N., Wang, L., Dong, H., Zhao, Y., and Jiang, L. (2012) Electrospun porous structure fibrous film with high oil adsorption capacity. *ACS Applied Materials & Interfaces*, **4**, 3207-3212.
80. Sai, H., Fu, R., Xing, L., Xiang, J., Li, Z., Li, F., and Zhang, T. (2015) Surface modification of bacterial cellulose aerogels' web-like skeleton for oil/water separation. *ACS Applied Materials & Interfaces*, **7**, 7373-7381.

81. Si, Y., Fu, Q., Wang, X., Zhu, J., Yu, J., Sun, G., and Ding, B. (2015) Superelastic and superhydrophobic nanofiber-assembled cellular aerogels for effective separation of oil/water emulsions. *ACS Nano*, **9**, 3791-3799.
82. Jiang, F., and Hsieh, Y.-L. (2014) Amphiphilic superabsorbent cellulose nanofibril aerogels. *Journal of Materials Chemistry A*, **2**, 6337-6342.
83. Zheng, Q., Cai, Z., and Gong, S. (2014) Green synthesis of polyvinyl alcohol (PVA)–cellulose nanofibril (CNF) hybrid aerogels and their use as superabsorbents. *Journal of Materials Chemistry A*, **2**, 3110-3118.
84. Lin, J., Tian, F., Shang, Y., Wang, F., Ding, B., Yu, J., and Guo, Z. (2013) Co-axial electrospun polystyrene/polyurethane fibres for oil collection from water surface. *Nanoscale*, **5**, 2745-2755.
85. Lin, J., Ding, B., Yang, J., Yu, J., and Sun, G. (2012) Subtle regulation of the micro-and nanostructures of electrospun polystyrene fibers and their application in oil absorption. *Nanoscale*, **4**, 176-182.
86. Zhao, J., Xiao, C., and Xu, N. (2013) Evaluation of polypropylene and poly (butylmethacrylate-co-hydroxyethylmethacrylate) nonwoven material as oil absorbent. *Environmental Science and Pollution Research*, **20**, 4137-4145.
87. Xue, C.-H., Ji, P.-T., Zhang, P., Li, Y.-R., and Jia, S.-T. (2013) Fabrication of superhydrophobic and superoleophilic textiles for oil–water separation. *Applied Surface Science*, **284**, 464-471.
88. Liu, X., Ge, L., Li, W., Wang, X., and Li, F. (2014) Layered double hydroxide functionalized textile for effective oil/water separation and selective oil adsorption. *ACS Applied Materials & Interfaces*, **7**, 791-800.
89. Zhang, J., and Seeger, S. (2011) Polyester materials with superwetting silicone nanofilaments for oil/water separation and selective oil absorption. *Advanced Functional Materials*, **21**, 4699-4704.
90. Wu, L., Zhang, J., Li, B., and Wang, A. (2014) Magnetically driven super durable superhydrophobic polyester materials for oil/water separation. *Polymer Chemistry*, **5**, 2382-2390.
91. Li, X., Wang, M., Wang, C., Cheng, C., and Wang, X. (2014) Facile immobilization of Ag nanocluster on nanofibrous membrane for oil/water separation. *ACS Applied Materials & Interfaces*, **6**, 15272-15282.
92. Tai, M. H., Gao, P., Tan, B. Y. L., Sun, D. D., and Leckie, J. O. (2014) Highly efficient and flexible electrospun carbon–silica nanofibrous membrane for ultrafast gravity-driven oil–water separation. *ACS Applied Materials & Interfaces*, **6**, 9393-9401.
93. Zhang, W., Shi, Z., Zhang, F., Liu, X., Jin, J., and Jiang, L. (2013) Superhydrophobic and superoleophilic PVDF membranes for effective separation of water-in-oil emulsions with high flux. *Advanced Mate-*

rials, **25**, 2071-2076.

94. Chen, P.-C., and Xu, Z.-K. (2013) Mineral-coated polymer membranes with superhydrophilicity and underwater superoleophobicity for effective oil/water separation. *Scientific Reports*, **3**.
95. Chakrabarty, B., Ghoshal, A., and Purkait, M. (2008) Ultrafiltration of stable oil-in-water emulsion by polysulfone membrane. *Journal of Membrane Science*, **325**, 427-437.
96. Zhang, C., Li, P., and Cao, B. (2015) Electrospun microfibrous membranes based on PIM-1/POSS with high oil wettability for separation of oil–water mixtures and cleanup of oil soluble contaminants. *Industrial & Engineering Chemistry Research*, **54**, 8772-8781.
97. Zhou, X., Zhang, Z., Xu, X., Guo, F., Zhu, X., Men, X., and Ge, B. (2013) Robust and durable superhydrophobic cotton fabrics for oil/water separation. *ACS Applied Materials & Interfaces*, **5**, 7208-7214.
98. Huang, M., Si, Y., Tang, X., Zhu, Z., Ding, B., Liu, L., Zheng, G., Luo, W., and Yu, J. (2013) Gravity driven separation of emulsified oil–water mixtures utilizing in situ polymerized superhydrophobic and superoleophilic nanofibrous membranes. *Journal of Materials Chemistry A*, **1**, 14071-14074.
99. Tang, X., Si, Y., Ge, J., Ding, B., Liu, L., Zheng, G., Luo, W., and Yu, J. (2013) In situ polymerized superhydrophobic and superoleophilic nanofibrous membranes for gravity driven oil–water separation. *Nanoscale*, **5**, 11657-11664.
100. Shang, Y., Si, Y., Raza, A., Yang, L., Mao, X., Ding, B., and Yu, J. (2012) An in-situ polymerization approach for the synthesis of superhydrophobic and superoleophilic nanofibrous membranes for oil–water separation. Nanoscale 4, 7847-7854.
101. Chen, W., Su, Y., Zheng, L., Wang, L., and Jiang, Z. (2009) The improved oil/water separation performance of cellulose acetate-graft-polyacrylonitrile membranes. *Journal of Membrane Science*, **337**, 98-105.
102. Obaid, M., Barakat, N. A., Fadali, O., Motlak, M., Almajid, A. A., and Khalil, K. A. (2015) Effective and reusable oil/water separation membranes based on modified polysulfone electrospun nanofiber mats. *Chemical Engineering Journal*, **259**, 449-456.
103. Extrand, C. (2002) Water contact angles and hysteresis of polyamide surfaces. *Journal of Colloid and Interface Science*, **248**, 136-142.
104. Extrand, C. (2004) Contact angles and their hysteresis as a measure of liquid-solid adhesion. *Langmuir*, **20**, 4017-4021.
105. Zhu, Y., Zhang, F., Wang, D., Pei, X. F., Zhang, W., and Jin, J. (2013) A novel zwitterionic polyelectrolyte grafted PVDF membrane for thoroughly separating oil from water with ultrahigh efficiency. *Journal of Materials Chemistry A*, **1**, 5758-5765.
106. Xiang, Y., Liu, F., and Xue, L. (2015) Under seawater superoleophobic PVDF membrane inspired by polydopamine for efficient oil/sea

water separation. Journal of Membrane Science 476, 321-329.

107. Li, H.-J., Cao, Y.-M., Qin, J.-J., Jie, X.-M., Wang, T.-H., Liu, J.-H., and Yuan, Q. (2006) Development and characterization of anti-fouling cellulose hollow fiber UF membranes for oil–water separation. *Journal of Membrane Science*, **279**, 328-335.
108. Raza, A., Ding, B., Zainab, G., El-Newehy, M., Al-Deyab, S.S., and Yu, J. (2014) In situ cross-linked superwetting nanofibrous membranes for ultrafast oil–water separation. *Journal of Materials Chemistry A*, **2**, 10137-10145.
109. Chen, W., Peng, J., Su, Y., Zheng, L., Wang, L., and Jiang, Z. (2009) Separation of oil/water emulsion using Pluronic F127 modified polyethersulfone ultrafiltration membranes. *Separation and Purification Technology*, **66**, 591-597.
110. Zhang, W., Zhu, Y., Liu, X., Wang, D., Li, J., Jiang, L., and Jin, J. (2014) Salt-induced fabrication of superhydrophilic and underwater superoleophobic PAA-g-PVDF membranes for effective separation of iil-in-water emulsions. *Angewandte Chemie International Edition*, **53**, 856-860.
111. Li, J.-J., Zhou, Y.-N., and Luo, Z.-H. (2015) Smart fiber membrane for pH-induced oil/water separation. *ACS Applied Materials & Interfaces*, **7**, 19643-19650.
112. Zhang, W., Shi, L., Ma, R., An, Y., Xu, Y., and Wu, K. (2005) Micellization of thermo-and pH-responsive triblock copolymer of poly (ethylene glycol)-b-poly (4-vinylpyridine)-b-poly (N-isopropylacrylamide). *Macromolecules*, **38**, 8850-8852.
113. Mendrek, S., Mendrek, A., Adler, H. J., Dworak, A., and Kuckling, D. (2009) Synthesis and characterization of pH sensitive poly (glycidol)-b-poly (4-vinylpyridine) block copolymers. *Journal of Polymer Science Part A: Polymer Chemistry*, **47**, 1782-1794.
114. Escale, P., Rubatat, L., Derail, C., Save, M., and Billon, L. (2011) pH sensitive hierarchically self-organized bioinspired films. *Macromolecular Rapid Communications*, **32**, 1072-1076.
115. Geng, Z., Guan, S., Jiang, H.-M., Liu, Z.-W., and Jiang, L. (2014) pH-sensitive wettability induced by topological and chemical transition on the self assembled surface of block copolymer. *Chinese Journal of Polymer Science*, **32**, 92-97.
116. Zhang, Y., Wei, S., Liu, F., Du, Y., Liu, S., Ji, Y., Yokoi, T., Tatsumi, T., and Xiao, F.-S. (2009) Superhydrophobic nanoporous polymers as efficient adsorbents for organic compounds. *Nano Today*, **4**, 135-142.
117. Atta, A. M., El-Hamouly, S. H., AlSabagh, A. M., and Gabr, M. M. (2007) Crosslinked poly (octadecene-alt-maleic anhydride) copolymers as crude oil sorbers. *Journal of Applied Polymer Science*, **105**, 2113-2120.
118. Avila, A. F., Munhoz, V. C., de Oliveira, A. M., Santos, M. C., Lacerda, G.

R., and Goncalves, C. P. (2014) Nano-based systems for oil spills con trol and cleanup. *Journal of Hazardous Materials*, **272**, 20-27.

119. Gu, J., Jiang, W., Wang, F., Chen, M., Mao, J., and Xie, T. (2014) Facile removal of oils from water surfaces through highly hydrophobic and magnetic polymer nanocomposites. *Applied Surface Science*, **301**, 492-499.
120. Chen, M., Jiang, W., Wang, F., Shen, P., Ma, P., Gu, J., Mao, J., and Li, F. (2013) Synthesis of highly hydrophobic floating magnetic polymer nanocomposites for the removal of oils from water surface. *Applied Surface Science*, **286**, 249-256.
121. Song, C., Ding, L., Yao, F., Deng, J., and Yang, W. (2013) β-Cyclodextrin-based oil-absorbent microspheres: Preparation and high oil absorbency. *Carbohydrate Polymers*, **91**, 217-223.
122. Mao, J., Jiang, W., Gu, J., Zhou, S., Lu, Y., and Xie, T. (2014) Synthesis of P (St-DVB)/Fe_3O_4 microspheres and application for oil removal in aqueous environment. *Applied Surface Science*, **317**, 787-793.
123. Huang, J. (2009) Treatment of phenol and p-cresol in aqueous solution by adsorption using a carbonylated hypercrosslinked polymeric adsorbent. *Journal of Hazardous Materials*, **168**, 1028-1034.
124. Caetano, M., Valderrama, C., Farran, A., and Cortina, J. L. (2009) Phenol removal from aqueous solution by adsorption and ion exchange mechanisms onto polymeric resins. *Journal of Colloid and Interface Science*, **338**, 402-409.
125. Senel, S., Kara, A., Alsancak, G., and Denizli, A. (2006) Removal of phenol and chlorophenols from water with reusable dye-affinity hollow fibers. *Journal of Hazardous Materials*, **138**, 317-324.
126. An, F., and Gao, B. (2008) Adsorption of phenol on a novel adsorption material PEI/SiO_2. *Journal of Hazardous Materials*, **152**, 1186-1191.
127. An, F., Gao, B., and Feng, X. (2008) Adsorption and recognizing ability of molecular imprinted polymer MIP-PEI/SiO_2 towards phenol. *Journal of Hazardous Materials*, **157**, 286-292.
128. Shao, D., Hu, J., Chen, C., Sheng, G., Ren, X., and Wang, X. (2010) Polyaniline multiwalled carbon nanotube magnetic composite prepared by plasma-induced graft technique and its application for removal of aniline and phenol. *The Journal of Physical Chemistry C*, **114**, 21524-21530.
129. Zeng, X., Fan, Y., Wu, G., Wang, C., and Shi, R. (2009) Enhanced adsorption of phenol from water by a novel polar post-crosslinked polymeric adsorbent. *Journal of Hazardous Materials*, **169**, 1022-1028.
130. Li, A., Zhang, Q., Zhang, G., Chen, J., Fei, Z., and Liu, F. (2002) Adsorption of phenolic compounds from aqueous solutions by a water-compatible hypercrosslinked polymeric adsorbent. *Chemosphere*, **47**, 981-989.

131. Ming, Z. W., Long, C. J., Cai, P. B., Xing, Z. Q., and Zhang, B. (2006) Synergistic adsorption of phenol from aqueous solution onto polymeric adsorbents. *Journal of Hazardous Materials*, **128**, 123-129.
132. Zhang, W., Du, Q., Pan, B., Lv, L., Hong, C., Jiang, Z., and Kong, D. (2009) Adsorption equilibrium and heat of phenol onto aminated polymeric resins from aqueous solution. *Colloids and Surfaces A: Physicochemical and Engineering Aspects*, **346**, 34-38.
133. Pan, B., Du, W., Zhang, W., Zhang, X., Zhang, Q., Pan, B., Lv, L., Zhang, Q., and Chen, J. (2007) Improved adsorption of 4-nitrophenol onto a novel hyper-cross-linked polymer. *Environmental Science & Technology*, **41**, 5057-5062.
134. Pan, B., Zhang, X., Zhang, W., Zheng, J., Pan, B., Chen, J., and Zhang, Q. (2005) Adsorption of phenolic compounds from aqueous solution onto a macroporous polymer and its aminated derivative: isotherm analysis. *Journal of Hazardous Materials*, **121**, 233-241.
135. Li, A., Zhang, Q., Chen, J., Fei, Z., Long, C., and Li, W. (2001) Adsorption of phenolic compounds on Amberlite XAD-4 and its acetylated derivative MX-4. *Reactive and Functional Polymers*, **49**, 225-233.
136. Pan, B., Pan, B., Zhang, W., Zhang, Q., Zhang, Q., and Zheng, S. (2008) Adsorptive removal of phenol from aqueous phase by using a porous acrylic ester polymer. *Journal of Hazardous Materials*, **157**, 293-299.
137. Juang, R.-S., and Shiau, J.-Y. (1999) Adsorption isotherms of phenols from water onto macroreticular resins. *Journal of Hazardous Materials*, **70**, 171-183.
138. Zeng, X., Yu, T., Wang, P., Yuan, R., Wen, Q., Fan, Y., Wang, C., and Shi, R. (2010) Preparation and characterization of polar polymeric adsorbents with high surface area for the removal of phenol from water. *Journal of Hazardous Materials*, **177**, 773-780.
139. Wagner, K., and Schulz, S. (2001) Adsorption of phenol, chlorophenols, and dihydroxybenzenes onto unfunctionalized polymeric resins at temperatures from 294.15 K to 318.15 K. *Journal of Chemical & Engineering Data*, **46**, 322-330.
140. Li, H., Xu, M., Shi, Z., and He, B. (2004) Isotherm analysis of phenol adsorption on polymeric adsorbents from nonaqueous solution. *Journal of Colloid and Interface Science*, **271**, 47-54.
141. Dursun, A. Y., and Kalayci, C. S. (2005) Equilibrium, kinetic and thermodynamic studies on the adsorption of phenol onto chitin. *Journal of Hazardous Materials*, **123**, 151-157.
142. Kawabata, N., and Ohira, K. (1979) Removal and recovery of organic pollutants from aquatic environment. 1. Vinylpyridine-divinylbenzene copolymer as a polymeric adsorbent for removal and recovery of phenol from aqueous solution. *Environmental Science & Technology*, **13**, 1396-1402.
143. Fowkes, F. M., Kaczinski, M. B., and Dwight, D. W. (1991) Character-

ization of polymer surface sites with contact angles of test solutions. 1. Phenol and iodine adsorption from methylene iodide onto PMMA films. *Langmuir*, **7**, 2464-2470.

144. Burleigh, M. C., Markowitz, M. A., Spector, M. S., and Gaber, B. P. (2002) Porous polysilsesquioxanes for the adsorption of phenols. *Environmental Science & Technology*, **36**, 2515-2518.
145. Huang, J., Jin, X., and Deng, S. (2012) Phenol adsorption on an N-methylacetamide-modified hypercrosslinked resin from aqueous solutions. *Chemical Engineering Journal*, **192**, 192-200.
146. Gupta, T., Pradhan, N. C., and Adhikari, B. (2003) Separation of phenol from aqueous solution by pervaporation using HTPB-based polyurethaneurea membrane. *Journal of Membrane Science*, **217**, 43-53.

6

Resorcinol-Formaldehyde Cryogel Nanocomposites for Oil Spill Clean Up

6.1 Introduction

Oil spills have resulted in catastrophic effects on aquatic ecosystems over the past years [1]. Therefore, there is a continuous requirement to develop ideal absorbent materials exhibiting low cost, environment favorability, high efficiency, high absorption capacity, excellent selectivity and recyclability. Variety of methods have been employed to clean the oil spills such as bioremediation [2], mechanical collection [3], *in-situ* burning [4], chemical dispersants [5] and absorbent materials [6]. Among these methods, sorbent materials have proven to be the most desirable option for the removal of oil spills from water bodies. In this category, some materials including wool fibers [7], organoclays [8], activated carbon [9], zeolites [10], and alkyl acrylate copolymers [11] have been employed to clean the oil spill. However, in practical applications, usage of these materials is limited due to their low absorption capacities, environmental incompatibilities and poor recyclability. To overcome these challenge, various advanced materials based on graphene [12], carbon nanotubes [13], mesh films [14], aerogels [15], sponges [16], foams [17], membranes [18] and fibers [19] have been developed in the past few years for the separation of oil from water.

Cryogels have large and well-defined interconnected porous heterogeneous structure [20,21]. Free-radical cryo-polymerization technique is generally employed to prepare cryogels from polymeric gel precursors below the solvent freezing point, as freeze-drying process maintains the interconnected structure of the resultant cryogel [22]. The properties of cryogels such as pore dimensions, pore wall thickness, mechanical toughness and elasticity depend on the conditions of preparation, amount of solvent and precursors, temperature gradients, cooling rate and other conditions, etc. [23,24]. In this study, the preparation of three-dimensional carbonaceous monolithic cryogel nanocomposites from resorcinol-formaldehyde

Naman Arora, Muthukumaraswamy R. Vengatesan and Vikas Mittal, Khalifa University of Science and Technology, Abu Dhabi, UAE

(R-F) resin precursor with graphene oxide (GO), phosphazene nanotubes (PZTs) and cobalt ferrite nanoparticles is reported. The nanocomposites exhibited porosity and high surface area, paving the way to utilize the nanomaterials effectively as a sorbent material at the macroscopic level to remove oil from oil/water mixture. During the formation of R-F based cryogel, divalent metal ions (such as Co^{2+} and Fe^{2+}) perform as both cross-linker and catalyst due to their binding ability towards oxygen groups [25] which leads to crosslinking between graphene oxide and R-F skeletons, thus, enhancing the strength of cryogel and also substituting Na_2CO_3 as a catalyst. Subsequently, the cryogel is converted into carbonaceous monoliths with excellent porosity and hydrophobicity *via* pyrolysis. During the carbonization process, divalent metal ions (Co^{2+} and Fe^{2+}) are converted into cobalt ferrite nanoparticles [26] which help in increasing the surface roughness and surface area of the carbonaceous monolithic cryogel, thus, enhancing the sorption capacity [27].

Graphene has a two-dimensional structure with hydrophobic properties and tends to aggregate due to van der Waals interaction between neighboring sheets. Reduction in specific surface area of graphene occurs due to the aggregation of sheets which is not beneficial for the adsorption of organic contaminants. The adsorption of molecules onto the sorbents is basically determined by hydrogen bonding, $\pi - \pi$ interactions, van der Waals interactions and electrostatic interactions [28,29]. Several methods have been adopted to reduce the sheet staking of the graphene. In particular, for this purpose, one-dimensional nanomaterials like carbon nanofibers [30], CNTs [31], sepiolite clay [32], etc. have been employed. Therefore, in this study, in order to attain high specific surface area of sorbent, graphene sheets have been modified with phosphazene nanotubes so as to avoid the stacking of graphene sheets. Phosphazene is a versatile class of hybrid organic-inorganic material with alternating phosphorus and nitrogen atoms. These materials have been used as membrane materials, biomaterials, optical materials, flame retardant material, etc. [33-37]. Phosphazene nanotubes containing hydroxyl groups are synthesized through one-pot reaction with simple process, high yield, controllable morphology, and low cost [38], which provides an optimal cost advantage for manufacturing of CNTs [39].

In specific, this work reports graphene-PZTs based carbonaceous monolithic cryogel containing cobalt ferrite nanoparticles as an efficient sorbent. Crude oil and vegetable oil were selected as model pollutants for evaluating the oil sorption performance under laboratory conditions.

6.2 Materials and Methods

6.2.1 Materials

Natural graphite powder (10 mesh), resorcinol, 2-nitrophenol (Merck), formaldehyde (37%, Merck), H_2SO_4 (95-97%, Merck), H_3PO_4 (85%, Merck), $KMnO_4$ (Eurolab), HCl (37%, Panreac), hydrogen peroxide (35%, Merck), $CoCl_2.6H_2O$ (Sigma-Aldrich), iron(III) nitrate nanohydrate (≥98%, Sigma-Aldrich), phosphonitrilic chloride trimer (99 %, Sigma-Aldrich), bisphenol S (Sigma-Aldrich), triethylamine (≥99%, Sigma Aldrich) and tetrahydrofuran were used for material synthesis. Crude oil and vegetable oil were procured locally.

6.2.2 GO Preparation

GO was prepared by using the improved Tour's synthesis method [40]. For this, graphite (10 g) was added into the mixture of H_2SO_4 (540 ml) and H_3PO_4 (60 ml) in a molar ratio of 9:1. The mixture was allowed to stir for 30 min, followed by gradual addition of $KMnO_4$ (56 g) and continuous stirring for 72 h. The reaction was initiated by pouring H_2O_2 slowly and allowed to stir for 8 h until the mixture color turned yellow. Produced graphene oxide was washed with 1 M HCl solution, until no sulfate ions were detected, followed by washing with excess of distilled water to remove chloride ions. The GO solution was subsequently poured in Teflon petri-dish for freeze drying.

6.2.3 Preparation of PZTs

3 g of phosphonitrilic chloride trimer was dissolved in 550 ml of tetrahydrofuran (THF) and the mixture was sonicated for 20 min. Another mixture containing triethylamine (TEA) (7.8 ml) and THF (50 ml) was prepared, followed by its dropwise addition in the mixture of phosphonitrilic chloride trimer and THF under ultra-sonication at 40 °C. Bisphenol S (6.5 g) was added after 30 min to the same mixture followed by the sonication for another 7 h. The final solution was filtered, and the precipitates were washed with the mixture of

THF and distilled water (1:1). The resulting product was dried under vacuum at 50 °C overnight.

6.2.4 Synthesis of Graphene-PZT based R-F Cryogel Containing Cobalt Ferrite Nanoparticles

GO exfoliation was carried out under ultra-sonication (2 h) by adding GO (0.2 g) into deionized water (100 ml). Later, phosphazene nanotubes (0.2 g) were added into the GO suspension followed by stirring for 2 h. $CoCl_2.6H_2O$ (0.0025 M) and $Fe(NO_3)_3.9H_2O$ (0.0025 M) were added as Co^{2+} and Fe^{2+} sources. Resorcinol (2.48 g) and formaldehyde (3.65 g) were then added in the molar ratio of 1:2. After continuous stirring for 4 h to form a homogeneous solution, the mixture was cast into a glass mold. It was kept at 85 °C for curing until a red-brown colored hydrogel was obtained. The hydrogel was washed with excess of distilled water to remove the residual ions, followed by a lyophilization process for 5 d at a temperature of -90 °C under vacuum to obtain a GO-PZT based R-F cryogel. The cryogel was carbonized for 2 h at 700 °C at a heating rate of 10 °C/min under N_2 atmosphere and cooled naturally in a quartz tube furnace. Finally, GO-PZTs based R-F cryogel containing cobalt ferrite nanoparticles ($CoFe_2O_4$/graphene-PZTs/R-F cryogel or resultant cryogel) was obtained. To investigate the influence of PZTs and cobalt ferrite nanoparticles on the formation of 3-D carbonaceous monolith material, the graphene/R-F based cryogel containing cobalt ferrite nanoparticles ($CoFe_2O_4$/graphene/R-F) was prepared using the same methodology. Graphene-R-F based cryogel (graphene/R-F/Na_2CO_3) was also synthesized using Na_2CO_3 catalyst by replacing Co^{2+} and Fe^{2+} ions for the gelation process.

6.3 Adsorbent Characterization

Wide-angle X-ray diffraction (WAXRD) patterns of the samples were collected using analytical powder (X'Pert PRO) diffractometer with instrument parameters of 40 kV and CuKα (1.5406 Å) radiation in reflection mode. The samples were step-scanned between 5 and 60° at a step size of $2\theta = 0.02°\ s^{-1}$.

Scanning electron microscopy (SEM) images and EDX analysis data were used to obtain morphology of the sample, using FEI Quanta, FEG250 (USA) microscope at accelerating voltages of 10-20 kV. FEI transmission electron microscope (TEM, Tecnai G20) at 200kV was used to obtain the morphology of $CoFe_2O_4$ /graphene/PZTs/R-F nanocomposites. The samples for TEM analysis were prepared by dispersing 0.5 mg of adsorbent in 20 ml of dimethylformamide under sonication at room temperature for 10 min. Two drops of suspension were poured onto a 400 mesh copper grid covered with thin lacey carbon film and dried in air.

Nitrogen adsorption-desorption isotherms were obtained on a Micromeritics ASAP 2010 volumetric adsorption analyzer. Prior to each adsorption measurement, the samples were degassed for 16 h at 200 °C, under vacuum (P < 5-10 mbar). Specific surface area of the nanocomposites was determined from the linear part of the BET equation. The pore volume was calculated using the BET plot corresponding to the amount of nitrogen adsorbed at the last adsorption point (P/P_0 = 0.99). Pore size distribution was estimated using the Barrett-Joyner-Halenda (BJH) method.

6.4 Oil Sorption Studies by Cryogels

Oil can be removed from aqueous media by different phenomena such as adsorption, absorption or both (sorption). Herein, sorption process has been employed to remove the oil from water. In the oil sorption experiments, the sorption capacities of graphene/R-F/Na_2CO_3, $CoFe_2O_4$/graphene/R-F and $CoFe_2O_4$/graphene/PZTs/R-F for crude oil and vegetable oil were measured. Initial weight of different monolithic cryogels was measured as $M_\circ$ (before immersing them into 200 ml glass bowl containing 100 ml of oil/water mixture). Oil/ water mixture was prepared using 30 ml of oil and 70 ml of water. The cryogels were allowed to remove the oil in the oil/water mixture at room temperature for an interval of 15 min. After being saturated with oil, the soaked cryogels were taken out and the weight was recorded as M_e. The following equation was used to determine the oil sorption capacities of the as-prepared cryogels:

$$Oil\ absorption\ capacity\left(\frac{g}{g}\right) = \frac{(M_e - M_\circ)}{M_\circ}$$

In order to study the recyclability of the cryogels, *n*-hexane was used to remove the oil from the monolithic cryogels by soaking them in *n*-hexane for several hours. After extracting the oil from the cryogels completely, the gels were taken out and heated to 65 °C for the evaporation of residual *n*-hexane. Subsequently, the developed cryogels were employed for the next sorption-desorption cyclic analysis.

6.5 Results and Discussion

6.5.1 Structure and Morphology of Adsorbent

Figure 6.1a shows the image of GO-PZTs/R-F based cryogel containing divalent metal salts of Co^{2+} and Fe^{2+}. After the carbonization process, the GO-PZTs/R-F based cryogel turned into black colored carbonaceous monolithic cryogel ($CoFe_2O_4$/graphene-PZTs/R-F), represented in Figure 6.1b. The cryogels were synthesized by curing the initial suspensions and further lyophilization of the corresponding hydrogels in order to eliminate the water inside, which resulted in dried monoliths/cryogels with desirable shape, depending on the glass molds. It can be noticed that GO-PZTs/R-F was a homogeneous monolith with yellow-brownish color, which was obtained when the divalent metal salts of Co^{2+} and Fe^{2+} were used to promote the gelation process of GO/R-F. The Co^{2+} and Fe^{2+} ions were chelated by the functional groups (negatively charged) of GO sheets and R-F, acting as both catalyst and crosslinking agents at the same time. After the carbonization of GO-PZTs/R-F cryogel, divalent metal ions (Co^{2+} and Fe^{2+}) and GO sheets were converted into cobalt ferrite nanoparticles and graphene sheets, respectively, along with PZTs turning into carbon fibers. As a result, $CoFe_2O_4$/graphene/PZTs/R-F) carbonaceous monolith was obtained.

Figure 6.1 (a) Digital image of GO/PZTs/R-F based cryogel containing divalent metal salt of Co^{2+} and Fe^{2+}, and (b) digital image of $CoFe_2O_4$/graphene/PZTs/R-F based cryogel.

The diffraction pattern of the GO/PZTs/R-F is presented in Figure 6.2a. A diffraction hump was observed in the range of ($2\theta = 20°$ to $33°$). The diffraction pattern of $CoFe_2O_4$/graphene/PZTs/R-F is depicted in Figure 6.2b, where the weak diffraction peak can be noticed at $2\theta = 26.2°$, attributed to the (002) plane of carbon. This indicated graphitization of amorphous carbon catalyzed by transition metals [41] and changes in interlayer spacing [42]. Strong diffraction peaks at $2\theta = 43°$ and $52.5°$ were attributed to (400) and (422) crystal planes of cubic spinel structure of cobalt ferrite nanoparticles (*JCPDS card No: 22-1086*) and a small diffraction peak at $2\theta = 45°$ corresponded to (110) plane of bcc Fe-Co [43]. Thus, it can be suggested that after the calcination at 700 °C, Co^{2+} and Fe^{2+} were completely reduced to zero valence state and cobalt ferrite nanoparticles were generated. This confirmed the achievement of intended crystalline state of the materials, which could be effectively controlled through the employed experimental protocols.

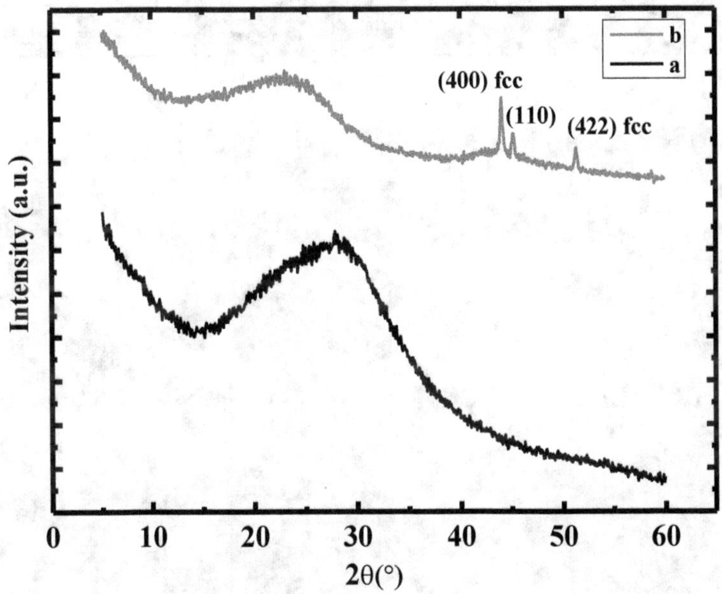

Figure 6.2 XRD patterns of (a) GO/PZTs/R-F, and (b) $CoFe_2O_4$/graphene/PZTs/R-F.

Figure 6.3 demonstrates the scanning electron micrographs of the resultant adsorbent ($CoFe_2O_4$/graphene/PZTs/R-F). It was observed that cobalt ferrite nanoparticles were present in a uniformly

Figure 6.3 SEM image of $CoFe_2O_4$/graphene/PZTs/R-F cryogel at (a)10 μm and (b) 20 μm magnifications.

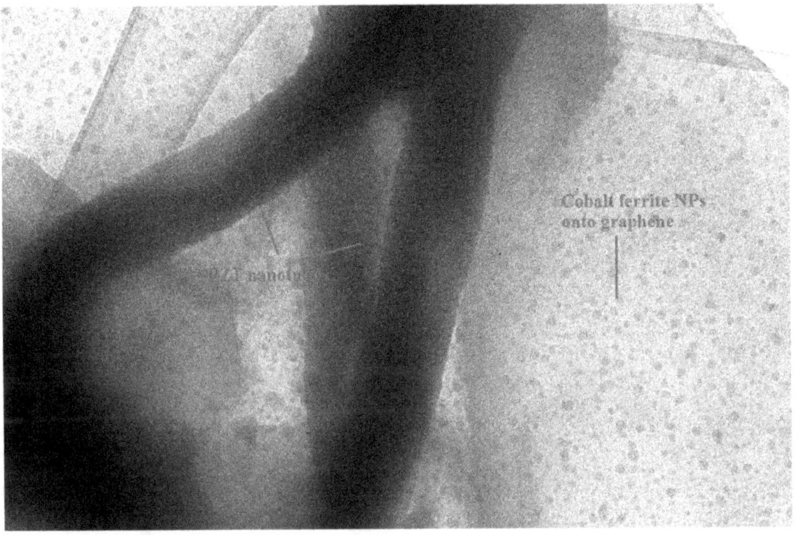

Figure 6.4 TEM image of $CoFe_2O_4$/graphene/PZTs/R-F cryogel.

defined interconnected porous structure. As suggested earlier, the process of gelation was facilitated by the metal ions (Co^{2+} and Fe^{2+}) which acted as cross-linkers, provided active sites for the assembly of graphene layers [44], and facilitated the formation of corrugations in the resultant cryogel, therefore, forming a well-defined interconnected porous structure. For gaining further insights into the cryogel morphology, TEM image of the resultant adsorbent is shown in Figure 6.4. It was observed that carbonaceous matrix contained carbon nanofibers resulting from PZTs which helped in avoiding the stacking of graphene sheets, thus, enhancing the surface area available for sorption. Additionally, cobalt ferrite nanoparticles were observed to be distributed uniformly onto graphene sheets. The nanoparticle might improve nano-scale roughness of the graphene surface, which can further increase the effective surface area for the sorption process [27]. Determination of the surface area and pore volume of adsorbent was carried out by N_2 adsorption-desorption measurements at 73 K, along with pore size distribution determination based on BJH desorption method. The textural properties were studied for cryogel nanocomposites such as (i) graphene/R-F/Na_2CO_3, (ii) $CoFe_2O_4$/graphene/R-F and (iii) $CoFe_2O_4$/graphene/PZTs/R-F (Table 6.1).

From Table 6.1, it was observed that the presence of PZTs along with cobalt ferrite nanoparticles led to the achievement of higher

BET surface area (629.5 m^2 g^{-1}) and pore radius (19.01 Å), as compared to the cryogels without PZTs and $CoFe_2O_4$ NPs, and cryogel with $CoFe_2O_4$ NPs but without PZTs.

Table 6.1 Textural properties of graphene/R-F/Na_2CO_3, $CoFe_2O_4$/graphene/R-F and $CoFe_2O_4$/graphene/PZTs/R-F based cryogels.

Sample	Surface area (m^2/g)	Pore volume (cc/g)	Pore radius (Å)
Graphene/R-F	324.3	0.54	7.05
$CoFe_2O_4$/graphene/R-F	478.7	0.42	7.12
$CoFe_2O_4$/graphene/PZTs/R-F	629.5	0.61	19.01

6.5.2 Hydrophobicity and Oil Sorption Capacity of Cryogels

With the combination of hydrophobic materials and multiple scale of surface roughness, enhanced hydrophobicity can be achieved [45]. The incorporation of intrinsic hydrophobic graphene sheets, carbon nanofibers of PZTs and cobalt ferrite nanoparticles into the R-F matrix, thus, provided the micro-nano scale roughness in combination with interconnected micro-porous structure of the R-F cryogel [45]. Other studies have also reported the use of nanoparticles for creating functional surfaces with hierarchical structures to improve the wettability of the substrate [27,46]. Figure 6.5 demonstrates the hydrophobic behavior of the $CoFe_2O_4$/graphene/PZTs/R-F cryogel. From the visual images, it was observed that the oil droplet was quickly sorbed by the cryogel, while the water droplet achieved quasi-spherical shape on the surface of the cryogel with a water contact angle (WCA) of an obtuse angle. In a typical experiment for the measurement of oil sorption capacity, the monolithic carbonaceous cryogels were kept in contact with a layer of crude oil floating on the surface of water. Cryogels repelled the water phase and sorbed the oil within a time interval of 15 min due to their hydrophobic property, as shown in Figure 6.6, exhibiting excellent properties of oil sorbent for oil spill clean up.

Many studies have reported the oil sorption process to be driven by the van der Waals forces, hydrophobic interactions between the oil and sorbent, pore morphology and viscosity of the oil [47-50]. To

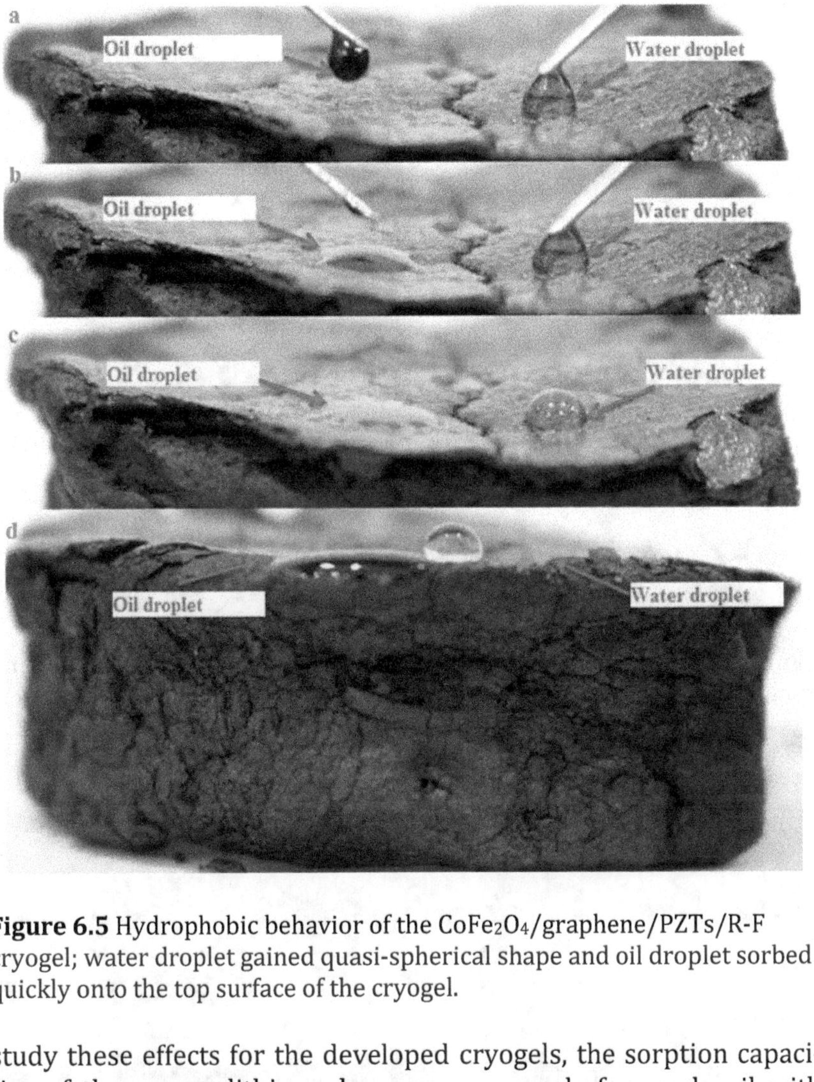

Figure 6.5 Hydrophobic behavior of the $CoFe_2O_4$/graphene/PZTs/R-F cryogel; water droplet gained quasi-spherical shape and oil droplet sorbed quickly onto the top surface of the cryogel.

study these effects for the developed cryogels, the sorption capacities of three monolithic carbonaceous cryogels for crude oil with higher viscosity and vegetable oil with lower viscosity are compared in Figure 6.7. As mentioned above, oil viscosity played an important role in the sorption process. Thus, the penetration of the oil into the network of the cryogels is facilitated by the lower viscosity of oil, resulting in a higher sorption capacity. On the other hand, diffusion of oil into the pores of the sorbent is inhibited by the high viscosity of oil, thus, resulting in a low sorption capacity [48]. In other words, this behavior can also be related as a function of time, indicating the sorption rate of the cryogel for vegetable oil is faster than the crude

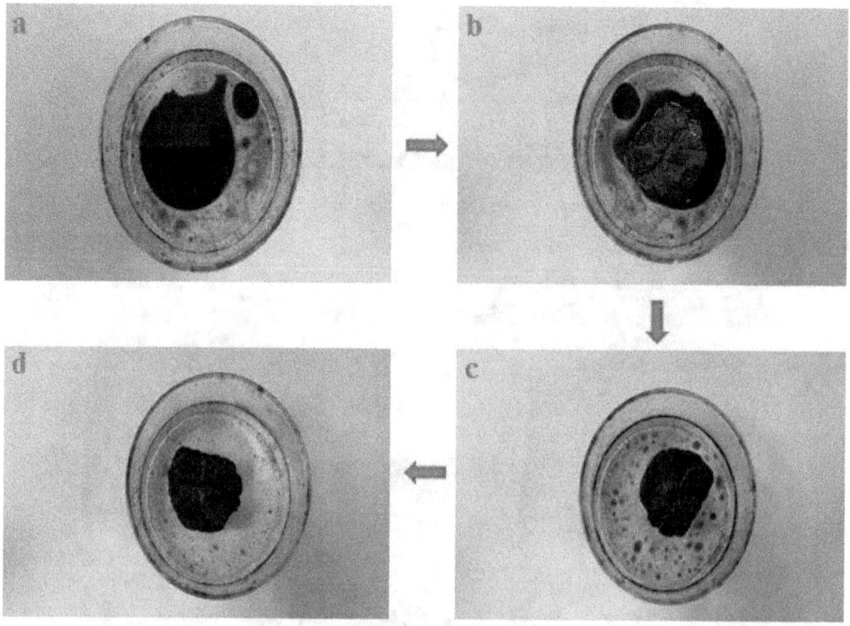

Figure 6.6 Oil sorption experiment by using $CoFe_2O_4$/graphene/PZTs/R-F cryogel for the time interval of 15 min.

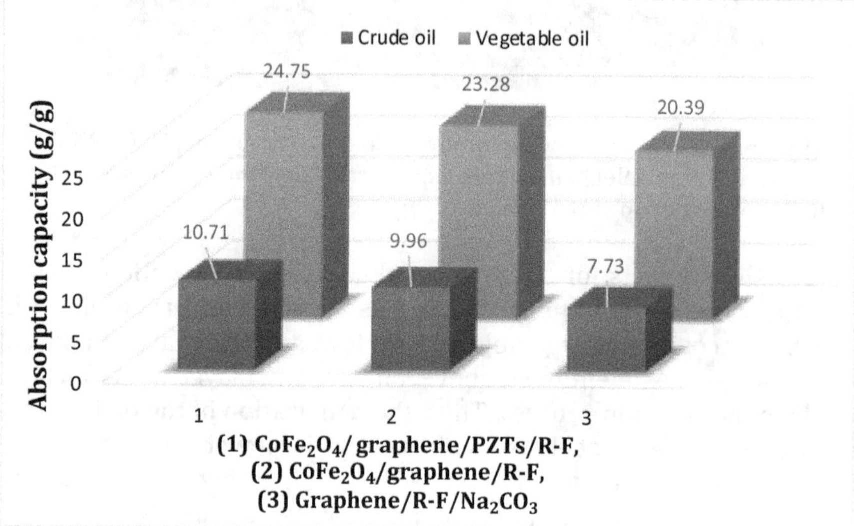

Figure 6.7 Sorption capacities of the three monolithic carbonaceous cryogels for crude oil and vegetable oil.

oil due to the high viscosity of the crude oil which delays the wicking rate of the oil onto the surface of the sorbent [50]. Therefore, the sorption capacity of the monolithic carbonaceous cryogels was higher for vegetable oil (low viscosity) as compared to crude oil (high viscosity). The removal of oil by graphene based sorbents has been reported to be based on van der Waals interactions and is related to the specific surface area of the sorbent [50,51]. It is observed from Figure 6.7 that the cryogels achieved high sorption capacity (10-25 times of its own weight) for crude oil and vegetable oil. The lower sorption capacity of graphene/R-F/Na_2CO_3 based cryogel could be attributed to low surface area, poor porosity and stacking of graphene sheets (adsorption sites) which may reduce the adsorption sites for oil adsorption [26]. After the incorporation of metal ions (Co^{2+} and Fe^{2+}) in case of $CoFe_2O_4$/graphene/R-F based cryogel, where metal ions acted as cross-linkers and provided active sites for the assembly of graphene layers to achieve well-defined interconnected porous structure, improved oil sorption capacity was achieved. The conversion of metal ions into cobalt ferrite NPs after the calcination process helped in increasing the surface roughness of sorbent to enhance the hydrophobic property, which increased the oil sorption capacity. The high sorption capacity of $CoFe_2O_4$/graphene/PZTs/R-F based cryogel was attributed to high surface area, excellent porosity and the presence of cobalt ferrite NPs along with the carbon nanofibers of PZTs which avoided the stacking of graphene sheets by maintaining the considerable gap between two graphene sheets, hence, increasing the surface area to provide the maximum oil adsorption sites. The sorption capacities of $CoFe_2O_4$/graphene/PZTs/R-F, $CoFe_2O_4$/graphene/R-F and graphene/R-F/Na_2CO_3 cryogels for crude oil and vegetable oil were determined as 10.71, 9.96, and 7.73 g/g, respectively and 24.75, 23.28, and 20.39 g/g, respectively. The recyclability of the $CoFe_2O_4$/graphene/PZTs/R-F cryogel for crude oil and vegetable oil is also presented in Figure 6.8. It was noticed that the recyclability of the resultant cryogel did not deteriorate even after five cycles of sorption-desorption process, implying a good recycling performance.

6.6 Conclusion

In this study, a well-defined interconnected porous structure cryogel ($CoFe_2O_4$/graphene/PZTs/R-F) has been successfully developed

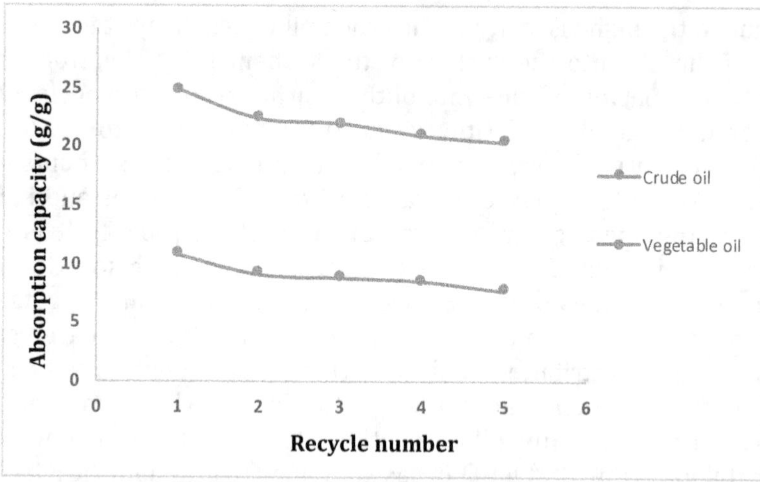

Figure 6.8 Recyclability of the $CoFe_2O_4$/graphene/PZTs/R-F cryogel for five cycles of sorption-desorption.

via freeze drying method, followed by carbonization in order to generate advanced materials to remove crude oil (high viscosity) and vegetable oil (low viscosity) from the oil/water mixture. The incorporation of PZTs into the monolithic carbonaceous cryogel enhanced the surface area of the sorbent by avoiding the stacking of the graphene sheets embedded in the R-F matrix. The combination of graphene, cobalt ferrite nanoparticles and carbon nanofibers of phosphazene tubes provided the micro-nano scale roughness in combination with a well-defined interconnected porous structure with good hydrophobic properties. The resultant cryogel exhibited good hydrophobic properties with water contact angle of obtuse angle. The sorption capacities of the resultant cryogel for crude oil and vegetable oil were higher than the sorption capacities of $CoFe_2O_4$/graphene/R-F and graphene/R-F/Na_2CO_3 cryogels. The recyclability of the $CoFe_2O_4$/graphene/PZTs/R-F cryogel did not deteriorate even after five cycles of the sorption-desorption process. Based on its promising performance, the developed cryogel is a promising sorbent for large quantity removal of various oils in different oil spill areas.

References

1. Chin, S. F., Romainor, A. N. B., and Pang, S. C. (2014) Fabrication of

hydrophobic and magnetic cellulose aerogel with high oil absorption capacity. *Materials Letters,* **115**, 241-243.

2. Zahed, M. A., Aziz, H. A., Isa, M. H., Mohajeri, L., and Mohajeri, S. (2010) Optimal conditions for bioremediation of oily seawater. *Bioresource technology,* **101**(24), 9455-9460.
3. Broje, V., and Keller, A. A. (2006) Improved mechanical oil spill recovery using an optimized geometry for the skimmer surface. *Environmental Science & Technology,* **40**(24), 7914-7918.
4. Buist, I., Potter, S., Nedwed, T., and Mullin, J. (2011) Herding surfactants to contract and thicken oil spills in pack ice for in situ burning. *Cold Regions Science and Technology,* **67**(1), 3-23.
5. Kujawinski, E. B., Kido Soule, M. C., Valentine, D. L., Boysen, A. K., Longnecker, K., and Redmond, M. C. (2011) Fate of dispersants associated with the Deepwater Horizon oil spill. *Environmental Science & Technology,* **45**(4), 1298-1306.
6. Xue, Z., Cao, Y., Liu, N., Feng, L., and Jiang, L. (2014) Special wettable materials for oil/water separation. *Journal of Materials Chemistry A,* **2**(8), 2445-2460.
7. Radetic, M. M., Jocic, D. M., Jovancic, P. M., Petrovic, Z. L., and Thomas, H. F. (2003) Recycled wool-based nonwoven material as an oil sorbent. *Environmental Science & Technology,* **37**(5), 1008-1012.
8. Moazed, H., and Viraraghavan, T. (2005) Removal of oil from water by bentonite organoclay. *Practice Periodical of Hazardous, Toxic, and Radioactive Waste Management,* **9**(2), 130-134.
9. Ahmad, A., Sumathi, S., and Hameed, B. (2005) Residual oil and suspended solid removal using natural adsorbents chitosan, bentonite and activated carbon: A comparative study. *Chemical Engineering Journal,* **108**(1), 179-185.
10. Cui, J., Zhang, X., Liu, H., Liu, S., and Yeung, K. L. (2008), Preparation and application of zeolite/ceramic microfiltration membranes for treatment of oil contaminated water. *Journal of Membrane Science,* **325**(1), 420-426.
11. Jang, J., and Kim, B. S. (2000) Studies of crosslinked styrene–alkyl acrylate copolymers for oil absorbency application. I. Synthesis and characterization. *Journal of applied polymer Science,* **77**(4), 903-913.
12. Iqbal, M. Z., and Abdala, A. A. (2013) Oil spill cleanup using graphene. *Environmental Science and Pollution Research,* **20**(5), 3271-3279.
13. Gui, X., Zeng, Z., Lin, Z., Gan, Q., Xiang, R., Zhu, Y., Cao, A., and Tang, Z. (2013) Magnetic and highly recyclable macroporous carbon nanotubes for spilled oil sorption and separation. *ACS Applied Materials & Interfaces,* **5**(12), 5845-5850.
14. Wang, C.-F., Tzeng, F.-S., Chen, H.-G., and Chang, C.-J. (2012) Ultra-

violet-durable superhydrophobic zinc oxide-coated mesh films for surface and underwater–oil capture and transportation. *Langmuir*, **28**(26), 10015-10019.

15. Reynolds, J. G., Coronado, P. R., and Hrubesh, L. W. (2001) Hydrophobic aerogels for oil-spill clean up–Synthesis and characterization. *Journal of Non-Crystalline Solids*, **292**(1), 127-137.
16. Zhu, Q., Pan, Q., and Liu, F., Facile removal and collection of oils from water surfaces through superhydrophobic and superoleophilic sponges. *The Journal of Physical Chemistry C*, **115**(35), 17464-17470.
17. Zhang, X., Li, Z., Liu, K., and Jiang, L. (2013) Bioinspired multifunctional foam with self-cleaning and oil/water separation. *Advanced Functional Materials*, **23**(22), 2881-2886.
18. Yuan, J., Liu, X., Akbulut, O., Hu, J., Suib, S. L., Kong, J., and Stellacci, F. (2008) Superwetting nanowire membranes for selective absorption. *Nature Nanotechnology*, **3**(6), 332-336.
19. Zhu, H., Qiu, S., Jiang, W., Wu, D., and Zhang, C. (2011) Evaluation of electrospun polyvinyl chloride/polystyrene fibers as sorbent materials for oil spill cleanup. *Environmental Science & Technology*, **45**(10), 4527-4531.
20. Bibi, N. S., Singh, N. K., Dsouza, R. N., Aasim, M., and Fernandez-Lahore, M. (2013) Synthesis and performance of megaporous immobilized metal-ion affinity cryogels for recombinant protein capture and purification. *Journal of Chromatography A*, **1272**, 145-149.
21. Zheng, S., Wang, T., Liu, D., Liu, X., Wang, C., and Tong, Z. (2013) Fast deswelling and highly extensible poly (N-isopropylacrylamide)-hectorite clay nanocomposite cryogels prepared by freezing polymerization. *Polymer*, **54**(7), 1846-1852.
22. Sahiner, N., Seven, F., and Al-lohedan, H. (2015) Superporous cryogel-M (Cu, Ni, and Co) composites in catalytic reduction of toxic phenolic compounds and dyes from wastewaters. *Water, Air, & Soil Pollution*, **226**(4), 1-12.
23. Reichelt, S., Becher, J., Weisser, J., Prager, A., Decker, U., Moeller, S., Berg, A., and Schnabelrauch, M. (2014) Biocompatible polysaccharide-based cryogels. *Materials Science and Engineering: C*, **35**, 164.
24. Petrov, P., Petrova, E., Tchorbanov, B., and Tsvetanov, C. B. (2007) Synthesis of biodegradable hydroxyethylcellulose cryogels by UV irradiation. *Polymer*, **48**(17), 4943-4949.
25. Park, S., Lee, K.-S., Bozoklu, G., Cai, W., Nguyen, S. T., and Ruoff, R. S. (2008) Graphene oxide papers modified by divalent ions-enhancing mechanical properties via chemical cross-linking. *ACS Nano*, **2**(3), 572-578.
26. Wei, G., Miao, Y.-E., Zhang, C., Yang, Z., Liu, Z., Tjiu, W. W., and Liu, T., Ni-doped graphene/carbon cryogels and their applications as

versatile sorbents for water purification. *ACS Applied Materials & Interfaces*, **5**(15), 7584-7591.

27. Huang, M., Si, Y., Tang, X., Zhu, Z., Ding, B., Liu, L., Zheng, G., Luo, W., and Yu, J. (2013) Gravity driven separation of emulsified oil–water mixtures utilizing in situ polymerized superhydrophobic and superoleophilic nanofibrous membranes. *Journal of Materials Chemistry A*, **1**(45), 14071-14074.
28. Zhang, S., Shao, T., Bekaroglu, S. S. K., and Karanfil, T. (2009) The impacts of aggregation and surface chemistry of carbon nanotubes on the adsorption of synthetic organic compounds. *Environmental Science & Technology*, **43**(15), 5719-5725.
29. Pan, B., and Xing, B., Adsorption mechanisms of organic chemicals on carbon nanotubes. *Environmental Science & Technology*, **42**(24), 9005-9013.
30. Fan, Z.-J., Yan, J., Wei, T., Ning, G.-Q., Zhi, L.-J., Liu, J.-C., Cao, D.-X., Wang, G.-L., and Wei, F. (2011) Nanographene-constructed carbon nanofibers grown on graphene sheets by chemical vapor deposition: high-performance anode materials for lithium ion batteries. *ACS Nano*, **5**(4), 2787-2794.
31. Yang, S.-Y., Chang, K.-H., Tien, H.-W., Lee, Y.-F., Li, S.-M., Wang, Y.-S., Wang, J.-Y., Ma, C.-C. M., and Hu, C.-C. (2011) Design and tailoring of a hierarchical graphene-carbon nanotube architecture for supercapacitors. *Journal of Materials Chemistry*, **21**(7), 2374-2380.
32. Vengatesan, M., Singh, S., Stephen, S., Prasanna, K., Lee, C., and Mittal, V. (2017) Facile synthesis of thermally reduced graphene oxide-sepiolite nanohybrid via intercalation and thermal reduction method. *Applied Clay Science*, **135**, 510-515.
33. Allcock, H. R., and Ambrosio, A. M., Synthesis and characterization of pH-sensitive poly (organophosphazene) hydrogels. *Biomaterials*, **17**(23), 2295-2302.
34. Allcock, H. R., Bender, J. D., Chang, Y., McKenzie, M., and Fone, M. M. (2003) Controlled refractive index polymers: Polyphosphazenes with chlorinated-and fluorinated-, aryloxy-and alkoxy-side-groups. *Chemistry of Materials*, **15**(2), 473-477.
35. Allcock, H. R., Kellam, E. C., and Morford, R. V. (2001) Gel electrolytes from co-substituted oligoethyleneoxy/trifluoroethoxy linear polyphosphazenes. *Solid State Ionics*, **143**(3), 297-308.
36. Lu, S.-Y., and Hamerton, I. (2002) Recent developments in the chemistry of halogen-free flame retardant polymers. *Progress in Polymer Science*, **27**(8), 1661-1712.
37. Allcock, H. R. (2006) Recent developments in polyphosphazene materials science. *Current Opinion in Solid State and Materials Science*, **10**(5), 231-240.
38. Zhu, L., Xu, Y., Yuan, W., Xi, J., Huang, X., Tang, X., and Zheng, S. (2006) One-pot synthesis of poly(cyclotriphosphazene-co-4,4′-

sulfonyldiphenol) nanotubes via an in situ template approach. *Advanced Materials*, **18**(22), 2997-3000.

39. Wang, P., Cao, M., Wang, C., Ao, Y., Hou, J., and Qian, J. (2014) Kinetics and thermodynamics of adsorption of methylene blue by a magnetic graphene-carbon nanotube composite. *Applied Surface Science*, **290**, 116-124.
40. Marcano, D. C., Kosynkin, D. V., Berlin, J. M., Sinitskii, A., Sun, Z., Slesarev, A., Alemany, L. B., Lu, W., and Tour, J. M. (2010) Improved synthesis of graphene oxide. *ACS Nano*, **4**(8), 4806-4814.
41. Maldonado-Hodar, F., Moreno-Castilla, C., and Perez-Cadenas, A. (2004) Surface morphology, metal dispersion, and pore texture of transition metal-doped monolithic carbon aerogels and steam-activated derivatives. *Microporous and Mesoporous Materials*, **69**(1), 119-125.
42. Chen, W., and Yan, L. (2011) In situ self-assembly of mild chemical reduction graphene for three-dimensional architectures. *Nanoscale*, **3**(8), 3132-3137.
43. Wu, A., Yang, X., and Yang, H. (2013) Magnetic properties of carbon-encapsulated Fe–Co alloy nanoparticles. *Dalton Transactions*, **42**(14), 4978-4984.
44. Tang, Z., Shen, S., Zhuang, J., and Wang, X. (2010) Noble-metal-promoted three-dimensional macroassembly of single-layered graphene oxide. *Angewandte Chemie*, **122**(27), 4707-4711.
45. Nguyen, D. D., Tai, N.-H., Lee, S.-B., and Kuo, W.-S. (2012) Superhydrophobic and superoleophilic properties of graphene-based sponges fabricated using a facile dip coating method. *Energy & Environmental Science*, **5**(7), 7908-7912.
46. Rafiee, J., Rafiee, M. A., Yu, Z. Z., and Koratkar, N. (2010) Superhydrophobic to superhydrophilic wetting control in graphene films. *Advanced Materials*, **22**(19), 2151-2154.
47. Gui, X., Li, H., Wang, K., Wei, J., Jia, Y., Li, Z., Fan, L., Cao, A., Zhu, H., and Wu, D. (2011) Recyclable carbon nanotube sponges for oil absorption. *Acta Materialia*, **59**(12), 4798-4804.
48. Nguyen, S. T., Feng, J., Le, N. T., Le, A. T., Hoang, N., Tan, V. B., and Duong, H. M., Cellulose aerogel from paper waste for crude oil spill cleaning. *Industrial & Engineering Chemistry Research*, **52**(51), 18386-18391.
49. Wang, J., Zheng, Y., and Wang, A. (2013) Coated kapok fiber for removal of spilled oil. *Marine Pollution Bulletin*, **69**(1), 91-96.
50. Lin, J., Shang, Y., Ding, B., Yang, J., Yu, J., and Al-Deyab, S. S. (2012) Nanoporous polystyrene fibers for oil spill cleanup. *Marine Pollution Bulletin*, **64**(2), 347-352.
51. Zhao, J., Ren, W., and Cheng, H.-M. (2012) Graphene sponge for efficient and repeatable adsorption and desorption of water contaminations. *Journal of Materials Chemistry*, **22**(38), 20197-20202.

7

Magnetic Polymer Nanocomposites for Environmental Applications

7.1 Introduction

Needless to say, nanomaterials are one of the most vigorously studied subject areas in the recent years. Nanocomposites are a class of such nanomaterials generated by synergistic combination of the components and have dimension of at least one of the constituents below 100 nm. The combination of the constituent materials in the composite results in improved properties. Thus, there has been a plethora of studies which have been carried out to develop a large variety of nanocomposites for different applications [1].

Polymer nanocomposites are a class of materials where polymer acts as the matrix which is incorporated with different type of nanomaterials, thus, improving the properties of the matrix dramatically. Generally, polymer nanocomposites exhibit enhanced stiffness, strength, gas barrier, flame retardance and thermal resistance as compared to pure polymer matrices. Due to significantly enhanced property profiles, polymer nanocomposites have been developed for various applications in the field of microelectronics, biomedicine, automobiles, etc. [2].

Magnetic nanomaterials are functional materials which exhibit different types of magnetism like ferromagnetism, paramagnetism, superparamagnetism, etc. The pure and alloyed forms of Fe, Ni, Co are the most studied magnetic nanomaterials. These have found uses in various fields like information storage, magnetic refrigeration systems, contrast enhancement in magnetic resonance imaging, magnetic carriers for drug targeting, catalysis, etc. [3]. Due to their nanosize, these materials tend to agglomerate swiftly, thus, leading to a reduction in the active surface area. In order to overcome this limitation, various types of coatings or surface functionalities are incorporated with these materials which make these materials more stable and attractive for various applications [4]. In addition, in particulate form, these materials do not fit many application requirements,

Liyamol Jacob and Vikas Mittal, Khalifa University of Science and Technology, Abu Dhabi, UAE

where the use of a continuous material or matrix is required. The nanoparticles, if not stabilized/immobilized through organic surface coatings, are also prone to be released in environment, water streams. Thus, to overcome these challenges, the use of polymers as a matrix or surface modifier for the magnetic nanomaterials has been of significant interest. The polymer matrix loaded with (or coated on) magnetic nanomaterials can respond to external magnetic fields, thus, exhibiting potential for a wide range of applications. Also, interaction between the external and internal magnetic field of the magnetic material can bring about a lot of changes in the host material (or organic coating), like induction of magneto-mechanical forces which can be utilized for changing the shape of the material or inducing its movement. Such properties can be utilized for biomedical applications, actuators, etc. In addition, the magnetic materials can work as nano-heaters responding to external alternating magnetic field. This property can be coupled with thermoresponsive polymers to generate advanced materials [5]. In the recent years, magnetic polymer nanocomposites have been studied for application as drug delivery agents, recovering agents in photocatalysts, negative contrast agents in imaging, etc. Synthesis of magnetic polymer composites can be achieved through various methods like plasma polymerization, melt and solution blending and co-precipitation [6].

This review focuses on the recent advances in the synthesis and properties of the functional magnetic polymer nanocomposite materials, along with summarizing their cutting-edge environmental applications.

7.2 Magnetic Polymer Nanocomposites

Magnetic properties shown by nanomaterials are size dependent to a large extent. Especially, the materials which have size comparable to the magnetic domains are of great interest. Unlike other nanomaterials, magnetic nanomaterials are expressed as small identical molecules which have magnetic property. As mentioned earlier, the utility of magnetic nanoparticles can be improved by forming composites with different polymers either by incorporating the particles in the polymers or by modifying the surface of the particles by polymers. There are numerous ways to achieve such hybridization of magnetic materials with polymers, such as ex-situ, in-situ, melt blending, solution blending, etc. [7]. The properties a magnetic material should have for the formation of a composite are same as for regular polymer composites such as: 1) easily availability, 2) cost-effectiveness, 3) reproducibility, 4) feasibility for large scale synthesis, 5) low toxicity, 6) no undesirable reactions with the matrix [8].

7.2.1 Fe based Magnetic Polymer Nanocomposites

Fe based magnetic materials have been commonly utilized for the fabrication of polymer nanocomposites with magnetic properties. Fernandez *et al.* [9] synthesized thermochromic thin films of spin crossover (SCO) polymer $[Fe(NH_2trz)_3](BF_4)$ using different organic polymers as hosts. The formation of these materials could be confirmed by the color changes associated with SCO phenomenon. The hydrophobic polymer hosts were characterized by large particles of SCO material, while hydrophilic polymers exhibited more dispersed nanocrystals with fibrous structures in the polymer matrix, which also affected their colorimetric transitions. The fibrous materials exhibited color transition at room temperature, while the bigger particles showed transitions at lower temperature. This was caused not only by the size of the particles, but also by the nature of solvent and polymer interactions during the fabrication of polymer nanocomposite. The color values vs. temperature exhibited a linear relation, thus, avoiding the uncertainty associated with hysteresis loop [9].

Electrocaloric ferroelectric polymer nanocomposites which can operate at high and average electric fields at room temperature were developed by Zhang *et al.* [10]. A low-cost emulsion polymerization was also utilized for the synthesis of magnetic polymer nanocomposites with hydrophobic and superoleophilic surface [11]. The composites were characterized by properties like unsinkability, thermal stability and corrosion resistance. The polymer nanocomposite was used as an oil adsorbent which could be controlled by applying external magnetic field. Repetitive use of the material for oil removal did not affect the surface properties, thus, confirming the development of a functional material which could be readily used for practical applications without any obvious contamination.

In-situ polymerization technique was used by Kalita *et al.* [12] for the fabrication of shape memory polymer composites using Mesua ferrea L. seed oil based hyperbranched polyurethane (HBPU) and Fe_3O_4. The polymer composite was characterized by good dispersity of the magnetic nanoparticles in the polymer matrix and strong interfacial attraction forces between the matrix and filler particles. The material was observed to exhibit shape memory properties, which also enhanced with Fe_3O_4 content in the composite. Under microwave atmosphere, the polymer had full shape fixity and recoverability.

Biocompatible conductive polyfuran and superparamagnetic Fe_3O_4 particles was used to fabricate Hb immobilized biosensor via in-situ polymerization technique [13]. The composite was used for the detection of H_2O_2 from biological samples. Xiong *et al.* [14] synthesized Fe_3O_4/Ag@nanofibrillated cellulose (NFC) nanocomposites which could change from highly porous aerogel to sheet architecture by actuating using a small magnet. The biocompatible and antibacterial polymer composite was observed to be efficient for the removal of 4-nitrophenol. In the composite, cellulose sheet acted both as a biocompatible substrate and reducing agent for silver ions.

Super paramagnetic iron oxide nanoparticles were generated by using double hydrophilic graft copolymers poly(acrylic acid)-g-poly(N-isopropylacrylamide) (PAA-g-PNIPAM) as a water soluble coating [15]. The preparation of the polymeric coating was realized in three different steps, viz. i) a well-defined hydrophobic PtBA-based backbone, poly(tert-butyl 2-((2-chloropropanoyloxy)-methyl)acrylate)-co-poly(tert-butyl acrylate), ($PtBCPMA_{19}$-co-$PtBA_{18}$), was prepared through RAFT copolymerization of a new trifunctional acrylic monomer, tert-butyl 2-((2-chloropropanoyloxy)methyl)acrylate and tert-butyl acrylate; ii) taking this backbone as a macroinitiator to initiate SET-LRP of N-isopropylacrylamide resulted in well-defined (poly(tert-butyl 2-((2-chloropropanoyloxy)methyl)-acrylate)-co-poly(tert-butyl acrylate))-g-poly(N-isopropylacrylamide) ((PtBCPMA-co-PtBA)-g-PNIPAM) amphiphilic graft copolymers with relatively narrow polydispersities ($M_w/M_n \leq 1.31$); iii) handling (PtBCPMA-co-PtBA)-g-PNIPAM in acidic conditions afforded PAA-g-PNIPAM graft copolymers. The polymer acted as a stabilizer during the preparation of superparamagnetic iron oxide particles, with size controllability.

Oxidative chemical polymerization was utilized for the synthesis of polyaniline (PANI)/cobalt-manganese ferrite, (PANI)/$Co_{0.5}Mn_{0.5}Fe_2O_4$ nanocomposite by Sozeri *et al.* [16]. Aniline monomer was polymerized in presence of ammonium peroxydisulfate (APS). The $CoFe_2O_4$ nanoparticles were synthesized using a microwave assisted method. PANI/$Co_{0.5}Mn_{0.5}Fe_2O_4$ nanocomposite exhibited superparamagnetic behavior with a blocking temperature higher than 300 K. Surface spin disorder resulted in the low magnetic saturation of the composite as compared to $CoFe_2O_4$ particles. Unsaturated magnetization could be obtained by temperature decrease which was characterized by increase in coercivity and antiferromagnetic interactions. In another study, ferromagnetic Fe_3O_4 nanoparticles were synthesized using a surfactant-free sonochemical reaction at room temperature without using any inert atmosphere [17]. The particles were integrated into poly(vinyl alcohol), which, in turn, significantly enhanced the thermal stability and flame retardancy of the polymer.

Removal of toxic Pb^{2+}, $Cd^{2+}$and Ni^{2+} from water sources was realized through the use of carboxymethyl-β-cyclodextrin (CM-β-CD) polymer modified Fe_3O_4 nanoparticles (CDpoly-MNPs) by Zayed *et al.* [18]. The magnetic material exhibited selective adsorption of all types of toxic ions from water with a high affinity towards the adsorption of Pb^{2+}. Chitosan based nanocomposite films were also synthesized on the magnetic α-Fe_2O_3 nanoparticles using hydrothermal method [19]. The magnetic nanoparticles exhibited rhombohedral structures and no phase change was observed by the use of chitosan molecules. The material exhibited excellent saturation magnetization and had ferromagnetic properties.

7.2.2 Ni based Magnetic Polymer Nanocomposites

Ni has been used to synthesize a variety of polymer magnetic nanocomposites with the induction of different types of magnetization into the polymer matrix [20-29]. For instance, Figure 7.1 shows the schematic of a phenolic biosensor prepared from the composite of polydopamine (PDA)-laccase (Lac)-nickel nanoparticle loaded carbon nanofibers (NiCNFs) [20]. Yoonessi *et al.* [30] also used in-situ magnetic field solvent method for the synthesis of oriented hybrid nickel tethered graphene polyimide resin nanocomposites with different degrees of orientation. The magnetization of the composite was observed to be maximum along the direction of magnetic field and minimum in the direction perpendicular to the magnetic field, indicating superparamagnetic orientation of Ni nanoparticles. The DC electrical conductivity of the film could be improved by casting it in the presence of a magnetic field. The nanocomposite exhibited actuation in static magnetic field.

In-situ chemical oxidative copolymerization was utilized to synthesize polystyrene@polypyrrole@nickel (PS@PPy@Ni) with core-shell architecture by Li *et al.* [31]. The Ni nanoparticles were deposited on to the surface of the synthesized polymer sheet using activation-electroless plating methods. The coating of Ni around PS@P(Py-

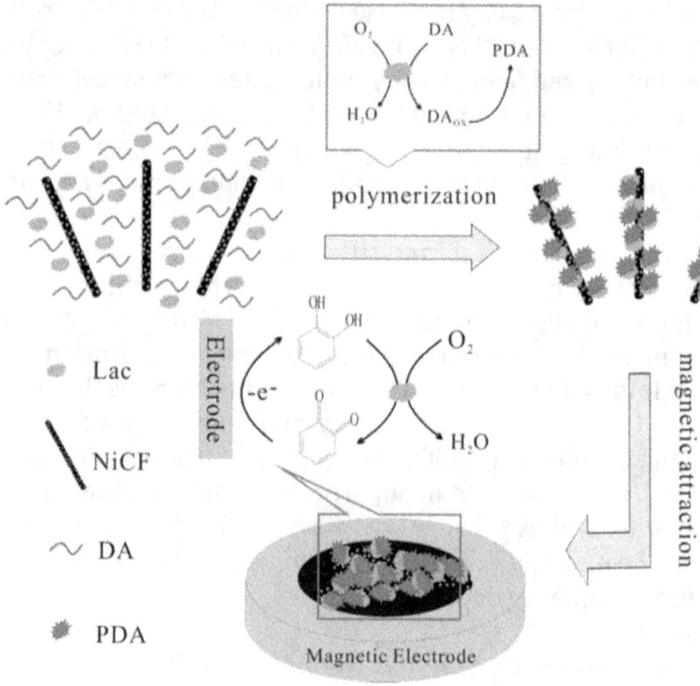

Figure 7.1 Representation of phenolic biosensor prepared from the composite of polydopamine (PDA)-laccase (Lac)-nickel nanoparticle loaded carbon nanofibers (NiCNFs). Reproduced from Reference 20 with permission from American Chemical Society.

PyCOOH) spheres induced ferromagnetism in the material. The material could be used for effective electromagnetic absorption with maximum reflection loss of around -20.06 dB at 10.69 GHz. Figure 7.2 also shows the scanning electron micrographs of the various hybrid materials.

Highly electrically conducting magnetic graphene-Ni polyimide nanocomposites were also synthesized by Yoonessi *et al.* [32]. Increasing the amount of magnetic nanoparticles resulted in an enhancement of saturation magnetization. Nanocomposite exhibited a bimodal nanoparticle distribution throughout the polymer layer and could be actuated in external static magnetic field. In-situ polymerization with the help of cetyltrimethylammonium bromide (CTAB) as a surfactant formed polyaniline-$NiFe_2O_4$ nanofiber composites at water and ionic liquid interphase by Baykal *et al.* [33]. The ionic liquid 1-butyl-3-methyl-imidazolium bromide [BMIM]Br was observed to have significant influence on the structure, morphology, conductivity and the magnetization of the resulting nanocomposite. In general, the introduction of ionic liquids has a positive impact on the conductivity, while a negative impact on the magnetization.

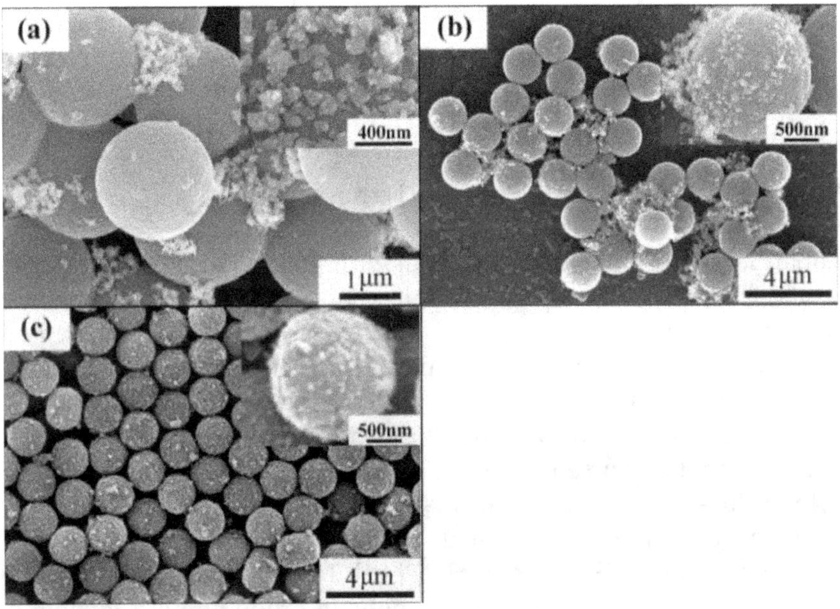

Figure 7.2 SEM images of (a) PS@P(Py-PyCOOH$_0$)@Ni, (b) PS@P(Py- Py-COOH$_{25}$)@Ni and (c) PS@P(Py-PyCOOH$_{50}$)@Ni core/shell microspheres. Reproduced from Reference 31 with permission from American Chemical Society.

Spatially distributed silicon rubber polymer nanocomposites were synthesized with varying the concentration of Ni in the polymer matrix by Boudenne *et al.* [34]. The creation of 1D Nickel structures contributed to higher electric and heat transportability. The spatial distribution did not have any effect on the specific heat capacity of the polymer. The 1D distribution also facilitated an increment of almost 100% thermal conductivity compared to the other orientation of Ni nanoparticles in the polymer matrix. This configuration also facilitated a reduction in percolation threshold from 25% to 2% (vol%) of Ni nanoparticles. In another study, field cooled magnetization of pure Ni ferrite powders and Ni ferrite nanocomposites synthesized using solution casting method exhibited shape memory properties [35]. Increasing particle concentration in the composites resulted in improved shape memory properties. Henkel plots confirmed improvement in the dipolar properties with the increase in particle concentration. However, the increase in cluster size negatively influenced the memory effects. Polydimethylsiloxane (PDMS) magnetic elastomer matrix filled with Ni fiber and Ni particles were also synthesized by speed mixing and curing at ambient temperature conditions [36]. Due to the use of magnetic field during the curing process, the Ni particles and fibers exhibited a distributed alignment throughout the polymer matrix.

7.2.3 Co based Magnetic Polymer Nanocomposites

Cobalt ferrite is a magnetic material with a high Curie temperature, which has found wide ranging applications in biomedical and environmental applications. It has characteristics like high coercivity, medium saturation magnetization, high anisotropy constant, chemical stability, mechanical hardness and electrical insulation [37-39].

The co-assembly of amphiphilic copolymers of PAA20-b-PS280 and oleic acid-stabilized magnetic iron oxide nanoparticles using tetrahydrofuran, N,N-dimethylformamide and water resulted in the synthesis of a well-defined, magnetic shell cross-linked knedel-like nanoparticles (MSCKs) (Figure 7.3) [40]. The material could be used as sequestering agent for hydrocarbons in crude oil with magnetic

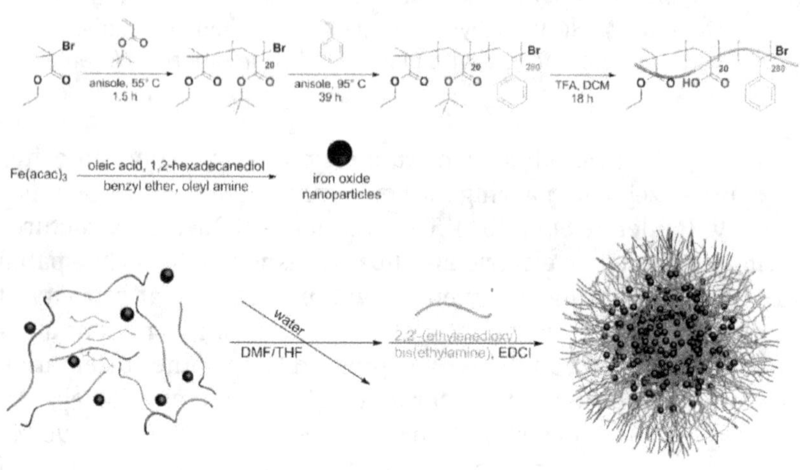

Figure 7.3 Schematic representation of the generation of magnetic shell cross-linked (MSCK) nanoparticles. Reproduced from Reference 40 with permission from American Chemical Society.

responsivity. The generated material in aqueous environments could remove pollutant hydrocarbons from water. The developed material has the potential to be used in the bulk recovery phase at oil spill sites.

The particle agglomeration in solution casting method for polymer magnetic nanomaterial synthesis was overcome by using the nanoparticles in their dry particulate state avoiding the use of solvents [41]. In another study, biocomaptible poly(vinyl alcohol) (PVA) was used for the surface modification of cobalt ferrite nanoparticles ($CoFe_2O_4$ NPs) synthesized via combustion method [42]. The PVA coating transformed the nature of nanoparticles from hydrophobic to hydrophilic. The polymer coated magnetic material exhibited less toxicity and high magnetization with a ferromagnetic behavior compared to the uncoated material.

7.3 Environmental Applications

In the recent years, there has been a significant interest for the use of polymer coated magnetic nanomaterials (or magnetic polymer nanocomposites) for various environmental applications. These materials have been used in areas like pollutant removal from water, oil spill removal, biomedical applications, etc. It is important to gain insights about the interactions of these magnetic nanomaterials with the environment to tune their applications in diverse fields.

7.3.1 Oil Spill Removal

Magnetic polymer nanocomposites have significant potential for use in oil spill removal. These materials are economical, reliable, relatively easy to synthesize and reusable which make them ideal for oil spill removal. The major challenge in the use of materials for the removal of oil is their fast distribution and collection which can be overcome by using magnetic materials as their performance can be efficiently controlled by the application of external magnetic field [43-46]. In a recent study, a well-defined interconnected porous structure cryogel based on resorcinol formaldehyde ($CoFe_2O_4$/graphene/PZTs/R-F) was prepared via freeze drying method, followed by carbonization in order to generate advanced materials to remove crude oil (high viscosity) and vegetable oil (low viscosity) from the oil/water mixture. The sorption capacities of the resultant cryogel for crude oil and vegetable oil were higher than the sorption capacities of $CoFe_2O_4$/graphene/R-F and graphene/R-F/Na_2CO_3 cryogels. The cryogel was responsive to external magnetic field and exhibited good hydrophobic properties with water contact angle of obtuse angle. Figure 7.4 also demonstrates the hydrophobic behavior of the cryogel. From the visual images, it was observed that the oil droplet was quickly sorbed by the cryogel, while the water droplet achieved quasi-spherical shape on the surface of the gel.

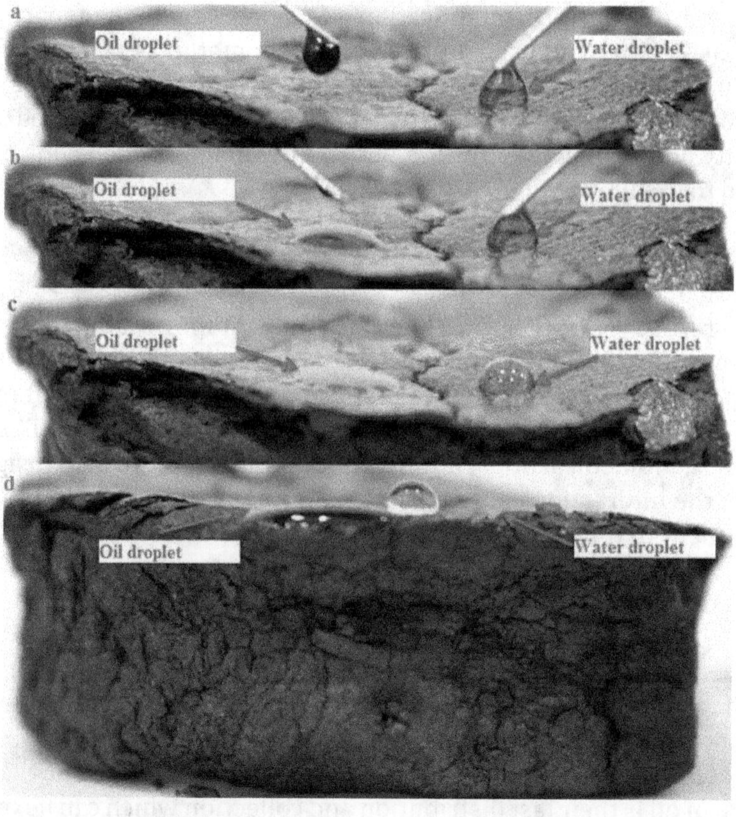

Figure 7.4 Hydrophobic behavior of the $CoFe_2O_4$/graphene/PZTs/R-F cryogel; water droplet gained quasi-spherical shape and oil droplet sorbed quickly onto the top surface of the cryogel. Reproduced from Reference 47.

In another study, superparamagnetic iron oxide nanoparticles functionalized on polyurethane foams were used to efficiently remove oil spills from aqueous environment by Calcagnile *et al.* [48]. The composite was also functionalized with sub-micrometer polytetrafluoroethylene to improve its efficiency. The use of solvent free electrostatic polytetrafluoroethylene particle deposition method rendered the polyurethane foams with water repellent and oil absorbing properties which were otherwise hydrophobic and oleophobic. The use of both types of nanoparticles enhanced the rate of oil absorption using polyurethane foams. In addition to the oil absorbing and water repelling properties, the composite exhibited magnetic responsivity (Figure 7.5). The material has the potential to be used in large area oil spills because the foam easily floats on the surface of water which can be moved using the help of a magnet, thus, cleaning the entire area of water underneath.

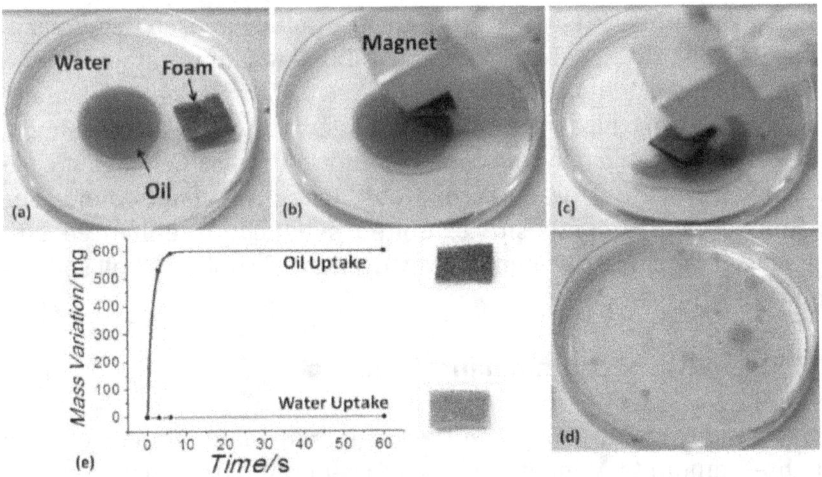

Figure 7.5 (a) Dye-colored oil on the surface of the water, (b,c) removal of the oil by the magnetically driven PU/NPs/PTFE sample, (d) cleaned water surface, (e) graph showing the variation of the weight of the foam with time. Reproduced from Reference 48 with permission from American Chemical Society.

A low-cost emulsion polymerization technique was utilized for the synthesis of superoleophilc and highly hydrophobic magnetic polymer nanocomposites which can remove oil from the surface of water [49]. The developed material had additional properties such as floatability, chemical stability, corrosion resistance and thermal stability, which further reinforced its potential for commercial applications. The material could absorb oil three times its weight with efficiently repelling water and could be retracted from the water surface by using a magnet. It had excellent reusability and the oil from the nanocomposite could be effectively removed using ultrasonication. The protein wastes from the leather industries were also utilized in another study for generating high stability magnetic collagen nanocomposites with superparamagnetic iron oxide nanoparticles [50]. The composite exhibited selective absorption of oils and was trackable by the application of external magnetic field. By heat treatment, the material could be converted into bi-functional graphitic nanocarbon, which proved its environmental sustainability.

Natural rubber's innate ability to resist oil has been utilized for generating oil spill removal materials [51]. Ecofriendly epoxidized natural rubber with magnetite nanoparticles was used for the absorption of gasoline. The epoxidization was used with the aim of retaining the rubber's elasticity and for the adhesion of the nanoparticles. 5% epoxidized natural rubber exhibited maximum stability with absorption capability even after many days of immersion. The material could absorb oil seven times its weight without losing its ability, even after repeated use. The composite formation using magnetic nanoparticles made it easily recoverable using a magnetic field. The magnetic rubber composite exhibited significantly enhanced efficiency as compared to commercially used butyl rubber for oil spill removal.

7.3.2 Removal of Contaminants from Water

Mercury is a highly toxic material for humans and animals even in slightest amounts. Magnetic composite generated from magnetic iron oxide nanoparticles (M-MIONPs) modified with 2-mercaptobenzothiazole were proved to be efficient in the removal of Pb(II) ions from water [52]. The material was observed to be efficient, fast, cost effective and reliable for the removal of toxic mercury ions from aquatic environments. The material was able to remove mercury ions within a short span of 4 min with almost 100% efficiency as compared to non-modified iron oxide nanoparticles.

A nanocomposite composed of multiwalled carbon nanotubes, poly(1-glycidyl-3-methylimidazolium chloride) and ferroferric oxide was used to remove anionic azo dyes orange II, sunset yellow FCF and amaranth [53]. Adsorption kinetics and isotherms followed pseudo second order kinetics and Langmuir isotherm respectively and the adsorption was affected by the pH of the system. The adsorption and separation processes were enhanced by the introduction of the polymer and magnetic nanomaterial together with carbon nanotubes. In another study, radical addition fragmentation chain transfer (RAFT) polymerization using RAFT agent functionalized iron oxide nanoparticles as the chain transfer agent was used to synthesize molecular imprinted core shell magnetic beads [54]. The beads had a polymer film with a thickness of about 22 nm, and exhibited thermal stability, magnetic responsivity and spherical shape. The material was utilized for the selective removal of bisphenol A and could be easily recovered using the application of an external magnetic field. The material also exhibited reusability and highly stable performance without much loss in efficiency up to 5 repeated cycles.

Polyethersulfone (PES) nanocomposite membrane containing polyaniline/iron(II, III) oxide (PANI/Fe_3O_4) nanoparticles synthesized using phase inversion method was utilized for Cu(II) ion removal from aqueous environment [55]. Different nanoparticle loading was tested for optimum removal efficiency. 0.1% loading of nanoparticles was observed to be most efficient due to the nanoparticle occupancy at the superficial pores of the polymer membrane during synthesis. The material exhibited a relatively complex adsorption mechanism with Redlich-Peterson model isotherm. The membrane was also observed to be highly reusable. In another study, in-situ polymerization of polypyrrole monomer with magnetic FE_3O_4 nanoparticles was used to synthesize Fe_3O_4 coated polypyrrole membrane which was efficient in the removal of Cr(VI) ions from water [56]. 100% adsorption was observed for 200 mg/L Cr(VI) aqueous solution at a low pH of 2. The removal was attributed to the ion exchange and reduction on the surface of the nanocomposite. The removal efficiency was negatively affected by an increase in pH of the aqueous solution. The adsorption followed pseudo second order kinetics and it could be fitted with a Langmuir adsorption isotherm. The material exhibited an endothermic reaction during the adsorption, which was spontaneous in nature. The material could be reused for at least two repeated cycles without much loss in the original capacity.

As mentioned earlier, carboxymethyl-β-cyclodextrin (CM-β-CD) polymer modified Fe_3O_4 nanoparticles were utilized for the selective removal of Pb^{2+}, Cd^{2+} and Ni^{2+} ions from aqueous solution [18]. The adsorption capability was dependent on temperature, pH and ionic strength of the aqueous solution. The adsorption exhibited pseudo second order kinetics and could be fitted with Langmuir isotherm. The polymer grafting enhanced the adsorption ability because of the hydroxyl and carboxyl groups from the polymer backbone complexing with the metal ions. The authors also explored the desorption and reusability of the developed materials.

Similar to earlier study, Arora *et al.* [57] investigated the potential application of cobalt ferrite doped graphene/phosphazene nanotubes (PZTs) - resorcinol-formaldehyde (R-F) cryogel nanocomposites as a new magnetic field responsive adsorbent to remove 2-nitrophenol from aqueous solution [47]. Adsorption kinetics and thermodynamics parameters ((ΔH°, ΔG° and ΔS°) were examined from experimental data, which indicated that the adsorption process was spontaneous and endothermic, which followed a pseudo second order model. The maximum adsorption capacity reached 92.37 $mg.g^{-1}$ at initial 2-nitrophenol concentration of 600 $mg.L^{-1}$ and 303 K. The experimental adsorption data was also fitted using Langmuir, Freundlich and Temkin isotherm models and the adsorption data was observed to fit well with Freundlich isotherm model (Figure 7.6).

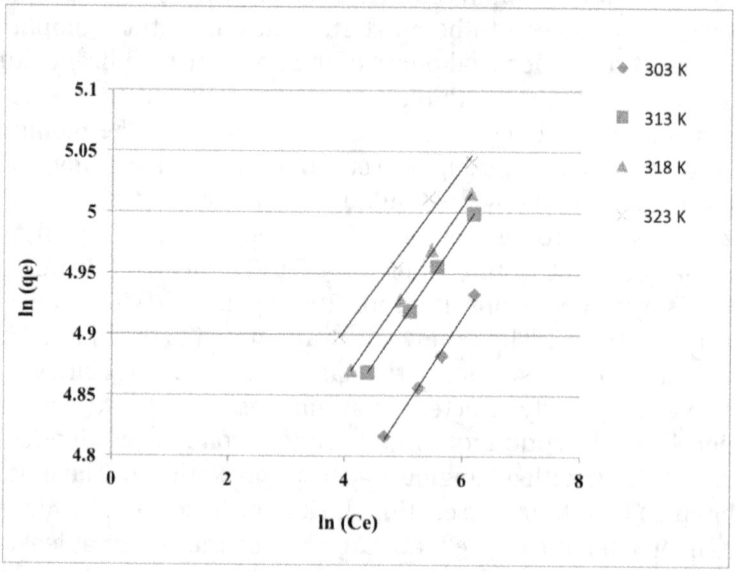

Figure 7.6 Freundlich isotherm for adsorption of 2-nitrophenol (600 $mg.L^{-1}$) onto magnetic adsorbent at 303, 313, 318 and 323 K. Reproduced from Reference 57.

7.3.3 Biocatalysts and Biological Host Systems

Lin *et al.* [58] reported the synthesis of enzyme nanogels with magnetic nanoparticle encapsulation. The synthesis was carried out as follows: i) synthesis of the aminated, acryloylated or carboxylated magnetic nanoparticles, ii) binding the oppositely charged enzymes on to the surface of magnetic nanoparticles using electrostatic interaction, iii) in-situ polymerization using acrylamide and methylenebis(acrylamide) as monomer and crosslinker respectively. The dissociation of enzyme from the polymer surface could be prevented by effectively controlling the pH of the system and its catalytic activity could be modified by covalent linkage. The polyacrylamide gel network was able to provide hydrophilic confinement and improve the thermal stability of the enzyme.

Copolymerization of styrene and maleic anhydride in the presence of magnetite seeds was carried out by Qiu *et al.* [59] to synthesize Fe_3O_4/poly(styrene-co-maleic anhydride) core-shell composite microspheres. The affinity between magnetite nanoparticles and monomers was improvised by encapsulating magnetite nanoparticles with polyethylene glycol. It was observed that the amount of surface anhydride, dimension of microspheres and percentage of magnetite present in the composite was influenced by the comonomer ratio, magnetite particles and solvent. Covalent binding led to the immobilization of free α-amylase to the reactive surfaces containing anhydride groups. The material exhibited a high magnetic immobilized enzyme and binding capacity with an activity recovery of 47.2%. Zhang *et al.* [60] reported the microwave solvothermal method for the rapid synthesis of Fe_3O_4 nanospheres with uniform dimensions. Chemical oxidative polymerization of pyrrole in the presence of magnetic nanospheres were carried out to synthesize Fe_3O_4/polypyrrole (PPy) composite nanospheres with well-defined core/shell structures. The coated shell exhibited an average thickness of 25 nm. Subsequently, gold nanoparticles were immobilized on to the surface of the polymer nanocomposite to generate Fe_3O_4/PPy/Au core/shell/shell structure. The composite exhibited excellent electrocatalytic and biocatalytic activity towards bio-species like ascorbic acid. The magnetic particles were controllable by the application of an external magnetic field and exhibited excellent conductivity. Figure 7.7 shows the micrographs of the generated materials.

β-cyclodextrin-polyurethane polymer coated Fe_3O_4 magnetic nanoparticle composite was prepared as solid-liquid phase transfer catalyst and molecular host system for nucleophilic substitution reactions [61]. The polymeric nanocomposite material was able to catalyze the nucleophilic substitution reaction of benzyl halides with thiocyanate, azide, cyanide and acetate anions in aqueous environment. The recovery and reuse of the developed material was achieved by the application of an external magnetic field. In addition, no toxic by-products were formed during the reaction. The nanoparticles exhibited superparamagnetic properties with a mean diameter of 59 nm.

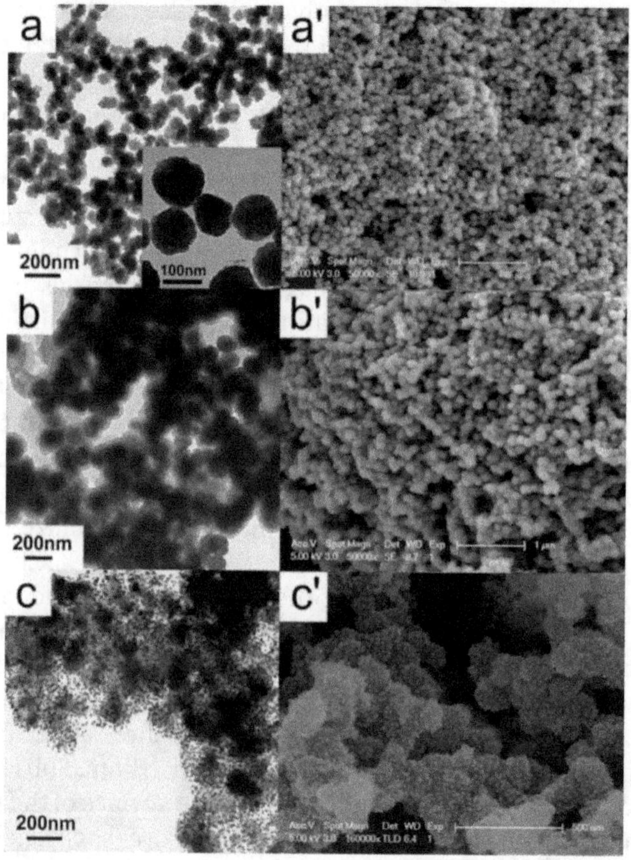

Figure 7.7 (a-c) transmission and (a′-c′) scanning electron images of (a, a′) Fe_3O_4, (b, b′) Fe_3O_4/PPy, and (c, c′) Fe_3O_4/PPy/Au NPs. Reproduced from Reference 60 with permission from American Chemical Society.

In another study, the efficiency of immobilized enzymes could be enhanced by using polymer nanocomposite microspheres (PNCMs) as solid supports, which reduced diffusional limitation and increased the surface area per mass unit [62]. Fe_3O_4/SiO_2-g-poly(PSStNa) was used as a solid support for the immobilization of pectin enzyme by covalent interaction. The material exhibited an enhanced storage capacity and activity over a wide range of temperature and pH. By using

Michaelis-Menten equation, the catalytic activity of the enzyme immobilized polymer magnetic nanocomposite was studied. The separation and recovery could be achieved using an external magnetic field and the method reduced the cost of enzyme immobilization considerably, thus, confirming the potential for a large variety of practical applications.

7.3.4 Gas and Humidity Sensors

Nanocomposites of polypyrrole and iron oxide have been successfully used as humidity sensors. In one such study, the composite was formed using a simultaneous gelation and polymerization process [63,64]. The pellet shaped nanocomposite was utilized for humidity and gas sensing studies. The humidity sensitivity of the as prepared composite was improved by increasing the polypyrrole content. The gas sensitivity studies conducted on CO_2, N_2 and CH_4 gases concluded that the composite exhibited a linear relationship between pressure and sensitivity.

Free standing ultrathin nanocomposite films were also synthesized based on conducting polymers like poly(3,4-ethylenedioxythiophene):polystyrene sulfonate (PEDOT:PSS) with embedded iron oxide nanoparticles [65]. Spin coating was utilized for the synthesis of the nanocomposite films in conjunction with a release technique to detach them from the substrate. The films were tested for their humidity sensing properties and the sensitivity was observed to increase with increase in nanoparticle concentration. Figure 7.8 also demonstrates the change in the resistance of the nanocomposite film during humidity step-change experiments. The response was observed to be fast and effectively matched to the changes in humidity. In addition, the resistance of the sensor was observed to return to the original value after the completion of the test.

In another study, polyaniline/γ-ferric oxide (PANi/γ-Fe_2O_3) nanocomposite films were synthesized by Sen *et al.* [66] for room temperature liquefied petroleum gas (LPG) sensing. The gas sensitivity of the nanocomposite was improved by its large surface area which enhanced the adsorption of gas molecules. The response time was as low as 60 s for the composite and the maximum response was observed for 200 ppm LPG concentration. The sensing mechanism of the nanocomposite was attributed to the change in the depletion region of the p–n junction, which was formed between polymer and magnetic nanoparticles.

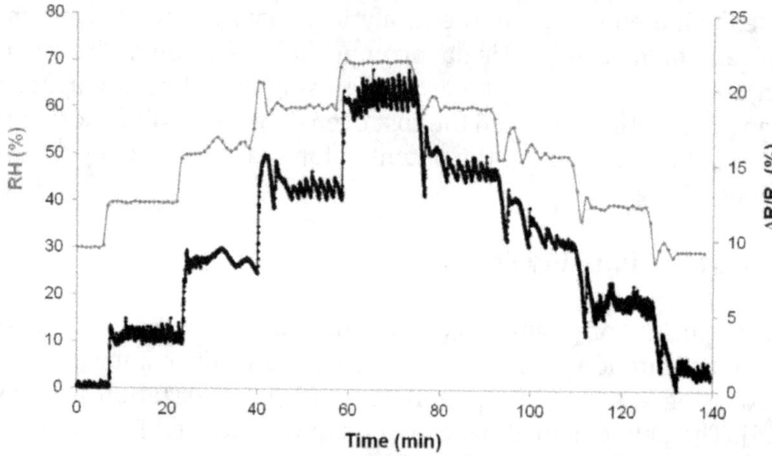

Figure 7.8 The change in the resistance (dark curve) of a PEDOT:PSS/NP nanofilm during humidity step-change experiments (light curve) in the range of 30-70% RH. Reproduced from Reference 65 with permission from American Chemical Society.

7.3.5 Biosensors

Nanocomposites formed by the conjugation of polymers and nanomaterials have found significant applications in biosensing field due to enhanced surface area and tunable surface properties. Many synthesis methods have been utilized for the generation of polymeric nanocomposites, especially magnetic polymer nanocomposites, for biosensing applications [67-69].

Estrone imprinted polymer nanocomposite with Fe_3O_4 coating was synthesized using semi-covalent imprinting method [70]. Firstly, silicone-estrone monomer complex was prepared using 3-(triethoxysilyl)propyl isocyanate and estrone, where the template could be linked to the silica coating on the iron oxide core through a thermally reversible bond. The material exhibited high saturation magnetization and specific recognition. The developed magnetic polymer nanocomposites were successfully utilized for the specific biochemical separation of estrone. In another study, emulsion polymerization was utilized for the synthesis of biocompatible poly(p-phenylenediamine) (PpPDA)-Fe_3O_4 nanocomposite which was further immobilized with hemoglobin (Hb) molecules for sensing of H_2O_2 [71]. The material exhibited good sensitivity and reproducibility towards the detection of H_2O_2. The catalytic reduction current of H_2O_2 was observed to be proportional to its concentration. The high affinity of Hb on PpPDA@Fe_3O_4 nanocomposite was also confirmed by the value of Michaelis–Menten constant (0.088 mM).

Biosensing nanocomposite polymer films based on polypyrrole were also synthesized by immobilizing magnetic nanoparticles on polypyrrole surface under appropriate magnetic field [72]. Streptavidin labeled magnetic particles were used to functionalize NPs assembly with biotin-Fab fragment K47 antibody. The biosensor was used to detect atrazine concentrations. The nanocomposite was observed to be fast, simple and accurate for the detection and exhibited a low concentration detection limit of 5 ng/ml. These properties make these materials attractive candidates for atrazine biosensing for practical and commercial applications. Zhang *et al.* [73] developed Fe_3O_4/polydopamine/Au composite nanoparticles with core shell structure by the polymerization of dopamine on Fe_3O_4 core. The changes in reaction time and reactant concentrations affected the polydopamine shell. The reduction of Au^{3+} between the core/shell and $HAuCl_4$ solution resulted in numerous satellite structures of Au nanoparticles on the surface of the Fe_3O_4/polydopamine composite. The material exhibited excellent magnetic responsivity. It was also observed to be specific to the detection and removal of bovine hemoglobin. These properties confirm the potential use of these materials for protein removal in proteomic analyses.

7.3.6 Other Applications

Along with the above-mentioned applications, polymer magnetic nanomaterials have found applications in other advanced areas such as drug delivery [74-84], magnetic resonance (MR) imaging [85-94], tissue engineering [95-103], etc. Their specificity, responsivity towards magnetic environment as well as low toxicity are the most important characteristics driving their use in such applications. For instance, Lutz *et al.* [85] reported the synthesis of a well-defined copolymer poly(oligo(ethylene glycol) methacrylate-co-methacrylic acid) P(OEGMA-co-MAA) as a novel water-soluble biocompatible coating for superparamagnetic iron oxide nanoparticles. The authors reported that the diameter of the nanoparticles was controlled in the range 10-25 nm, which was dependent on the initial copolymer concentration. The magnetic nanomaterials were observed to have long term colloidal stability in physiological buffer. These materials could be studied in vivo by MR imaging. On injecting in rats, the nanoparticles were not observed to accumulate in the liver during the first 2 h, and exhibited maximum accumulation after 6 h, thus, indicating a prolonged circulation in bloodstream. Figure 7.9 presents the schematic of the preparation of copolymer as well as copolymer modified iron oxide nanoparticles.

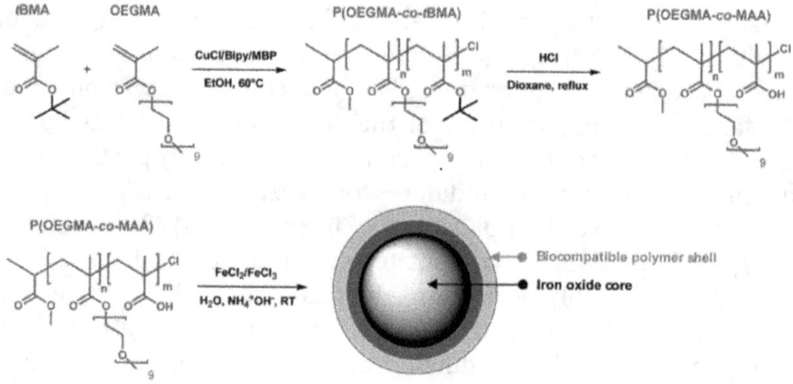

Figure 7.9 Schematic of the preparation of copolymer as well as copolymer modified iron oxide nanoparticles. Reproduced from Reference 85 with permission from American Chemical Society.

In another study, Berret *et al.* [89] also reported controlled clustering of superparamagnetic nanoparticles using block copolymers (Figure 7.10) [89]. The complexes were generated using 6.3 nm diameter superparamagnetic nanoparticles (maghemite γ-Fe_2O_3) and cationic-neutral copolymers such as poly(trimethylammonium ethylacrylate methyl sulfate)-b-poly-(acrylamide) with varying molecular weights of the blocks. The authors confirmed through magnetic resonance spin-echo measurements that the developed polymer-nanoparticle complexes exhibited a better contrast in magnetic resonance imaging as compared to individual nanoparticles.

Arora *et al.* [104] also reported the synthesis of cobalt ferrite doped graphene polyacrylamide cryogel nanocomposites synthesized by employing in situ-polymerization technique for bio-separation processes. In the formation of polyacrylamide based cryogel, divalent metal ions (such as Co^{2+} and Fe^{2+}) performed both as crosslinker and catalyst due to their ability of binding towards oxygen groups which led to crosslinking between graphene oxide and polyacrylamide skeleton, thus, enhancing the strength of the cryogel. The properties of cryogel such as pore dimensions, pore wall thickness, mechanical toughness and elasticity were affected by synthesis conditions, amount of solvent and precursors, temperature gradients, cooling rate and other conditions.

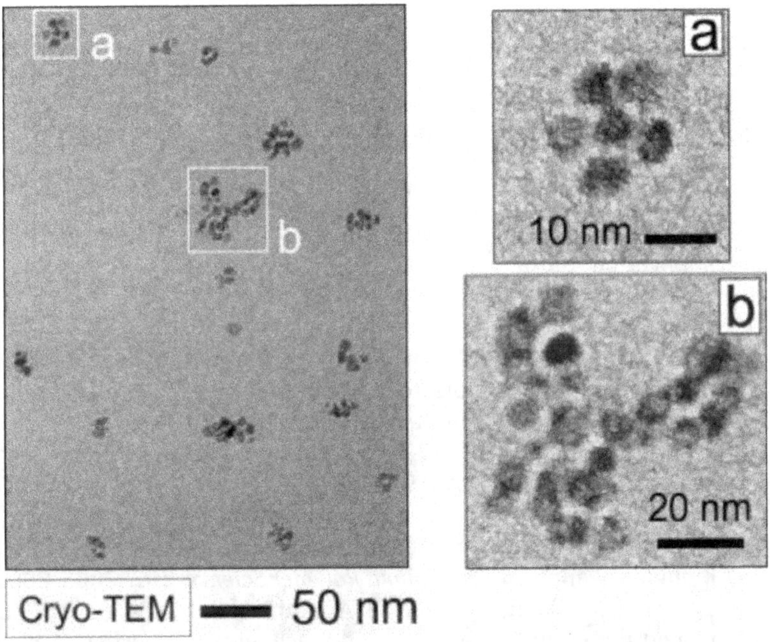

Figure 7.10 Cryo-transmission electron images of mixed aggregates obtained by complexation of PTEA(5K)-b-PAM(30K) and iron oxide nanoparticles. Reproduced from Reference 89 with permission from American Chemical Society.

7.4 Conclusions

Magnetic polymer nanocomposites (either nanoparticles embedded in polymer matrix or nanoparticles surface modified with polymer) are useful class of materials with functional characteristics. Polymer coating or grafting the nanoparticles to various polymer substrates avoids the usual agglomeration of nanoparticles. The magnetic polymer composites exhibit excellent improvement in properties as compared with individual components owing to the interactions between organic and inorganic frameworks. Specially, these nanomaterials have been utilized in various applications such as water purification, oil spill removal, tissue engineering, biocatalysis, biosensing, drug delivery, etc.

References

1. Smith, D. M., Baker Jr., J. R., and Simon, J. K. (2013) Applications of nanotechnology for immunology. *Nature Reviews Immunology,* **13**(8), 592-605.
2. Fischer, H. (2003) Polymer nanocomposites: from fundamental research to specific applications. *Materials Science and Engineering C,* **23**(6-8), 763-772.
3. Zhu, M., and Diao, G. (2011) Review on the progress in synthesis and application of magnetic carbon nanocomposites. *Nanoscale,* **3**, 2748-2767.
4. Liu, Y., Jiang, W., Wang, Y., Zhang, X. J., Song, D., and Li, F. S. (2009) Synthesis of Fe_3O_4/CNTs magnetic nanocomposites at the liquid–liquid interface using oleate as surfactant and reactant. *Journal of Magnetism and Magnetic Materials,* **321**(5), 408-412.
5. Behrens, S., and Appel, I. (2016) Magnetic nanocomposites. *Current Opinion in Biotechnology,* **39**, 89-96.
6. Kalia, S., Kango, S., Kumar, A., Haldorai, Y., Kumari, B., and Kumar, R. (2014) Magnetic polymer nanocomposites for environmental and biomedical applications. *Colloid Polymer Science,* **292**, 2025-2052.
7. Yoonessi, M., Gaier, J. R., Peck, J. A., and Meador, M. A. (2015) Controlled direction of electrical and mechanical properties in nickel tethered graphene polyimide nanocomposites using magnetic field. *Carbon,* **84,** 375-382.
8. Kainz, Q. M., and Reiser, O. (2014) Polymer- and Dendrimer-Coated Magnetic Nanoparticles as Versatile Supports for Catalysts, Scavengers, and Reagents. *Accounts of Chemical Research,* **47**(2), 667-677.
9. Fernandez, A. L., Cuellar, M. P., Herrera, J. M., Castillo, A. S., Pegalajar, M. D. C., Padilla, S. T., Colacio, E., and Vallvey L. F. C. (2014) Particle tuning and modulation of the magnetic/color synergy in Fe(II) spin crossover-polymer nanocomposites in a thermochromic sensor array. *Journal of Material Chemistry C,* **2**, 7292-7303.
10. Zhang, G., Li, Q., Gu, H., Jiang, S., Han, K., Gadinski, M. R., Haque, M. A., Zhang, Q., and Wang, Q. (2015) Ferroelectric polymer nanocomposites for room-temperature electrocaloric refrigeration. *Advanced Materials,* **27**, 1450-1454.
11. Gu, J., Jiang, W., Wang, F., Chen, M., Mao, J., and Xie, T. (2014) Facile removal of oils from water surfaces through highly hydrophobic and magnetic polymer nanocomposites. *Applied Surface Science,* **301,** 492-499.
12. Kalita, H., and Karak, N. (2013) Bio-based hyperbranched polyuret-

hane/Fe_3O_4 nanocomposites as shape memory materials. *Polymers for Advanced Technologies*, **24**, 819-823.

13. Baghayeri, M., Zare, E. N., and Lakouraj, M. M. (2014) Novel superparamagnetic PFu@Fe_3O_4 conductive nanocomposite as a suitable host for hemoglobin immobilization. *Sensors and Actuators B: Chemical*, **202**, 1200-1208.
14. Xiong, R., Lu, C., Wang, Y., Zhou, Z., and Zhang, X. (2013) Nanofibrillated cellulose as the support and reductant for the facile synthesis of Fe_3O_4/Ag nanocomposites with catalytic and antibacterial activity. *Journal of Material Chemistry A*, **1**, 14910-14918.
15. Jiang, X., Zhai, S., Jiang, X., Lu, G., and Huang, X. (2014) Synthesis of PAA-g-PNIPAM well-defined graft polymer by sequential RAFT and SET-LRP and its application in preparing size-controlled super-paramagnetic Fe_3O_4 nanoparticles as a stabilizer. *Polymer*, **55**(16), 3703-3712.
16. Sozeri, H., Kurtan, U., Topkaya, R., Baykal, A., and Toprak, M. S. (2013) Polyaniline (PANI)-$Co_{0.5}Mn_{0.5}Fe_2O_4$ nanocomposite: Synthesis, characterization and magnetic properties evaluation. *Ceramics International*, **39**(5), 5137-5143.
17. Ghanbari, D., Salavati-Niasari, M., and Ghasemi-Kooch, M. (2014) A sonochemical method for synthesis of Fe_3O_4 nanoparticles and thermal stable PVA-based magnetic nanocomposite. *Journal of Industrial and Engineering Chemistry*, **20**(6), 3970-3974.
18. Badruddoza, A. M., Bin Zakir Shawon, Z., Tay, W. J. D., Hidajat, K., and Uddin, M. S. (2013) Fe_3O_4/cyclodextrin polymer nanocomposites for selective heavy metals removal from industrial wastewater. *Carbohydrate Polymers*, **91**(1), 322-332.
19. Singh, J., Srivastava, M., Dutta, J., and Dutta, P. K. (2011) Preparation and properties of hybrid monodispersed magnetic α-Fe_2O_3 based chitosan nanocomposite film for industrial and biomedical applications. *International Journal of Biological Macromolecules*, **48**(1), 170-176.
20. Li, D., Luo, L., Pang, Z., Ding, L., Wang, Q., Ke, H., Huang, F., and Wei, Q. (2014) Novel phenolic biosensor based on a magnetic polydopamine-laccase-nickel nanoparticle loaded carbon nanofiber composite. *ACS Applied Materials and Interfaces*, **6**(7), 5144-5151.
21. Krupa, I., Cecen, V., Boudenne, A., Prokes, Jan., and Novak, I. (2013) The mechanical and adhesive properties of electrically and thermally conductive polymeric composites based on high density polyethylene filled with nickel powder. *Materials & Design*, **51**, 620-628.
22. Bora, C., Kalita, A., Das, D., Dolui, S. K. and Mukhopadhyay, P. K. (2014), Preparation of polyaniline/nickel oxide nanocomposites by liquid/liquid interfacial polymerization and evaluation of their electrical, electrochemical and magnetic properties. *Polymer Internatio-*

nal, **63**, 445-452.

23. Agarwal, S., Greiner, A., and Wendorff, J. H. (2013) Functional materials by electrospinning of polymers. *Progress in Polymer Science*, **38**(6), 963-991.
24. Sudagar, J., Lian, J., and Sha, W. (2013) Electroless nickel, alloy, composite and nano coatings - A critical review. *Journal of Alloys and Compounds*, **571**, 183-204.
25. He, S., Lu, C., Wang, G.-S., Wang, J.-W., Guo, H.-Y. and Guo, L. (2014), Synthesis and growth mechanism of white-fungus-like nickel sulfide microspheres, and their application in polymer composites with enhanced microwave-absorption properties. *ChemPlusChem*, **79**, 569-576.
26. Nandapure, B. I., Kondawar, S. B., Salunkhe, M. Y., and Nandapure, A. I. (2013) Magnetic and transport properties of conducting polyaniline/nickel oxide nanocomposites. *Advanced Materials Letters*, **4**(2), 134-140.
27. Zhai, W., Chen, Y., Ling, J., Wen, B., and Kim, Y.-W. (2014) Fabrication of lightweight, flexible polyetherimide/nickel composite foam with electromagnetic interference shielding effectiveness reaching 103 dB. *Journal of Cellular Plastics*, **50**(6), 537-550.
28. Romio, A. P., Rodrigues, H. H., Peres, A., Da Cas Viegas, A., Kobitskaya, E., Ziener, U., Landfester, K., Sayer, C. and Araujo, P. H. H. (2013) Encapsulation of magnetic nickel nanoparticles via inverse miniemulsion polymerization. *Journal of Applied Polymer Science*, **129**, 1426-1433.
29. Hui, N., Wang, S., Xie, H., Xu, S., Niu, S., and Luo, X. (2015) Nickel nanoparticles modified conducting polymer composite of reduced graphene oxide doped poly(3,4-ethylenedioxythiophene) for enhanced nonenzymatic glucose sensing. *Sensors and Actuators B: Chemical*, **221**(31), 606-613.
30. Yoonessi, M., Gaier, J. R., Peck, J. A., Meador, M. A. (2015) Controlled direction of electrical and mechanical properties in nickel tethered graphene polyimide nanocomposites using magnetic field. *Carbon*, **84**, 375-382.
31. Li, W., Qiu, T., Wang, L., Ren, S., Zhang, J., He, L., and Li, X. (2013) Preparation and electromagnetic properties of core/shell polystyrene@polypyrrole@Nickel composite microspheres. *ACS Applied Materials and Interfaces*, **5**(3), 883-891.
32. Yoonessi, M., Scheiman, D. A., Dittler, M., Peck, J. A., Ilavsky J., Gaier J. R., and Meador, M. A. (2013) High-temperature multifunctional magnetoactive nickel graphene polyimide nanocomposites. *Polymer*, **54**(11), 2776-2784.
33. Baykal, A., Günay, M., Toprak, M. S., and Sozeri, H. (2013) Effect of ionic liquids on the electrical and magnetic performance of polyaniline–nickel ferrite nanocomposite. *Materials Research Bulletin*, **48**,

378-382.

34. Boudenne, A., Mamunya, Y., Levchenko, V., Garnier, B., and Lebedev, E. (2015) Improvement of thermal and electrical properties of Silicone–Ni composites using magnetic field. *European Polymer Journal*, **63**, 11-19.
35. Malik, R., Sehdev, N., Lamba, S., Sharma, P., Makino, A., and Annapoorni, S. (2014) Magnetic memory effects in nickel ferrite/ polymer nanocomposites. *Applied Physics Letters*, **104**, 122407.
36. Zhu, L., Xie, D., Ma, J., Shao, J., and Shen, X. (2013) Fabrication of polydimethylsiloxane composites with nickel particles and nickel fibers and study of their magnetic properties. *Smart Materials and Structures*, **22** (4), 045015.
37. Amiri, S., and Shokrollahi, H. (2013) The role of cobalt ferrite magnetic nanoparticles in medical science. *Materials Science and Engineering C*, **33**(1), 1-8.
38. Wohlert, S., Tomkowicz, Z., Rams, M., Ebbinghaus, S. G., Fink, L., Schmidt, and M. U., Nather, C. (2014) Influence of the co-Ligand on the magnetic and relaxation properties of layered cobalt(II) thiocyanato coordination polymers. *Inorganic Chemistry*, **53**(16), 8298-8310.
39. Zeng, M.-H., Zheng Y., Tan, Y.-X., Zhang, W.-X., He, Y.-P., Kurmoo, M. (2014) Nanoporous cobalt(II) MOF exhibiting four magnetic ground states and changes in gas sorption upon post-synthetic modification. *Journal of American Chemical Society*, **136**(12), 4680-4688.
40. Pavia-Sanders, A., Zhang, S., Flores, J. A., Sanders, J. E., Raymond, J. E., and Wooley, K. L. (2013) Robust magnetic/polymer hybrid nanoparticles designed for crude oil entrapment and recovery in aqueous environments. *ACS Nano*, **7**(9), 7552-7561.
41. Khademolhosseini, F., and Chiao, M. (2013) Fabrication and patterning of magnetic polymer micropillar structures using a dry-nanoparticle embedding technique. *Journal of Microelectromechanical Systems*, **22**(1), 131-139.
42. Salunkhe, A. B., Khot, V. M., Thorat, N. D., Phadatare, M. R., Sathish, C. I., Dhawale, D. S. and Pawar, S. H. (2013) Polyvinyl alcohol functionalized cobalt ferrite nanoparticles for biomedical applications. *Applied Surface Science*, **264**, 598-604.
43. Zhu, Q., Tao, F., and Pan, Q. (2010) Fast and selective removal of oils from water surface via highly hydrophobic core–shell Fe_2O_3@C nanoparticles under magnetic field. *ACS Applied Materials and Interfaces*, **2**(11), 3141-3146.
44. Korhonen, J. T., Kettunen, M., Ras, R. H. A., and Ikkala, O. (2011) Hydrophobic nanocellulose aerogels as floating, sustainable, reusable, and recyclable oil absorbents. *ACS Applied Materials and Interfaces*, **3**(6), 1813-1816.

45. Chu, Y., and Pan, Q. (2012) Three-dimensionally macroporous Fe/C nanocomposites as highly selective oil-absorption materials. *ACS Applied Materials and Interfaces*, **4**(5), 2420-2425.
46. Kumar, A., Sharma, G., Naushad, M., and Thakur, S. (2015) SPION/β-cyclodextrin core–shell nanostructures for oil spill remediation and organic pollutant removal from waste water. *Chemical Engineering Journal*, **280**, 175-187.
47. Arora, N., Vengatesan, M. R., and Mittal, V. (2018) Resorcinol-formaldehyde cryogel nanocomposites for oil spill clean up. In: *Polymers in Oil and Gas Industry*, Mittal, V. (ed.), Central West Publishing, Australia, pp. 147-163.
48. Calcagnile, P., Fragouli, D., Bayer, I. S., Anyfantis, G. C., Martiradonna, L., Cozzoli, P. D., Cingolani, R., and Athanassios, A. (2012) Magnetically driven floating foams for the removal of oil contaminants from water. *ACS Nano*, **6**(6), 5413-5419.
49. Gu, J., Jiang, W., Wang, F., Chen, M., Mao, J., and Xie, T. (2014) Facile removal of oils from water surfaces through highly hydrophobic and magnetic polymer nanocomposites. *Applied Surface Science*, **301**, 492-499.
50. Thanikaivelan, P., Narayanan, N. T. Pradhan, B. K., and Ajayan, P. M. (2012) Collagen based magnetic nanocomposites for oil removal applications. *Scientific Reports*, **2**, 230.
51. Venkatanarasimhan, S., and Raghavachari, D. (2013) Epoxidized natural rubber–magnetite nanocomposites for oil spill recovery. *Journal of Material Chemistry A*, **1**, 868-876.
52. Parham, H., Zargar, B., and Shiralipour, R. (2012) Fast and efficient removal of mercury from water samples using magnetic iron oxide nanoparticles modified with 2-mercaptobenzothiazole. *Journal of Hazardous Materials*, **205-206**, 94-100.
53. Gao, H., Zhao, S., Cheng, X., Wang, X., and Zheng, L. (2013) Removal of anionic azo dyes from aqueous solution using magnetic polymer multi-wall carbon nanotube nanocomposite as adsorbent. *Chemical Engineering Journal*, **223**, 84-90.
54. Li, Y., Li, X., Chu, J., Dong, C., Qi, J., and Yuan, Y. (2010) Synthesis of core-shell magnetic molecular imprinted polymer by the surface RAFT polymerization for the fast and selective removal of endocrine disrupting chemicals from aqueous solutions. *Environmental Pollution*, **158**(6), 2317-2323.
55. Daraei, P., Madaeni, S. S., Ghaemi, N., Salehi, E., Khadivi, M. A., Moradian, R., and Astinchap, B. (2012) Novel polyethersulfone nanocomposite membrane prepared by PANI/Fe_3O_4 nanoparticles with enhanced performance for Cu(II) removal from water. *Journal of Membrane Science*, **415-416**, 250-259.
56. Bhaumik, M., Maity, A., Srinivasu, V. V., and Onyango, M. S. (2011) Enhanced removal of Cr(VI) from aqueous solution using polypyro-

ole/Fe_3O_4 magnetic nanocomposite. *Journal of Hazardous Materials*, **190**(1-3), 381-390.

57. Arora, N., Vengatesan, M. R., and Mittal, V. (2019) Cobalt ferrite doped graphene/phosphazene nanotubes (PZTs) - Resorcinol-Formaldehyde (R-F) cryogel nanocomposites: Synthesis and adsorption studies. In: *Porous Polymer Networks*, Mittal, V. (ed.), Central West Publishing, Australia, pp. 229-256.
58. Lin, M., Lu, D., Zhu, J., Yang, C., Zhang, Y. and Zheng, L. (2012) Magnetic enzyme nanogel (MENG): a universal synthetic route for biocatalysts. *Chemical Communications*, **48**, 3315-3317.
59. Qiu, G.-M., Zhu, B.-K. and Xu, Y.-Y. (2005) α-Amylase immobilized by Fe_3O_4/poly(styrene-co-maleic anhydride) magnetic composite microspheres: Preparation and characterization. *Journal of Applied Polymer Science*, **95**, 328-335.
60. Zhang, H., Zhong, X., Xu, J., and Chen, H. (2008) Fe_3O_4/polypyrrole/Au nanocomposites with core/shell/shell structure: synthesis, characterization, and their electrochemical properties. *Langmuir*, **24**(23), 13748-13752.
61. Kiasat, A. R., and Nazari, S. (2012) Magnetic nanoparticles grafted with β-cyclodextrin–polyurethane polymer as a novel nanomagnetic polymer brush catalyst for nucleophilic substitution reactions of benzyl halides in water. *Journal of Molecular Catalysis A: Chemical*, **365**, 80-86.
62. Lei, Z., Ren, N., Li, Y., Li, N., and Mu, B. (2009) Fe_3O_4/SiO_2-g-PSStNa polymer nanocomposite microspheres (PNCMs) from a surface-initiated atom transfer radical polymerization (SI-ATRP) approach for pectinase immobilization. *Journal of Agricultural and Food Chemistry*, **57**(4), 1544-1549.
63. Suri, K., Annapoorni, S., Sarkar, A. K., and Tandon, R. P. (2002) Gas and humidity sensors based on iron oxide–polypyrrole nanocomposites. *Sensors and Actuators B: Chemical*, **81**(2-3), 277-282.
64. Tandon, R. P., Tripathy, M. R., Arora, A. K., and Hotchandani, S. (2006) Gas and humidity response of iron oxide-polypyrrole nanocomposites. *Sensors and Actuators B: Chemical*, **114**(2), 768-773.
65. Taccola, S., Greco, F., Zucca, A., Innocenti, C., Fernandez, C. J., Campo, G., Sangregorio, C., Mazzolai, B., and Mattoli, V. (2013) Characterization of free-standing PEDOT:PSS/Iron oxide nanoparticle composite thin films and application as conformable humidity sensors. *ACS Applied Materials and Interfaces*, **5**(13), 6324-6332.
66. Sen, T., Shimpi, N. G., Mishra, S., and Sharma, R. (2014) Polyaniline/γ-Fe_2O_3 nanocomposite for room temperature LPG sensing. *Sensors and Actuators B: Chemical*, **190**, 120-126.
67. Wu, S., Wang, H., Tao, S., Wang, C., Zhang, L., Liu, Z., and Meng, C. (2011) Magnetic loading of tyrosinase-Fe_3O_4/mesoporous silica-core/shell microspheres for high sensitive electrochemical biosens-

ing. *Analytica Chimica Acta*, **686**(1-2), 81-86.

68. Oh, J. K., and Park, J. M. (2011) Iron oxide-based superparamagnetic polymeric nanomaterials: design, preparation, and biomedical application. *Progress in Polymer Science*, **36**(1), 168-189.
69. Kim, M. I., Ye, Y., Won, B. Y., Shin, S., Lee, J., and Park, H. G. (2011), A highly efficient electrochemical biosensing platform by employing conductive nanocomposite entrapping magnetic nanoparticles and oxidase in mesoporous carbon foam. *Advanced Functional Materials*, **21**, 2868-2875.
70. Wang, X., Wang, L., He, X., Zhang, Y., and Chen, L. (2009) A molecularly imprinted polymer-coated nanocomposite of magnetic nanoparticles for estrone recognition. *Talanta*, **78**(2), 327-332.
71. Baghayeri, M., Zare, E. N., and Lakouraj, M. M. (2014) A simple hydrogen peroxide biosensor based on a novel electro-magnetic poly(p-phenylenediamine)@Fe_3O_4 nanocomposite. *Biosensors and Bioelectronics*, **55**, 259-265.
72. Fredj, H. B., Helali, S., Esseghaier, C., Vonna, L., Vidal, L., and Abdelghani, A. (2008) Labeled magnetic nanoparticles assembly on polypyrrole film for biosensor applications. *Talanta*, **75**(3), 740-747.
73. Zhang, M., He, X., and Zhang, L. C. Y. (2010) Preparation of IDA-Cu functionalized core–satellite Fe_3O_4/polydopamine/Au magnetic nanocomposites and their application for depletion of abundant protein in bovine blood. *Journal of Materials Chemistry*, **20**, 10696-10704.
74. Singh, J., Srivastava, M., Dutta, J., and Dutta, P. K. (2011) Preparation and properties of hybrid monodispersed magnetic α-Fe_2O_3 based chitosan nanocomposite film for industrial and biomedical applications. *International Journal of Biological Macromolecules*, **48**(1), 170-176.
75. Perera, A. S., Zhang, S., Homer-Vanniasinkam, S., Coppens, M.-C., and Edirisinghe, M. (2018) Polymer-magnetic composite fibers for remote-controlled drug release. *ACS Applied Materials and Interfaces*, **10**(18), 15524-15531.
76. Xu, H., Cheng, L., Wang, C., Ma, X., Li, Y., and Liu, Z. (2011) Polymer encapsulated upconversion nanoparticle/iron oxide nanocomposites for multimodal imaging and magnetic targeted drug delivery. *Biomaterials*, **32**(35), 9364-9373.
77. Ma, X., Tao, H., Yang, K., Feng, L., Cheng, L., Shi, X., Li, Y., Guo, L., and Liu, Z. (2012) A functionalized graphene oxide-iron oxide nanocomposite for magnetically targeted drug delivery, photothermal therapy, and magnetic resonance imaging. *Nano Research*, **5**(3), 199-212.
78. Satarkar, N. S., and Hilt, J. Z. (2008) Magnetic hydrogel nanocomposites for remote controlled pulsatile drug release. *Journal of Contro-*

lled Release, **130**(3), 246-251.

79. Kim, J., Lee, J. E., Lee, J., Yu, J. H., Kim B. C., An, K., Hwang, Y., Shin, C.-H., Park, J.-G., Kim, J., and Hyeon, T. (2006) Magnetic fluorescent delivery vehicle using uniform mesoporous silica spheres embedded with monodisperse magnetic and semiconductor nanocrystals. *Journal of American Chemical Society*, **128**(3), 688-689.
80. Zhao, W., Gu, J., Zhang, L., Chen, H., and Shi, J. (2005) Fabrication of uniform magnetic nanocomposite spheres with a magnetic core/mesoporous silica shell structure. *Journal of American Chemical Society*, **127**(25), 8916-8917.
81. Asmatulu, R., Zalich, M. A., Claus, R. O., and Riffle, J. S. (2005) Synthesis, characterization and targeting of biodegradable magnetic nanocomposite particles by external magnetic fields. *Journal of Magnetism and Magnetic Materials*, **292**, 108-119.
82. Urbina, M. C., Zinoveva, S., Miller, T., Sabliov, C. M., Monroe, W. T., and Kumar, C. S. S. R. (2008) Investigation of magnetic nanoparticle-polymer composites for multiple-controlled drug delivery. *Journal of Physical Chemistry C*, **112**, 11102-11108.
83. Brazel, C. S. (2009) Magnetothermally-responsive nanomaterials: combining magnetic nanostructures and thermally-sensitive polymers for triggered drug release. *Pharmaceutical Research*, **26**(3), 644-656.
84. Zhao, W., Chen, H., Li, Y., Li, L., Lang, M., and Shi, J. (2008), Uniform rattle-type hollow magnetic mesoporous spheres as drug delivery carriers and their sustained-release property. *Advanced Functional Materials*, **18**, 2780-2788.
85. Lutz, J.-F., Stiller, S., Hoth, A., Kaufner, L., Pison, U., and Cartier, R. (2006) One-pot synthesis of PEGylated ultrasmall iron-oxide nanoparticles and their in vivo evaluation as magnetic resonance imaging contrast agents. *Biomacromolecules*, **7**(11), 3132-3138.
86. Lee, J.-H., Jun, Y.-w., Yeon, S.-I., Shin, J.-S., and Cheon, J. (2006) Dual-mode nanoparticle probes for high-performance magnetic resonance and fluorescence imaging of neuroblastoma. *Angewandte Chemie*, **118**, 8340-8342.
87. Li, K., Ding, D., Huo, D., Pu, K.-Y., Thao, N. N. P., Hu, Y., Li, Z. and Liu, B. (2012) Conjugated polymer based nanoparticles as dual-modal probes for targeted in vivo fluorescence and magnetic resonance imaging. *Advanced Functional Materials*, **22**, 3107-3115.
88. Baldock, C., Burford, R. P., Billingham, N., Wagner, G. S., Patval, S., Badawi, R. D., and Keevil, S. F. (1998) Experimental procedure for the manufacture and calibration of polyacrylamide gel (PAG) for magnetic resonance imaging (MRI) radiation dosimetry, *Physics in Medicine and Biology*, **43**(3), 695-702.
89. Berret, J.-F., Schonbeck, N., Gazeau, F., Kharrat, D. E. Sandre, O., Vacher, A., and Airiau, M. (2006) Controlled clustering of superpar-

amagnetic nanoparticles using block copolymers: design of new contrast agents for magnetic resonance imaging. *Journal of American Chemical Society*, **128**(5), 1755-1761.

90. Lee, J. E., Lee, N., Kim, H., Kim, J., Choi, S. H., Kim, J. H., Kim, T., Song, I. C., Park, S. P., Moon, W. K., and Hyeon, T. (2010) Uniform mesoporous dye-doped silica nanoparticles decorated with multiple magnetite nanocrystals for simultaneous enhanced magnetic resonance imaging, fluorescence imaging, and drug delivery. *Journal of American Chemical Society*, **132**(2), 552-557.
91. Sitharaman, B., Van Der Zande, M., Ananta, J. S., Shi, X., Veltien, A., Walboomers, X. F., Wilson, L. J., Mikos, A. G., Heerschap, A., and Jansen, J. A. (2010) Magnetic resonance imaging studies on gadonanotube-reinforced biodegradable polymer nanocomposites. *Journal of Biomedical Materials Research*, **93A**, 1454-1462.
92. Jun, Y.-W., Huh, Y.-M., Choi, J.-S., Lee, J.-H., Song, H.-T., Sungjun, S., Yoon, S., Kim, K.-S., Shin, J.-S., Suh, J.-S., and Cheon, J. (2005) Nanoscale size effect of magnetic nanocrystals and their utilization for cancer diagnosis via magnetic resonance imaging. *Journal of American Chemical Society*, **127**(16), 5732-5733.
93. Ai, H., Flask, C., Weinberg, B., Shuai, X.-T., Pagel, M. D., Farrell, D., Duerk, J., and Gao, J. (2005) Magnetite-loaded polymeric micelles as ultrasensitive magnetic-resonance probes. *Advanced Materials*, **17**, 1949-1952.
94. Lu, J., Ma, S., Sun, J., Xia, C., Liu, C., Wang, Z., Zhao, X., Gao, F., Gong, Q., Song, B., Shuai, X., Ai, H., and Gu, Z. (2009) Manganese ferrite nanoparticle micellar nanocomposites as MRI contrast agent for liver imaging. *Biomaterials*, **30**(15), 2919-2928.
95. Armentano, I., Dottori, M., Fortunati, E., Mattioli, S., and Kenny, J. M. (2010) Biodegradable polymer matrix nanocomposites for tissue engineering: A review. *Polymer Degradation and Stability*, **95**(11), 2126-2146.
96. Gloria, A., Russo, T., D'Amora, U., Zeppetelli, S., D'Alessandro, T., Sandri, M., Banobre-Lopez, M., Pineiro-Redondo, Y., Uhlarz, M. Tampieri, A., Rivas, J., Herrmannsdorfer, T., Dediu, V. A., Ambrosio, L., and De Santis, R. (2013) Magnetic poly(ε-caprolactone)/iron-doped hydroxyapatite nanocomposite substrates for advanced bone tissue engineering. *Journal of the Royal Society Interface*, DOI: https://doi.org/10.1098/rsif.2012.0833.
97. Pankhurst, Q. A., Connolly, J., Jones, S. K., and Dobson, J. (2003) Applications of magnetic nanoparticles in biomedicine. *Journal of Physics D: Applied Physics*, **36**, R167.
98. De Santis, R., Gloria, A., Russo, T., D'Amora, U., Zeppetelli, S., Dionigi, C., Sytcheva, A., Herrmannsdorfer, T., Dediu, V. and Ambrosio, L. (2011) A basic approach toward the development of nanocomposite magnetic scaffolds for advanced bone tissue engineering. *Jour-*

nal of Applied Polymer Science, **122**, 3599-3605.

99. Zhang, Y., Venugopal, J. R., El-Turki, A., Ramakrishna, S. Su, B., and Lim, C. T. (2008) Electrospun biomimetic nanocomposite nanofibers of hydroxyapatite/chitosan for bone tissue engineering. *Biomaterials*, **29**(32), 4314-4322.
100. Gupta, A. K., Naregalkar, R. R., Vaidya, V. D., and Gupta, M. (2007) Recent advances on surface engineering of magnetic iron oxide nanoparticles and their biomedical applications. *Nanomedicine*, **2**(1), 23-39.
101. Meenach, S. A., Hilt, J. Z., and Anderson, K. W. (2010) Poly(ethylene glycol)-based magnetic hydrogel nanocomposites for hyperthermia cancer therapy. *Acta Biomaterialia*, **6**(3), 1039-1046.
102. Tan, S. T., Wendorff, J. H., Pietzonka, C., Jia, Z. H., and Wang, G. Q. (2005) Biocompatible and biodegradable polymer nanofibers displaying superparamagnetic properties. *Chemphyschem*, **6**(8), 1461-5.
103. Sureshkumar, M., Siswanto, D. Y., and Lee, C.-K. (2010) Magnetic antimicrobial nanocomposite based on bacterial cellulose and silver nanoparticles. *Journal of Materials Chemistry*, **20**, 6948-6955.
104. Arora, N., and Mittal, V. (2018) Cobalt ferrite doped graphene/polyacrylamide cryogel nanocomposites: Synthesis and characterization studies. In: *Polymers and Polymer Nanocomposites*, Volume 2, Mittal, V. (ed.), pp. 47-54.

8

Cobalt Ferrite Doped Graphene/Phosphazene Nanotubes - Resorcinol-Formaldehyde Cryogel Nanocomposites

8.1 Introduction

A variety of molecules are used/generated during industrial processes which when released in environment cause negative effects on health and environment (due to toxicity, mutagenic and carcinogenic properties) [1,2]. 2-Nitrophenol is used in various industries such as steel foundry, insecticides, explosives, petrochemicals, herbicides, petroleum refineries, coke-oven, etc. [3]. Its release in water and soil can be controlled though its degradation, however, it is generally possible at the water and soil surface, and its degradation takes longer for lower soil depths and groundwater. Its exposure through inhalation, ingestion and skin contact causes tissue erosion, protein degeneration, paralysis of the central nervous system as well as damage to liver, kidney and pancreas [4]. Many techniques have been applied to remove phenolic derivatives, especially from aqueous media, such as membrane filtration [5], biological degradation [6], solvent extraction [7], photocatalytic degradation [8] and electrochemical oxidation [9]. Among these methods, adsorption is of the simplest and widely used treatment approaches due to its high efficiency and low cost. For this purpose, many adsorbents such as activated alumina [10], clay [11], chitosan [12], activated carbon [13], graphene [14] and multi-wall carbon nanotubes [15] have been employed. Among these, excellent adsorption capabilities have been shown by carbonaceous materials due to their pore size distribution, large surface area and controllable surface functional groups. Moreover, cryogels possess large and well-defined interconnected porous structures [16,17]. Free-radical cryo-polymerization technique is employed to prepare cryogels from polymeric gel precursors below the solvent freezing point, where freeze-drying process maintains the interconnected structure of the resultant cryogel [18]. The properties of cryogels, such as pore dimensions, pore wall thickness, mechanical toughness and elasticity, depend on preparation conditions, amount

Naman Arora, Muthukumaraswamy R. Vengatesan and Vikas Mittal, Khalifa University of Science and Technology, Abu Dhabi, UAE

of solvent and precursors, temperature gradients, cooling rate and other conditions [19,20]. Owing to their properties, cryogels form optimal substrates for adsorption processes in a large number of industrial fields.

In this study, we have prepared 3-D carbonaceous monolithic cryogel nanocomposites from resorcinol-formaldehyde (RF) resin precursor with graphene oxide (GO), phosphazene nanotubes (PZT) and cobalt ferrite nanoparticles. In the formation of RF based cryogel, divalent metal ions (such as Co^{2+} and Fe^{2+}) perform both as crosslinker and catalyst due to their binding ability towards oxygen groups [21], which leads to crosslinking between graphene oxide and RF skeletons, thus, enhancing the strength of whole cryogel. Subsequently, the cryogel is converted into carbonaceous monolith with excellent porosity and hydrophobicity via pyrolysis. During the carbonization process, the divalent metal ions (Co^{2+} and Fe^{2+}) are converted into cobalt ferrite nanoparticles [22], which act as the additional adsorption sites and increase the interaction with the negatively charged phenolic ions, thus, significantly enhancing the overall adsorption process [23].

A few research studies have reported graphene-CNTs composites as effective adsorbents [24], which have an excellent adsorption capacity towards methylene blue [25]. Phosphazenes represent a versatile class of hybrid organic-inorganic materials with alternating phosphorus and nitrogen atoms. These have been used as membranes, biomaterials, optical materials, flame retardants, etc. [26-30]. PZTs with hydroxyl groups are synthesized through one-pot reaction with simple process, high yield, controllable morphology and low cost [31].

Therefore, in this study, in order to attain the high specific surface area of adsorbent, graphene has been modified with phosphazene nanotubes as an approach to avoid the stacking of graphene sheets. Moreover, the development of graphene-PZTs based carbonaceous monolithic materials containing cobalt ferrite nanoparticles as an efficient adsorbent is first of its kind in the literature. 2-Nitrophenol was selected as a model pollutant for evaluating the adsorption performance under stringent laboratory conditions. The equilibrium, kinetic and thermodynamic studies have been carried out, along with structural characterization of the as-developed cryogel nanocomposites.

Overall, the cryogels developed in this study represent high potential of application in the diverse industrial fields.

8.2 Materials and Methods

8.2.1 Materials

Natural graphite powder (10 mesh), resorcinol, 2-nitrophenol (MERCK), formaldehyde (37%, MERCK), H_2SO_4 (95-97%, MERCK), H_3PO_4 (85%, MERCK), $KMnO_4$ (eurolab), HCl (37%, Panreac), hydrogen peroxide (35%, MERCK), $CoCl_2.6H_2O$ (Sigma-Aldrich), iron(III) nitrate nanohydrate (≥98%, Sigma-Aldrich), phosphonitrilic chloride trimer (99%, Sigma-Aldrich), bisphenol S (Sigma-Aldrich), triethylamine (TEA, ≥99%, Sigma Aldrich) and tetrahydrofuran (THF) were procured and used as received.

8.2.2 Preparation of GO

GO was prepared by using the improved Tour's synthesis method [32]. In this method, graphite (10 g) was added to a mixture of H_2SO_4 (540 mL) and H_3PO_4 (60 mL) in a molar ratio of 9:1. The mixture was stirred for 30 min, followed by the gradual addition of $KMnO_4$ (56 g) and continued stirring for 72 h. The reaction was continued by pouring H_2O_2 slowly and allowing it to stir for 8 h, until the mixture color turned yellow. Produced graphene oxide was washed with 1 M HCl solution until no sulfate ions are detected, followed by washing with excess distilled water to remove chloride ions. The GO solution was then poured into Teflon petri dishes for freeze drying.

8.2.3 Preparation of PZTs

Phosphonitrilic chloride trimer (3 g) was dissolved in 550 mL of THF, followed by sonication for 20 min. Another mixture containing TEA (7.8 mL) and THF (50 mL) was prepared, followed by its dropwise addition to the earlier mixture under ultra-sonication at 40 °C. After 30 min, bisphenol S (6.5 g) was added to the mixture, followed by sonication for another 7 h. The final solution was filtered, and the precipitates were washed with a THF-distilled water (1:1) mixture. The resulting product was dried overnight under vacuum at 50 °C.

8.2.4 Synthesis of Graphene-PZT based RF Cryogel Containing Cobalt Ferrite Nanoparticles

GO exfoliation was carried out under ultra-sonication by adding GO

(0.2 g) into deionized water (100 mL) for 2 h. Later, PZTs (0.2 g) were added to the GO solution followed by stirring for 2 h. $CoCl_2.6H_2O$ (0.0025 M) and $Fe(NO_3)_3.9H_2O$ (0.0025 M) were subsequently added as Co^{2+} and Fe^{2+} sources. Resorcinol (2.475 g) and formaldehyde (3.65 g) were then added in the molar ratio of 1:2. After continuous stirring for 4 h to form a homogeneous solution, the mixture was cast into a glass mold. It was kept at 85 °C for curing, until a red-brown colored hydrogel was obtained. The hydrogel was washed with excess distilled water to remove the residual ions. Followed by a lyophilization process for 5 days at a temperature of -89 °C and vacuum of 10 Pa, GO-PZT based RF cryogel was obtained. The cryogel was carbonized for 2 h at 700 °C with a heating rate of 10 °C/min under N_2 atmosphere and cooled naturally in a quartz tube furnace, thus, resulting in GO-PZTs based RF cryogel containing cobalt ferrite nanoparticles ($CoFe_2O_4$/Graphene/PZTs-RF). To investigate the influence of PZTs and cobalt ferrite nanoparticles on the formation of 3-D carbonaceous monolith material, graphene-RF based cryogel containing cobalt ferrite nanoparticles ($CoFe_2O_4$/Graphene-RF) was prepared using same methodology. Similarly, graphene-RF based cryogel (Graphene-RF-Na_2CO_3) was also synthesized using Na_2CO_3 catalyst by replacing Co^{2+} and Fe^{2+} ions for the gelation process.

8.2.5 Adsorbent Characterization

Wide-angle X-ray diffraction (WAXRD) patterns of the samples were collected using analytical powder (X' Pert PRO) diffractometer using instrument parameters of 20 A intensity, 40 kV and Cu Kα (1.5406 Å) radiation in reflection mode. The samples were step-scanned between 5 and 60° at a step size of $2\theta = 0.02° \ s^{-1}$. Scanning electron microscopy (SEM) and energy dispersive X-ray spectroscopy (EDX) analysis were used to analyze the morphology of the samples, using FEI Quanta, FEG250, USA at accelerating voltages of 10-20 kV. FEI transmission electron microscope (TEM, Tecnai G20) at 200 kV was used to obtain the morphology of $CoFe_2O_4$/Graphene/PZTs/RF nanocomposite. The samples for TEM analysis were prepared by dispersing 0.5 mg of adsorbent in 20 mL of dimethylformamide under sonication at room temperature for 10 min. Two drops of suspension were placed on a 400 mesh copper electron microscopy grid covered with thin carbon film and dried in air. The nitrogen adsorption-desorption isotherms were measured on a Micromeritics ASAP 2010 volumetric adsorption analyzer. Prior to the analysis, the samples were degassed for 16 h at 200 °C under vacuum (p < 5-10 mbar) in the degassing port. The specific surface area of the nanocomposites was determined from the linear part of the BET equation. The pore volume was calculated using the BET plot corresponding to the amount of nitrogen gas adsorbed at the last adsorption point (P/P_0= 0.99), and the pore size distribution of the nanocomposites was estimated using BJH method

8.2.6 Batch Adsorption Analysis

2-Nitrophenol stock solution (600 mg L^{-1}) was prepared by dissolving it in distilled water and further diluted as per the required concentration. The isotherms were determined by a batch equilibration technique at 303 K, 313 K, 318 K and 323 K. Adsorbent (0.05 g) and 2-nitrophenol solutions (15 mL) were added to 50 mL flasks for adsorption experiments. The flasks were sealed with glass stoppers and shaken at 130 rpm in a shaker for 24 h. Small amount of samples were withdrawn from each flask with the help of syringes to carry out the absorbance measurement at the absorbance maximum wavelength of 277 nm by a UV–visible spectrophotometer (Varian). The effect of pH on 2-nitrophenol adsorption was investigated in a pH range of 2-12 at 303 K. The pH of 2-nitrophenol solution was adjusted by using 0.1 M HNO_3 and 0.1 M NaOH. In dosage studies, amount of adsorbent varied between 0.02 g to 0.1 g (corresponding adsorbent to 2-nitrophenol ratio from 0.8 to 4) in 25 mL of 2-nitrophenol solution with concentration of 600 mg L^{-1} in a 50 mL conical flask shaken at 130 rpm, using different temperatures (303 K, 313 K and 323 K). Kinetics experiments were performed using identical procedure as equilibrium studies with 200, 400, 500 and 600 mg L^{-1} of 2-nitrophenol solution. A small amount of samples was taken at preset time intervals and the concentrations were measured at various time intervals. The uptake of the adsorbate at any time per gram of adsorbent (q_t, mg g^{-1}) was determined as follows:

$$q_t = \left(\frac{C_o - C_t}{m}\right) V$$

where, C_o and C_t are the initial concentration and concentration at time (t) respectively, V is the volume of solution and m is the mass of the adsorbent. The equilibrium adsorption capacity (q_e) was calculated by replacing C_t with the equilibrium concentration (C_e).

8.3 Results and Discussion

8.3.1 Structure and Morphology of the Adsorbent

Figure 8.1a shows the digital photographic image of GO-PZTs-RF based cryogel containing divalent metal salts of Co^{2+} and Fe^{2+}. After carbonization process, the GO-PZTs-RF based cryogel turned into black colored carbonaceous cryogel nanocomposite ($CoFe_2O_4$/Graphene/PZTs- RF), as presented in Figure 8.1b. As mentioned earlier, the cryogels were synthesized by curing the initial suspensions and further lyophilization of the corresponding hydrogels in order to eliminate the water molecules, which resulted in the dried monoliths/cryogels with desirable shapes, depending on the glass molds. GO/PZTs-RF was observed to be a homogeneous monolith with yellow-brownish color, which was obtained when the divalent metal salts of Co^{2+} and Fe^{2+} were used to promote the gelation process of GO-RF; Co^{2+} and Fe^{2+} ions are chelated by the functional groups (negatively charged) of GO sheets and RF, acting as both catalyst and crosslinking agents at the same time. After the carbonization of GO/PZTs-RF, divalent metal ions (Co^{2+} and Fe^{2+}) and GO were converted into cobalt ferrite nanoparticles and graphene, respectively, along with conversion of PZTs into carbon fibers. Thus, $CoFe_2O_4$/Graphene/PZTs-RF carbonaceous monolith was obtained.

Figure 8.1 Digital photographs of (a) GO-PZTs-RF based cryogel containing divalent metal salts of Co^{2+} and Fe^{2+} and (b) $CoFe_2O_4$/Graphene/PZTs-RF based cryogel.

The diffraction pattern of GO/PZTs-RF cryogel is presented in Figure 8.2a. A diffraction hump was observed in the range of $2\theta = 20°$ to

33°. The diffraction pattern of $CoFe_2O_4$/Graphene/PZTs-RF is depicted in Figure 8.2b, where the weak diffraction peak could be noticed at 2θ = 26.2°, indexing to the (002) plane of carbon. This showed the graphitization of amorphous carbon catalyzed by transition metal [33] and the change of the interlayer spacing [34]. Strong diffraction peaks at 2θ = 43° and 52.5° were respectively indexed to (400) and (422) crystal planes of cubic spinel structure of cobalt ferrite nanoparticles (*JCPDS card No: 22-1086*), and a small diffraction peak at 2θ = 45° corresponding to (110) plane was attributed to bcc Fe-Co [35]. Thus, it can be suggested that after the calcination at 700 °C, Co^{2+} and Fe^{2+} were completely reduced to zero valence state, thus, confirming the presence of cobalt ferrite nanoparticles.

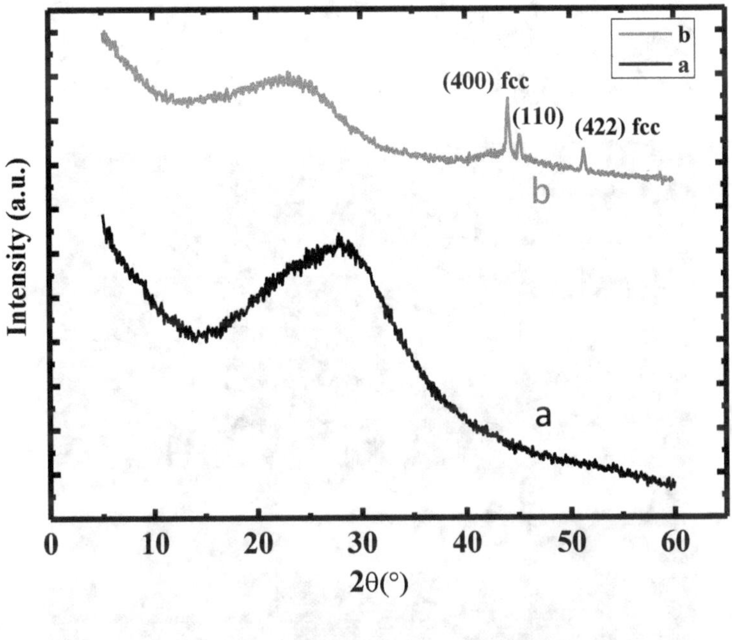

Figure 8.2 XRD patterns of (a) GO/PZTs-RF and (b) $CoFe_2O_4$/Graphene/PZTs-RF.

Figures 8.3a and b represent the SEM micrographs of the adsorbent ($CoFe_2O_4$/Graphene/PZTs-RF). It can be observed that cobalt ferrite nanoparticles were present in a well-defined interconnected porous structure. As discussed earlier, the gelation process was provided by the metal ions (Co^{2+} and Fe^{2+}), which behave as crosslinkers and provide active sites for the assembly of graphene layers [36], along with facilitating the formation of corrugations in the resultant

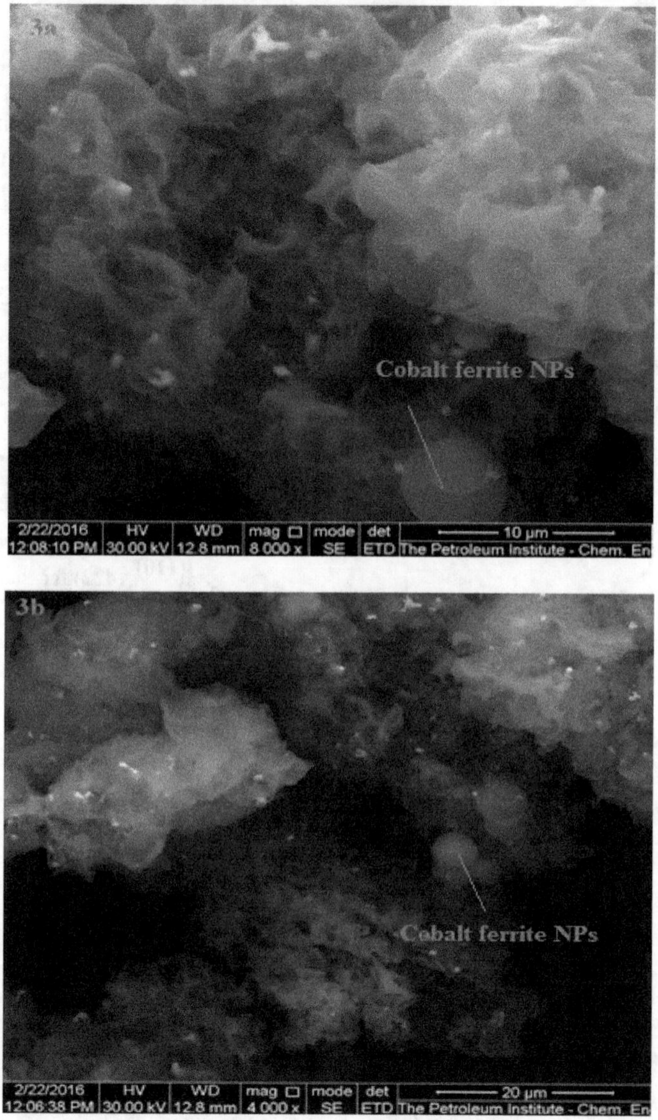

Figure 8.3 SEM images of $CoFe_2O_4$/Graphene/PZTs-RF cryogel at different magnifications.

cryogel, therefore, forming a well-defined interconnected porous structure. Determination of the surface area and pore volume of the adsorbent was carried out by N_2 adsorption-desorption measurements at 73 K, and the pore size distribution was determined based on BJH desorption method. The textural properties were studied for the three cryogel nanocomposites such as (i) Graphene-RF-Na_2CO_3, (ii) $CoFe_2O_4$/Graphene-RF and (iii) $CoFe_2O_4$/ Graphene/PZTs-RF (Table 8.1). From Table 8.1, it can be observed that the presence of PZTs

Table 8.1 Determination of surface area, pore radius and pore volume of samples by N_2 adsorption-desorption measurements

Sample	Surface area ($m^2\,g^{-1}$)	Pore volume ($cc\,g^{-1}$)	Pore radius (Å)
Graphene- RF-Na_2CO_3	324	0.54	7.05
Graphene/$CoFe_2O_4$-RF	479	0.42	7.12
Graphene/PZT/$CoFe_2O_4$-R	630	0.06	19.01

along with cobalt ferrite nanoparticles achieved better BET surface area (630 $m^2\,g^{-1}$) and pore radius (19.01 Å), as compared to cryogel without PZTs and $CoFe_2O_4$ NPs and cryogel with $CoFe_2O_4$ NPs but without PZTs. Thus, the increased BET surface area and better porosity of the resultant adsorbent can enhance the active adsorption sites for efficient adsorption process. For further investigation of morphology, TEM image of the resultant adsorbent is presented in Figure 8.4. It can be observed that the carbonaceous matrix contained carbon fibers of PZTs which helped in avoiding the stacking of graphene sheets, thus, contributing to enhanced surface area of the adsorbent.

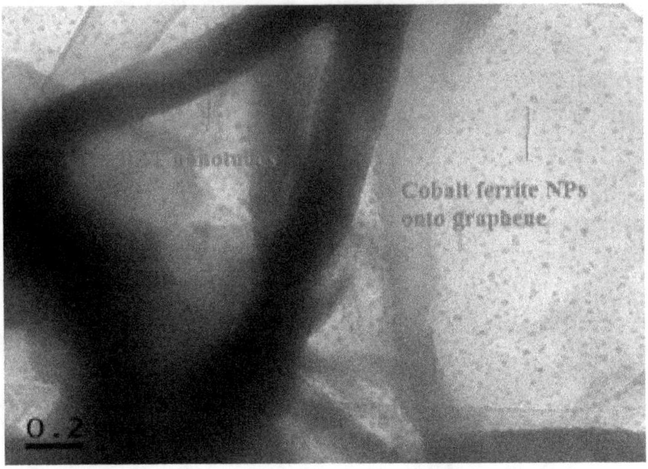

Figure 8.4 TEM image of $CoFe_2O_4$/Graphene/PZTs-RF cryogel.

Additionally, cobalt ferrite nanoparticles were uniformly placed onto graphene sheets which corresponded to additional active adsorption sites as well as an increase in the interaction with negatively charged 2-nitrophenol ions.

8.3.2 Effect of Various Parameters on Adsorption

In this section, the effect of adsorption parameters such as contact time, 2-nitrophenol concentration, dosage and pH on the adsorption capacity as well as removal efficiency from aqueous solution has been analyzed.

Time Effect

In Figure 8.5, the effect of C_o on q_e is presented as a function of contact time. A rapid adsorption was observed during the first 4 h. After 4 h, the rate of adsorption became almost stationary. The quick initial 2-nitrophenol adsorption occurred due to the presence of active adsorption sites provided by the combination of large surface of graphene layers and cobalt ferrite nanoparticles. With increasing time, the rate of uptake of 2-nitrophenol gradually reduced due to the diminished availability of remaining active adsorption sites. In Figure 8.6, the effect of C_o on equilibrium adsorption capacity (q_e) and

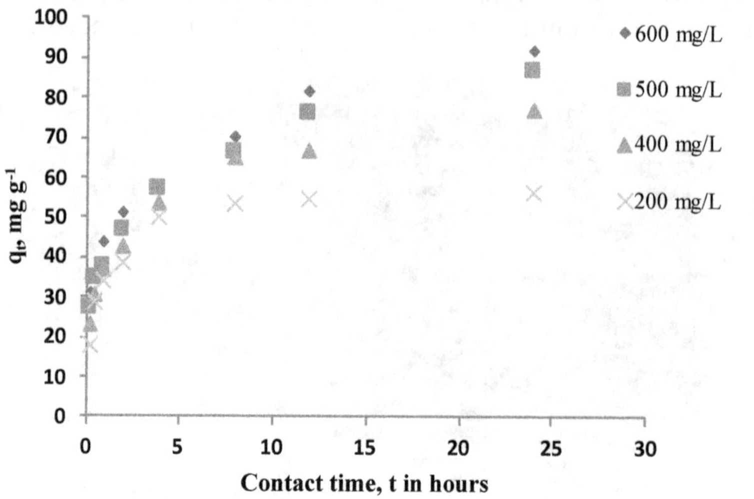

Figure 8.5 Effect of contact time on the adsorption capacity at 200, 400, 500 and 600 mg L^{-1}.

percentage (%) removal is presented. It can be observed that with increasing concentration (C_o, mg L^{-1}) of 2-nitrophenol, q_e increased and % removal decreased.

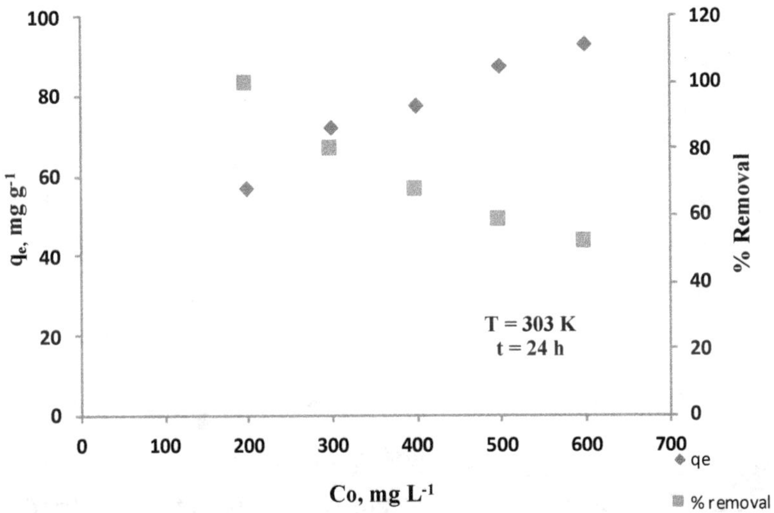

Figure 8.6 Effect of initial concentration (C_0) on equilibrium adsorption capacity (q_e) and percentage (%) removal.

Dosage Effect

The effect of adsorbent dose on the adsorption capacity (q_e) and percentage (%) removal of 2-nitrophenol (C_o = 600 mg L^{-1}, V = 25 mL) was studied at three different temperatures (303 K, 313K and 323 K), as shown in Figures 8.7 and 8.8, respectively. Amount of adsorbent varied between 0.02 g to 0.1 g (corresponding adsorbent to 2-nitrophenol ratio from 0.8 to 4). As the dosage ratio increased from 0.8 to 4, the equilibrium adsorption capacity (q_e) decreased from 138.75 to 123.49 mg L^{-1}, 148.25 to 130.17 mg L^{-1}, and 154.72 to 133.99 mg L^{-1} at 303 K, 313 K and 323 K, respectively. The percentage removal increased from 18.4 to 82.06%, 19.7 to 86.49% and 21.12 to 91.46% at 303 K, 313 K and 323 K, respectively. Due to the increased absolute adsorption surface, the percentage (%) removal of 2-nitrophenol increased with increasing adsorbent to 2-nitrophenol ratio [37], whereas the equilibrium adsorption capacity was reduced, which can be attributed to the decrease in total surface area of the adsorbent, increase in adsorbent dose (2-nitrophenol molecules get partially adsorbed at adsorption sites) and increase in diffusion path length of 2-nitrophenol molecules [37].

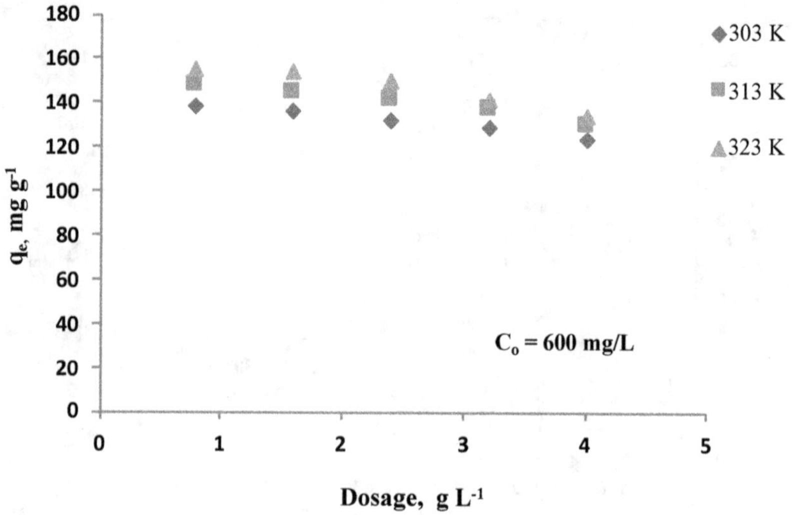

Figure 8.7 Effect of dosage on adsorption capacity (q_e) at 303, 313 and 323 K.

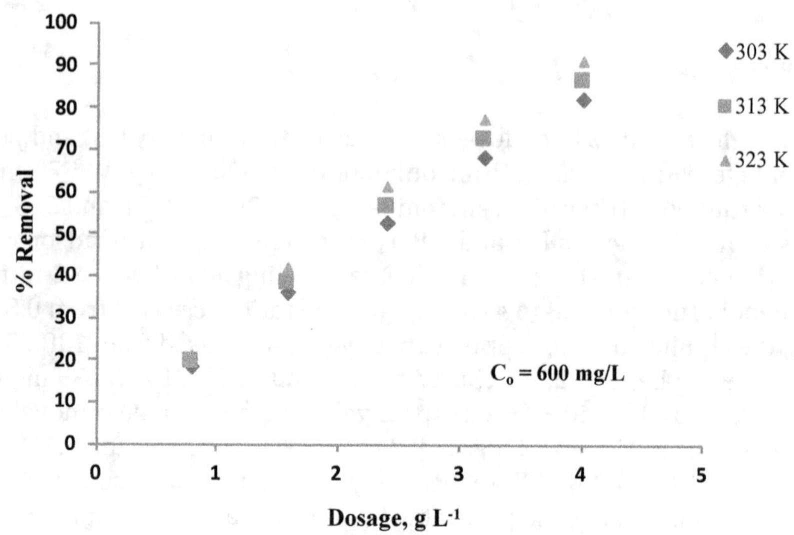

Figure 8.8 Effect of dosage on percentage (%) removal at 303, 313 and 323 K.

pH Effect

Industrial effluents may exhibit various pH which can limit the efficacy of an adsorbent. Therefore, it is very crucial to investigate the effect of pH effect on adsorption. The pH studies were carried out within the pH range of 2-12 by employing 0.05 g of adsorbent in 15 mL of 2-nitrophenol (600 mg L^{-1}) at 303 K. The effect of pH on the adsorption capacity (q_e) and the percentage (%) removal is shown in Figure 8.9. The adsorbent exhibited isoelectric point (point of zero charge) around pH = 4. Thus, it had positive surface charge at pH < 4, whereas the surface charge of adsorbent was negative at pH > 4. It is also worth mentioning about the pk_a value of 7.23 for 2-nitrophenol,

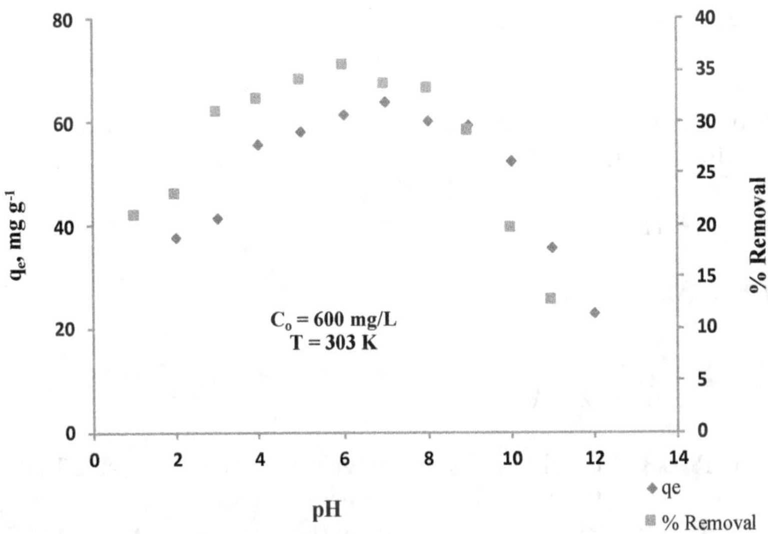

Figure 8.9 Effect of pH on equilibrium adsorption capacity (q_e) and percentage (%) removal.

as, below this pH, 2-nitrophenol existed in undissociated form, whereas it was in the dissociated form above this pH (contained constituent negatively charged ions) [38]. It can be observed that at initial lower pH (2-3), the adsorption capacity was observed to be low, which was attributed to the positive surface charge of adsorbent due to which there was an electrostatic repulsion between the adsorbent and 2-nitrophenol molecules, thus, further reducing the adsorbent-adsorbate interaction. As pH range shifted towards 4-7, the adsorption capacity increased, which might be due to the surface charge of adsorbent turning to be negative, thus, promoting the adsorbent-adsorbate interaction. As pH range shifted to 7-10, the adsorption capacity gradually decreased, as 2-nitrophenol started to dissociate into constituent negatively charged anions at pH > 7.23, thus, leading to a sort of electrostatic repulsion between the negative surface of graphene and negatively charged phenolate anions, which lowered the adsorption capacity [14,38].

8.3.3 Thermodynamic Analysis

Adsorption experiments were carried out at different temperatures (303 K, 313 K, 318 K and 323 K) to evaluate the effect on the adsorption process. To determine the thermodynamic parameters, such as change in enthalpy ($\Delta H°$), Gibbs free energy ($\Delta G°$) and entropy change ($\Delta S°$), the values of Langmuir isotherm constant (K_L) were obtained at four different temperatures. The thermodynamic parameters of adsorbent/2-nitrophenol system were calculated by using the following equations [39].

$$\Delta G° = -RT ln k_L$$

$$\Delta G° = \Delta H° - T\Delta S$$

$$\ln k_L = -\left(\frac{\Delta H°}{RT}\right) + \left(\frac{\Delta S°}{R}\right)$$

where, R (8.314 J mol^{-1} K^{-1}) is the universal gas constant and T (K) is the temperature. The thermodynamics parameters ($\Delta G°, \Delta H°$ and $\Delta S°$) were obtained from the slope and intercept of linear plot of ln k_L versus 1/T (Figure 8.10). The calculated values of thermodynamics parameters are listed in Table 8.2. The positive value of $\Delta S°$ (108.406 J $mol^{-1}k^{-1}$) signified the increased randomness or degree of freedom at the adsorbent/2-nitrophenol solution interface during the adsorption of 2-nitrophenol on adsorbent. The adsorption process was observed to be an endothermic reaction, shown by the positive value of $\Delta H°$ (27529.3 J mol^{-1}) [40]. The negative values of $\Delta G°$ signified the spontaneous nature of the adsorption process. As temperature increased from 303 K to 323 K, the value of $\Delta G°$ decreased from -5419.42 to -7612.88, which indicated that the adsorption process on the adsorbent surface was more favorable at higher temperature.

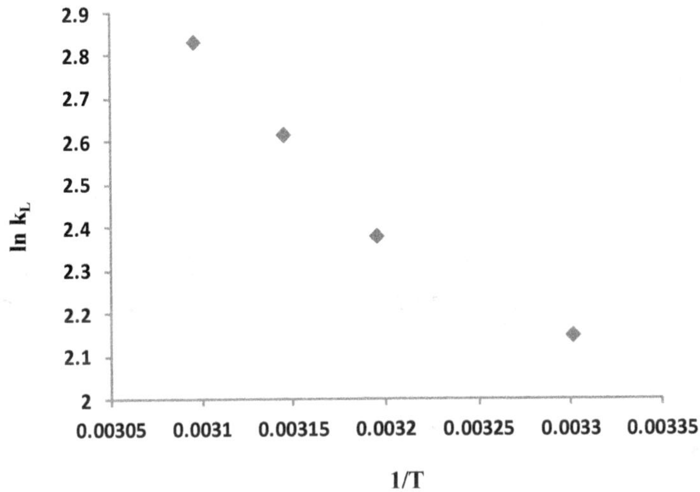

Figure 8.10 Plot of ln k_L versus 1/T for the evaluation of thermodynamic parameters for the adsorption of 2-nitrophenol.

Table 8.2 Thermodynamic parameters for the adsorption of 2-nitrophenol on adsorbent

$\Delta S°$ $Jmol^{-1}K^{-1}$	$\Delta H°$ $Jmol^{-1}$	$\Delta G°$ $Jmol^{-1}$			
		303 K	313 K	318 K	323 K
108.406	27529.3	-5419.42	-6192.24	-6920.66	-7612.88

8.3.4 Adsorption Kinetics

The equilibrium time and efficiency of the adsorption process was investigated by the kinetics of the adsorption process as well as by the kinetic models to examine the potential rate controlling step. Pseudo-first-order and pseudo-second-order models were used to investigate the kinetic data obtained from batch studies.

Pseudo-first-order Model or Lagergren's Equation

Pseudo-first-order model [41] is commonly used for the analysis of adsorption of solute from the liquid phase. Its linearized equation can be expressed as follows:

$$\log(q_e - q_t) = \log q_e - \frac{k_1}{2.303}t$$

where, q_t and q_e (mg g^{-1}) are the amounts of 2-nitrophenol adsorbed at time t and at equilibrium, respectively, and k_1 is the Lagergren rate constant of adsorption (min^{-1}). Plots of log (q_e - q_t) versus t are shown in Figure 8.11, the rate constant k_1 and equilibrium concentration q_e were determined from the slope and intercept by linear fitting, and their values are presented in Table 8.3.

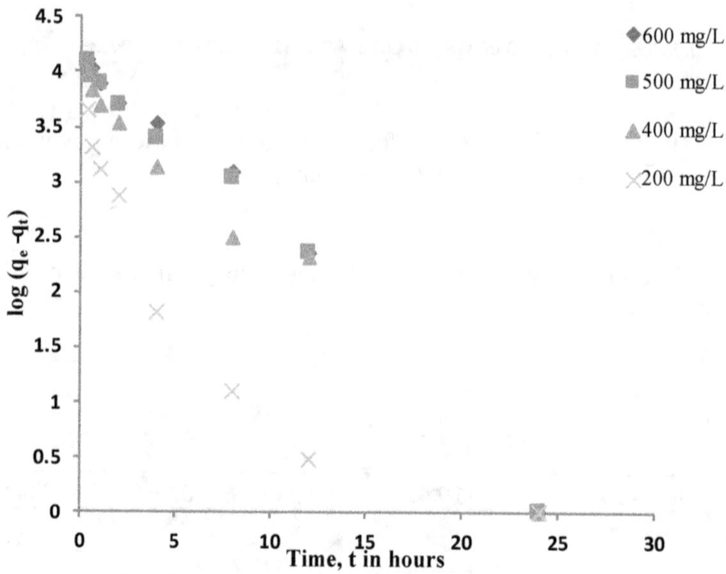

Figure 8.11 Pseudo-first-order model used for adsorption of 2-nitrophenol on adsorbent for various concentrations at 303 K.

Pseudo-second-order Model

The pseudo-second-order equation [42] is represented as follows:

$$\frac{t}{q_t} = \frac{1}{k_2 q_e^2} + \frac{t}{q_e}$$

where, k_2 is the pseudo-second-order rate constant of adsorption (g

mg^{-1} min^{-1}). The values of q_e and k_2 are calculated from the slope and intercept of the plot of t/q_t versus t, as shown in Figure 8.12. The corresponding kinetic parameters and determination coefficients are also mentioned in Table 8.3.

Table 8.3 Evaluated values from pseudo-first and second-order kinetic equations and intra-particle diffusion model at different concentrations (C_o)

			C_o, mg L^{-1}			
Model			200	400	500	600
Pseudo-first-order		$q_{e,exp.}$	56.56	77.01	86.7	92.37
		q_e calc.	20.96	49.44	61.20	64.62
		k_1	0.154	0.158	0.164	0.166
		R^2	0.822	0.987	0.988	0.987
Pseudo-second-order		$q_{e,exp.}$	56.56	77.01	86.7	92.37
		$q_{e,cal.}$	58.13	79.36	89.28	95.23
		k_2	0.025	0.0093	0.0071	0.0068
		R^2	0.999	0.994	0.99	0.989
Intra-particle diffusion model	Below $t_c^{1/2}$	K_{id-1}	19.32	19.25	19.02	18.12
		C_1	12.33	15.91	19.24	23.75
		R^2	0.949	0.98	0.989	0.975
	Above $t_c^{1/2}$	K_{id-2}	2.036	7.56	10.35	11.73
		C_2	47.15	40.73	37.35	36.99
		R^2	0.903	0.955	0.975	0.959

The lower determination coefficients (R^2 = 0.82-0.98) of plot log (q_e-q_t) versus t suggested that the pseudo-first-order model provided a poor fit of the experimental data, and there was a considerable difference between the calculated (q_e = 64.62 mg g^{-1} at 600 mg L^{-1}) and experimental (q_e = 92.37 mg g^{-1} at 600 mg L^{-1}) values of equilibrium capacity. On the other hand, pseudo-second-order model showed higher determinant coefficients (R^2) of plot t/q_t versus t which fitted the experimental data, and the calculated (q_e = 95.23 mg g^{-1} at 600 mg L^{-1}) and experimental (q_e = 92.37 mg g^{-1} at 600 mg L^{-1}) values of equilibrium capacity were observed to be very close to each other. The above results suggested that the adsorption process followed the pseudo-second-order model.

8.3.5 Sorption Mechanism

To explain the diffusion mechanism, kinetic results were studied using the intra-particle diffusion model. In general, adsorption process

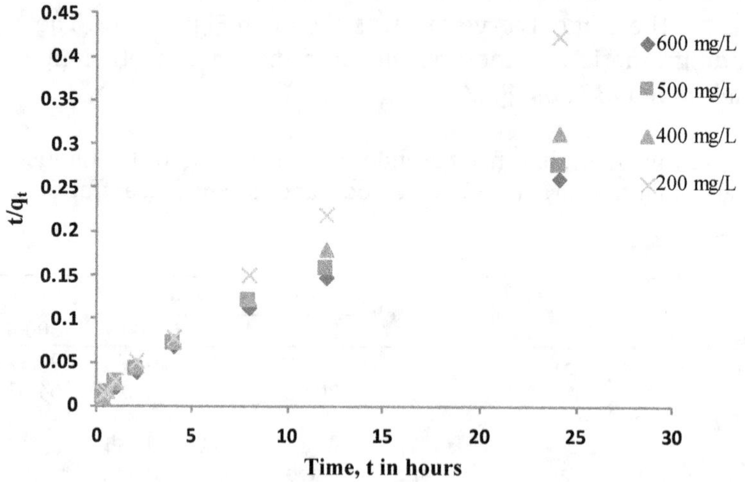

Figure 8.12 Pseudo-second-order model used for adsorption of 2-nitrophenol for various concentration at 303 K.

is governed by following steps: (a) instantaneous adsorption step or an external surface adsorption, (b) a slow adsorption step where intra-particle diffusion is the rate limiting mechanism, and (c) an extremely slow adsorption step due to the very low 2-nitrophenol concentration or the saturation of the adsorbent [43]. Weber and Morris model is a widely used intra-particle diffusion model which predicts the rate controlling step. It is stated that the adsorbate uptake varies with the square root of adsorption time [44]. The rate constant of intra-particle diffusion (k_{id}) at the stage 'i' was determined by using the following equation:

$$q_t = k_{id}\, t^{\frac{1}{2}} + C_i$$

where, q_t (mg g^{-1}) is the amount of 2-nitrophenol adsorbed at time t (min), k_{id} (mg g^{-1} $min^{-1/2}$) is the intra-particle diffusion rate constant and C is the intercept. A plot of q_t versus $t^{1/2}$ at different initial 2-nitrophenol concentrations is shown in Figure 8.13. The adsorption data is split into two regions, both fitted by a straight line. These lines intercept at a critical time, t_c. This multi-linearity characterization shows the presence of two or more steps in the adsorption process. Macropore diffusion dominates the section 1, where adsorption time is shorter than t_c. Section 1 resembles the external surface adsorption or instantaneous adsorption stage. Micropore diffusion dominates the section 2, where adsorption time is longer than t_c [45].

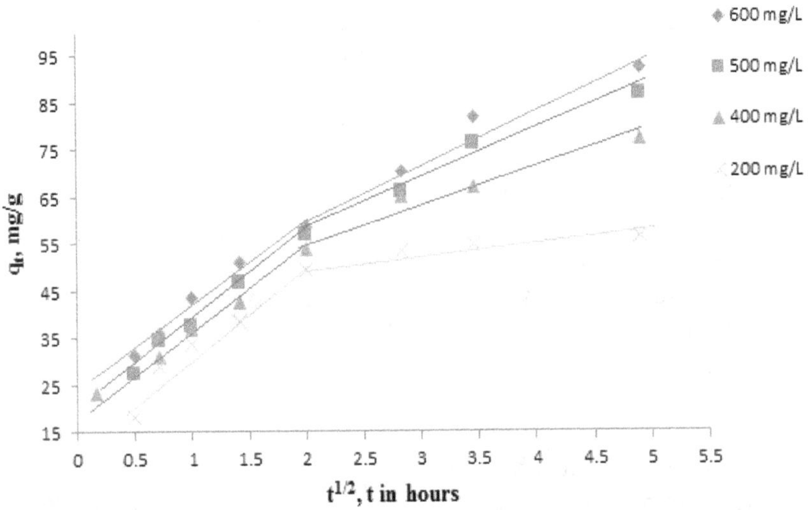

Figure 8.13 Intra-particle diffusion model plots for adsorption of 2-nitrophenol for various concentration at 303 K. The lines represent fitting of data.

It was noticed from Table 8.3 that the value of k_{id-1} (section 1) was higher than k_{id-2} (section 2) for all concentrations (200, 400, 500 and 600 mg L^{-1}) of 2-nitrophenol, which indicated that the macropore diffusion was dominant and the rate of micropore diffusion was the rate determining step [14].

The value of C resembles the thickness of the boundary layer. The larger value of C is related to the greater boundary layer diffusion effect [14]. It was observed from Table 8.3 that the value of C_2 was greater than C_1 for all concentration of 2-nitrophenol, which showed greater boundary layer effect at the micropore diffusion stage than at the macropore diffusion stage [46].

The two sections of the plot (q_t versus $t^{1/2}$) and their deviation from the origin for various 2-nitrophenol concentrations (200, 400, 500 and 600 mg L^{-1}) suggested that the intra-particle diffusion was not the only rate limiting step [14]. Moreover, the high initial adsorption rate, which was further followed by an almost stagnant rate before reaching the equilibrium, also explained that intra-particle diffusion was not the sole rate limiting step for the adsorption of 2-nitrophenol on adsorbent. Therefore, it can be suggested that there are several stages for the adsorption process, such as 2-nitophenol diffusion from the solution onto the surface of adsorbent, intra-particle diffusion, sorption of solution into the interior part and smaller mesopores of adsorbent, sorption of the 2-nitrophenol molecules into the capillaries to reach equilibrium, etc. [45].

8.3.6 Adsorption Isotherms

The distribution of the adsorbate molecules between the solid and liquid phases, when the adsorption reaches the equilibrium, is determined using the isotherm Langmuir, Freundlich and Temkin isotherms. These isotherms provide the information about the heterogeneity and homogeneity of the adsorbent surface.

The Langmuir isotherm [47] assumes that the adsorption of adsorbate occurs on a homogeneous surface by monolayer adsorption with well-defined and energetically equal sites as well as no interaction between the adsorbate molecules. The linear form of Langmuir isotherm is expressed as:

$$\frac{C_e}{q_e} = \frac{1}{q_{max}k_L} + \frac{1}{q_{max}}C_e$$

$$\frac{1}{q_e} = \left(\frac{1}{q_{max}k_L}\right)\left(\frac{1}{C_e}\right) + \frac{1}{q_{max}}$$

where, q_{max} is the maximum adsorption capacity of 2-nirophenol (mg g^{-1}), C_e is the equilibrium concentration (mg L^{-1}) and K_L is the Langmuir constant (L mg^{-1}). The Freundlich isotherm [48] assumes multilayer adsorption with non-uniform distribution of the affinities over the heterogeneous surface. The linear form of Freundlich isotherm is expressed as:

$$\ln q_e = \ln k_F + \frac{1}{n}\ln C_e$$

where, k_F (mg g^{-1}) is the theoretical value of maximum adsorption capacity and n is the Freundlich constant. The equation is characterized by the heterogeneity factor (n), thus, this isotherm is used to describe many heterogeneous systems. Temkin isotherm [49] has also been used to describe the adsorption system behavior on heterogeneous surfaces and can be expressed by the following equations:

$$q_e = \frac{RT}{b}\ln(k_T C_e)$$

$$q_e = \frac{RT}{b}\ lnk_T\ + \frac{RT}{b}\ lnC_e$$

where, T (K) is the temperature, k_T (L g^{-1}) is the equilibrium binding constant related to the maximum binding energy, b (kJ mol^{-1}) is the Temkin isotherm constant, $\frac{RT}{b}$ corresponds to heat of adsorption [50,51] and R (8.314 J K^{-1} mol^{-1}) is the gas constant. The equation represents the linear form of the Temkin isotherm. For Temkin isotherm, experimental adsorption data was used to obtain a linear relation by plotting q_e versus ln(C_e), as shown in Figure 8.14. The values of b and k_T (Temkin model parameters) are presented in Table 8.4.

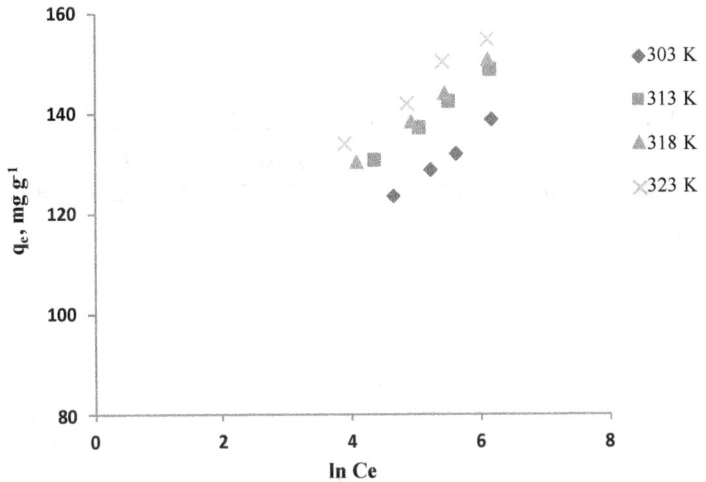

Figure 8.14 Temkin isotherm for the adsorption of 2-nitrophenol (600 mg L^{-1}) at 303, 313, 318 and 323 K.

In Langmuir isotherm, a linear relation was obtained by plotting $1/q_e$ versus $1/C_e$ at three different temperatures, as shown in Figure 8.15. The Langmuir model parameters (K_L and $q_{max,}$) could be obtained from the plot of $1/q_e$ versus $1/C_e$, as mentioned in Table 8.4. The theoretical values of maximum adsorption capacity (q_{max} = 140.84, 149.25, 151.51 and 153.84 mg g^{-1} at 303, 313, 318 and 323 K, respectively) were not very close to the experimental values (q_e = 92.37, 100.11, 103.8 and 107.13 mg g^{-1} at 303, 313, 318 and 323 K,

Table 8.4 Langmuir, Freundlich and Temkin isotherm parameters (C_o = 600 mg L^{-1}, t =24 h) at 303 K, 313 K, 318 K and 323 K

T (K)		Langmuir				Freundlich		
	q_{e}, exp mg g^{-1}	q_{max} mg g^{-1}	K_L Lg^{-1}	R_L	R^2	n	K_F	R^2
303	92.37	140.84	0.06	0.03	0.92	13.14	86.25	0.99
313	100.11	149.25	0.08	0.02	0.94	13.64	94.28	0.99
318	103.8	151.51	0.09	0.02	0.92	13.91	97.03	0.99
323	107.13	153.84	0.12	0.01	0.91	14.88	102.97	0.98

T(K)		Temkin		
	q_{e}, exp mg g^{-1}	b kJmol^{-1}	k_T	R^2
303	92.37	0.252	1.0077	0.990
313	100.11	0.255	1.0084	0.999
318	103.8	0.262	1.0088	0.998
323	107.13	0.277	1.0099	0.978

respectively), which showed that adsorption of 2-nitrophenol on the prepared adsorbent did not follow the Langmuir isotherm. The relation of Langmuir constant (k_L) to the separation factor (R_L) is represented by equation (13):

$$R_L = \frac{1}{1 + C_o k_L}$$

where, C_o is the initial concentration of 2-nitrophenol (600 mg g^{-1}). R_L provides a qualitative measure of the favorability of the adsorption process; $R_L > 1$ (unfavorable), $R_L = 1$ (linear), $0 < R_L < 1$ (favorable) and $R_L = 0$ (irreversible) [52]. The predicted values of R_L from Langmuir isotherm varied between 0.013 and 0.026, indicating the favorable behavior towards 2-nitrophenol adsorption. Moreover, with increasing temperature, the value of R_L decreased which showed that the process became more favorable at higher temperature.

Freundlich isotherm was also used to model the adsorption data. The values of n and K_F could be calculated by plotting ln(C_e) versus ln(q_e), as shown in Figure 8.16. K_F and n represent the capacity of adsorbent and favorability of the adsorption process, respectively. In this study, the value of n increased from 13.14 to 14.88 as temperature increased from 303 K to 323 K (Table 8.4), which confirmed the

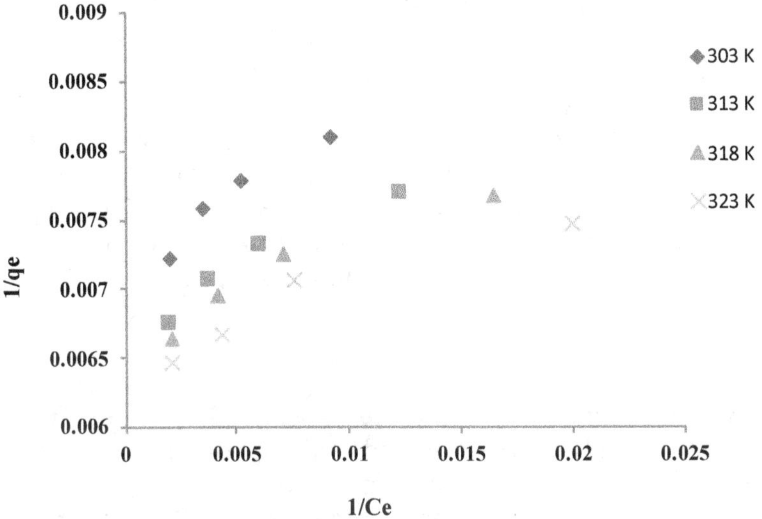

Figure 8.15 Langmuir isotherm for the adsorption of 2-nitrophenol (600 mg L^{-1}) on adsorbent at 303, 313, 318 and 323 K.

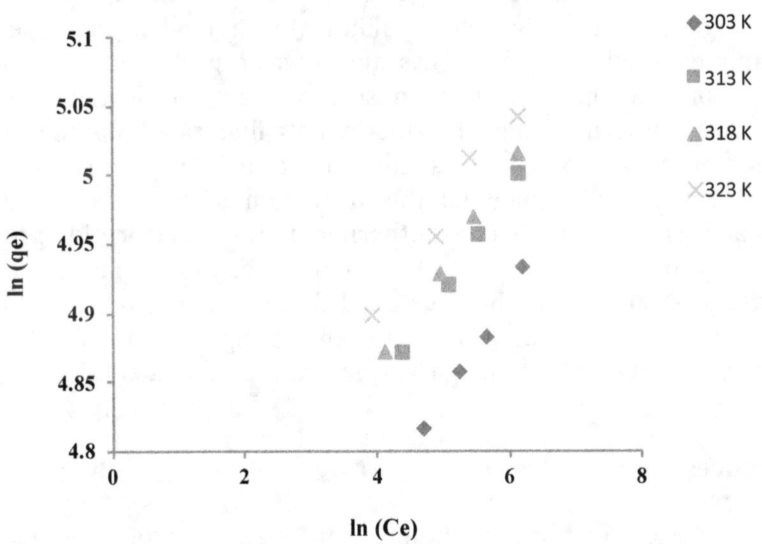

Figure 8.16 Freundlich isotherm for the adsorption of 2-nitrophenol (600 mg L^{-1}) on adsorbent at 303, 313, 318 and 323 K.

increase in favorability of the adsorption process. As n increased, the adsorption became more favorable (poor adsorption $n < 1$, difficult adsorption n = 1-2, good adsorption n > 2) [45]. Freundlich isotherm was found to be more suitable as evidenced from the greater values of regression coefficients (R^2) among three isotherm models. The theoretical values of maximum adsorption capacity (K_F = 86.254, 94.282, 97.028 and 102.97 mg g^{-1} at 303, 313, 318 and 323 K, respectively) were relatively very close to the experimental values (q_e = 92.37, 100.11, 103.8 and 107.13 mg g^{-1} at 303, 313, 318 and 323 K, respectively). This indicated that the adsorption of 2-nitrophenol on the prepared adsorbent followed the Freundlich isotherm.

8.4 Conclusions

The cryogel porous nanocomposite $CoFe_2O_4$/Graphene/PZTs-RF could be successfully synthesized and exhibited promising potential for the removal of 2-nitrophenol from aqueous solutions. Especially, the addition of PZTs with graphene containing $CoFe_2O_3$ nanoparticles effectively contributed to an increase in the specific surface area of final adsorbent, thus, increasing the active adsorption sites to adsorb the 2-nitrophenol molecules. The adsorption capacity and removal efficiency were affected by dosage ratio, solution pH and temperature. Langmuir, Freundlich and Temkin isotherm models were used to examine the adsorption process, and it was revealed that the adsorption of 2-nitrophenol on the adsorbent was well described by Freundlich isotherm model. The kinetics data illustrated that the adsorption process followed the pseudo-second-order model. Thermodynamics studies confirmed that the adsorption process was spontaneous and endothermic. The endothermic nature of adsorption process was also confirmed by the increase in adsorption capacity with increasing temperature. The developed three dimensional network porous cryogel nanocomposites represent strong potential for utilization as effective adsorbents for various industrial effluents.

References

1. Busca, G., Berardinelli, S., Resini, C., and Arrighi, L. (2008) Technologies for the removal of phenol from fluid streams: a short review of recent developments. *Journal of Hazardous Materials*, **160**(2), 265-288.
2. Jordan, W., van Barneveld, H., Gerlich, O., Kleine-Boymann, M., and Ullrich, J. (1991) Phenol. In: *Ullmann's Encyclopedia of Industrial Chemistry*, Wiley-VCH, Germany, doi:10.1002/14356007.a19_299.

3. Kumar, A., Kumar, S., Kumar, S., and Gupta, D. V. (2007) Adsorption of phenol and 4-nitrophenol on granular activated carbon in basal salt medium: equilibrium and kinetics. *Journal of Hazardous Materials*, **147**(1), 155-166.
4. Knop, A., and Pilato, L. A. (1985) *Phenolic Resins*, Springer, USA.
5. Park, Y., Skelland, A., Forney, L. J., and Kim, J.-H. (2006) Removal of phenol and substituted phenols by newly developed emulsion liquid membrane process. *Water Research*, **40**(9), 1763-1772.
6. Moussavi, G., Mahmoudi, M., and Barikbin, B. (2009) Biological removal of phenol from strong wastewaters using a novel MSBR. *Water Research*, **43**(5), 1295-1302.
7. Juang, R.-S., Huang, W.-C., and Hsu, Y.-H. (2009) Treatment of phenol in synthetic saline wastewater by solvent extraction and two-phase membrane biodegradation. *Journal of Hazardous Materials*, **164**(1), 46-52.
8. Sun, S., Wang, W., Zhang, L., and Shang, M. (2009) Visible light-induced photocatalytic oxidation of phenol and aqueous ammonia in flowerlike $Bi_2Fe_4O_9$ suspensions. *Journal of Physical Chemistry C*, **113**(29), 12826-12831.
9. Wu, Z., and Zhou, M. (2001) Partial degradation of phenol by advanced electrochemical oxidation process. *Environmental Science & Technology*, **35**(13), 2698-2703.
10. Roostaei, N., and Tezel, F. H. (2004) Removal of phenol from aqueous solutions by adsorption. *Journal of Environmental Management*, **70**(2), 157-164.
11. Froehner, S., Martins, R. F., Furukawa, W., and Errera, M. R. (2009) Water remediation by adsorption of phenol onto hydrophobic modified clay. *Water, Air, and Soil Pollution*, **199**(1-4), 107-113.
12. Li, J.-M., Meng, X.-G., Hu, C.-W., and Du, J. (2009) Adsorption of phenol, p-chlorophenol and p-nitrophenol onto functional chitosan. *Bioresource Technology*, **100**(3), 1168-1173.
13. Kumar, S., Zafar, M., Prajapati, J. K., Kumar, S., and Kannepalli, S. (2011) Modeling studies on simultaneous adsorption of phenol and resorcinol onto granular activated carbon from simulated aqueous solution. *Journal of Hazardous Materials*, **185**(1), 287-294.
14. Li, Y., Du, Q., Liu, T., Sun, J., Jiao, Y., Xia, Y., Xia, L., Wang, Z., Zhang, W., and Wang, K. (2012) Equilibrium, kinetic and thermodynamic studies on the adsorption of phenol onto graphene. *Materials Research Bulletin*, **47**(8), 1898-1904.
15. Arasteh, R., Masoumi, M.,; Rashidi, A., Moradi, L., Samimi, V., and Mostafavi, S. (2010) Adsorption of 2-nitrophenol by multi-wall carbon nanotubes from aqueous solutions. *Applied Surface Science*, **256**(14), 4447-4455.
16. Bibi, N. S., Singh, N. K., Dsouza, R. N., Aasim, M., and Fernandez-Lahore, M. (2013) Synthesis and performance of megaporous immob-

ilized metal-ion affinity cryogels for recombinant protein capture and purification. *Journal of Chromatography A*, **1272**, 145-149.

17. Zheng, S., Wang, T., Liu, D., Liu, X., Wang, C., and Tong, Z. (2013) Fast deswelling and highly extensible poly (N-isopropylacrylamide)-hectorite clay nanocomposite cryogels prepared by freezing polymerization. *Polymer*, **54**(7), 1846-1852.
18. Sahiner, N., Seven, F., and Al-lohedan, H. (2015) Superporous cryogel-M (Cu, Ni, and Co) composites in catalytic reduction of toxic phenolic compounds and dyes from wastewaters. *Water, Air & Soil Pollution*, **226**(4), 1-12.
19. Reichelt, S., Becher, J., Weisser, J., Prager, A., Decker, U., Moeller, S., Berg, A., and Schnabelrauch, M. (2014) Biocompatible polysaccharide-based cryogels. *Materials Science and Engineering C*, **35**,164-170.
20. Petrov, P., Petrova, E., Tchorbanov, B., and Tsvetanov, C. B. (2007) Synthesis of biodegradable hydroxyethylcellulose cryogels by UV irradiation. *Polymer*, **48**(17), 4943-4949.
21. Park, S., Lee, K.-S., Bozoklu, G., Cai, W., Nguyen, S. T., and Ruoff, R. S. (2008) Graphene oxide papers modified by divalent ions-enhancing mechanical properties via chemical cross-linking. *ACS Nano*, **2**(3), 572-578.
22. Wei, G., Miao, Y.-E., Zhang, C., Yang, Z., Liu, Z., Tjiu, W. W., and Liu, T. (2013) Ni-doped graphene/carbon cryogels and their applications as versatile sorbents for water purification. *ACS Applied Materials & Interfaces* **2013, 5**(15), 7584-7591.
23. Asmaly, H. A., Abussaud, B., Saleh, T. A., Gupta, V. K., and Atieh, M. A. (2015) Ferric oxide nanoparticles decorated carbon nanotubes and carbon nanofibers: from synthesis to enhanced removal of phenol. *Journal of Saudi Chemical Society*, **19**(5), 511-520.
24. Ai, L., and Jiang, J. (2012) Removal of methylene blue from aqueous solution with self-assembled cylindrical graphene-carbon nanotube hybrid. *Chemical Engineering Journal*, **192**, 156-163.
25. Wang, P., Cao, M., Wang, C., Ao, Y., Hou, J., and Qian, J. (2014) Kinetics and thermodynamics of adsorption of methylene blue by a magnetic graphene-carbon nanotube composite. *Applied Surface Science*, **290**, 116-124.
26. Allcock, H. R., and Ambrosio, A. M. (1996) Synthesis and characterization of pH-sensitive poly (organophosphazene) hydrogels. *Biomaterials*, **17**(23), 2295-2302.
27. Allcock, H. R., Bender, J. D., Chang, Y., McKenzie, M., and Fone, M. M. (2003) Controlled refractive index polymers: Polyphosphazenes with chlorinated-and fluorinated-, aryloxy-and alkoxy-side-groups. *Chemistry of Materials*, **15**(2), 473-477.
28. Allcock, H. R., Kellam, E. C., and Morford, R. V. (2001) Gel electrolytes from co-substituted oligoethyleneoxy/trifluoroethoxy linear poly-

phosphazenes. *Solid State Ionics*, **143**(3), 297-308.

29. Lu, S.-Y., and Hamerton, I. (2002) Recent developments in the chemistry of halogen-free flame retardant polymers. *Progress in Polymer Science*, **27**(8), 1661-1712.
30. Allcock, H. R. (2006) Recent developments in polyphosphazene materials science. *Current Opinion in Solid State and Materials Science*, **10**(5), 231-240.
31. Zhu, L., Xu, Y., Yuan, W., Xi, J., Huang, X., Tang, X., and Zheng, S. (2006) One-pot synthesis of Poly (cyclotriphosphazene-co-4, 4′-sulfonyldiphenol) nanotubes via an in situ template approach. *Advanced Materials*, **18**(22), 2997-3000.
32. Marcano, D. C., Kosynkin, D. V., Berlin, J. M., Sinitskii, A., Sun, Z., Slesarev, A., Alemany, L. B., Lu, W., and Tour, J. M. (2010) Improved synthesis of graphene oxide. *ACS Nano*, **4**(8), 4806-4814.
33. Maldonado-Hodar, F., Moreno-Castilla, C., and Perez-Cadenas, A. (2004) Surface morphology, metal dispersion, and pore texture of transition metal-doped monolithic carbon aerogels and steam-activated derivatives. *Microporous and Mesoporous Materials*, **69**(1), 119-125.
34. Chen, W., and Yan, L. (2011) In situ self-assembly of mild chemical reduction graphene for three-dimensional architectures. *Nanoscale*, **3**(8), 3132-3137.
35. Wu, A., Yang, X., and Yang, H. (2013) Magnetic properties of carbon-encapsulated Fe-Co alloy nanoparticles. *Dalton Transactions*, **42**(14), 4978-4984.
36. Tang, Z., Shen, S., Zhuang, J., and Wang, X. (2010) Noble-metal-promoted three-dimensional macroassembly of single-layered graphene oxide. *Angewandte Chemie International Edition*, **122**(27), 4707-4711.
37. Wang, H.-L., and Jiang, W.-F. (2007) Adsorption of dinitro butyl phenol (DNBP) from aqueous solutions by fly ash. *Industrial & Engineering Chemistry Research*, **46**(16), 5405-5411.
38. Karimi-Jashni, A., and Narbaitz, R. M. (1997) Impact of pH on the adsorption and desorption kinetics of 2-nitrophenol on activated carbons. *Water Research*, **31**(12), 3039-3044.
39. Pal, P., Banat, F., and AlShoaibi, A. (2013) Adsorptive removal of heat stable salt anions from industrial lean amine solvent using anion exchange resins from gas sweetening unit. *Journal of Natural Gas Science and Engineering*, **15**, 14-21.
40. Teng, H., and Hsieh, C.-T. (1998) Influence of surface characteristics on liquid-phase adsorption of phenol by activated carbons prepared from bituminous coal. *Industrial & Engineering Chemistry Research*, **37**(9), 3618-3624.
41. Lagergren, S. (1898) About the theory of so-called adsorption of soluble substances. *Kungliga Svenska Vetenskapsakademiens Handlin-*

gar, **24**(4), 1-39.

42. Reddad, Z., Gerente, C., Andres, Y., and Le Cloirec, P. (2002) Adsorption of several metal ions onto a low-cost biosorbent: kinetic and equilibrium studies. *Environmental Science & Technology*, **36**(9), 2067-2073.
43. Wu, F.-C., Tseng, R.-L., and Juang, R.-S. (2005) Comparisons of porous and adsorption properties of carbons activated by steam and KOH. *Journal of Colloid and Interface Science*, **283**(1), 49-56.
44. Weber, W. J., and Morris, J. C. (1963) Kinetics of adsorption on carbon from solution. *Journal of the Sanitary Engineering Division*, **89**(2), 31-60.
45. Iqbal, M. Z., and Abdala, A. A. (2013) Thermally reduced graphene: synthesis, characterization and dye removal applications. *RSC Advances*, **3**(46), 24455-24464.
46. Tarawou, T., and Young, E. (2015) Intraparticle and liquid film diffusion studies on the adsorption of Cu^{2+} and Pb^{2+} ions from aqueous solution using powdered cocoa pod (Theobroma cacao). *International Research Journal of Engineering and Technology*, **2**(8), 236-243.
47. Langmuir, I. (1916) The constitution and fundamental properties of solids and liquids. Part 1. Solids. *Journal of American Chemical Society*, **38**(11), 2221-2295.
48. Freundlich, U. (1906) über die adsorption in lusungen. *Zeitschrift für Physikalische Chemie*, **57**, 385-490.
49. Temkin, M., and Pyzhev, V. (1940) Recent modifications to Langmuir isotherms. *Acta Physiochim URSS*, **12**, 217- 225.
50. Bulut, E., Ozacar, M., and Sengil, I. A. (2008) Equilibrium and kinetic data and process design for adsorption of Congo Red onto bentonite. *Journal of Hazardous Materials*, **154**(1), 613-622.
51. Bulut, E., Ozacar, M., and Sengil, I. A. (2008) Adsorption of malachite green onto bentonite: equilibrium and kinetic studies and process design. *Microporous and Mesoporous Materials*, **115**(3), 234-246.
52. Krishna, R. H., and Swamy, A. V. V. S. (2012) Physico-chemical key parameters, Langmuir and Freundlich isotherm and Lagergren rate constant Studies on the removal of divalent nickel from the aqueous solutions onto powder of calcined brick. *International Journal of Engineering Research and Development*, **4**(1), 29-38.

9

DFT based Investigation of N_2O Adsorption on Graphene and Heteroatom-doped Graphene

9.1 Introduction

Graphene, the novel two-dimensional carbon based nanomaterial, has aroused great interest in numerous application areas including electronics [1, 2], energy storage [3-6], biomedicine [7, 8], catalysis [9], etc. The physicochemical properties of graphene such as large surface area (2630 m^2/g for single layer graphene), excellent electrical conductivity, low Johnson noise, and limited crystal defects along with the high mechanical characteristics [1,10-13] make graphene one of the most promising candidates for gas sensing applications [14-16]. Experimental research studies have shown that graphene can be used as a good sensing material to detect various gas molecules by measuring the change in the conductivity induced by the gas adsorbates, which act as either electron donors or acceptors [14,17]. Previous studies that focused on pure or intrinsic graphene (hereafter abbreviated as IG) showed weak physical adsorption of gas molecules on its surface [18,19]. It has been found that the gas sensing response of graphene towards gas molecules could be improved by using appropriate dopants [20,21].

The detection of N_2O, commonly known as laughing gas, is very important since increasing atmospheric concentrations of N_2O leads to global warming and stratospheric ozone-layer destruction [22-24]. N_2O is considered as the single greatest ozone-depleting substance and is expected to remain the dominant ozone-depleting substance throughout the 21st century [25]. The global warming potential of N_2O is 298 times higher than that of CO_2, with an atmospheric lifetime of about 120 years [26]. N_2O is estimated to contribute about 6% of the global warming effect due to green-house gases [27]. Significant efforts have been made to find an effective method for detecting N_2O gas. The sensing affinity of graphene or doped

Seba S. Varghese[a,b], Sundaram Swaminathan[c], Krishna K. Singh[b] and Vikas Mittal[a]
[a]Khalifa University of Science and Technology, Abu Dhabi, UAE; [b]Birla Institute of Technology and Science, UAE; [c]DIT University, India

graphene to N_2O may provide new insights into the development of future N_2O gas sensors for environmental monitoring and control.

In this study, the adsorption of N_2O on B-, N-, Si-, P-, Ga-, Cr- and Mn-doped graphene (hereafter abbreviated as BG, NG, SiG, PG, GG, CG and MG) sheets have been investigated by performing first-principles calculations based on density functional theory (DFT). The adsorption of N_2O on IG is also studied for comparison. The energetically favorable adsorption configurations, adsorption energies and charge transfers of the graphene-N_2O adsorbed systems were calculated and analyzed to exploit the potential application of graphene or doped graphene as N_2O gas sensors.

9.2 Computational Methods

All the calculations have been carried out using DFT framework implemented in ABINIT simulation package [28]. The exchange correlation functional of generalized gradient approximation (GGA) in the Perdew-Burke-Ernzerhof (PBE) form is employed for the study [29]. The interactions of the valence electrons with atomic core are represented using norm-conserving Troullier-Martins pseudopotentials [30]. The single-layer graphene sheet is modelled using a 4 × 4 graphene supercell with a single molecule of N_2O adsorbed onto it. Different doped graphenes (abbreviated as XG, where X corresponds to the dopant atom), are modelled by substituting a carbon (C) atom with the dopant atom (B, N, Si, P, Ga, Cr or Mn). To minimize the interactions between adjacent graphene layers, a space width of 16 Å is taken in the direction normal to the graphene plane. A planewave basis set with the converged kinetic energy cutoff value of 816 eV is used. The Brillouin zone is sampled using a 5 × 5 × 1 Monkhorst-Pack k-point [31]. Structural optimization has been performed for all considered systems using the Broyden-Fletcher-Goldfard-Shanno minimization (BFGS) [32] until the residual forces on each atom were smaller than 5×10^{-4} Hartree/Bohr.

The adsorption energy, E_{ad} of N_2O on graphene is calculated as:

$$E_{ad} = E_{total} - (E_{sheet} + E_{N_2O}) \quad (1)$$

where E_{total}, E_{sheet} and E_{N_2O} represent the energies of the relaxed graphene system with the adsorbed N_2O molecule, isolated graphene sheet and isolated N_2O molecule respectively. The charge transfer analysis has been carried out using Hirshfeld method [33].

9.3 Results and Discussion

The atomic geometries of isolated N_2O molecule and IG were first optimized. The structural parameters for N_2O molecule are determined with N-N bond length of 1.13 Å, N-O bond length of 1.19 Å and the N–N–O angle measures 180°, which are in good agreement with experimental values (N-N bond length of 1.1282 Å, N-O bond length of 1.1842 Å and N–N–O angle of 180^0 respectively). For graphene, the calculated C-C bond length of 1.42 Å (Figure 9.1) is observed to be in good agreement with the experimental value [34,35].

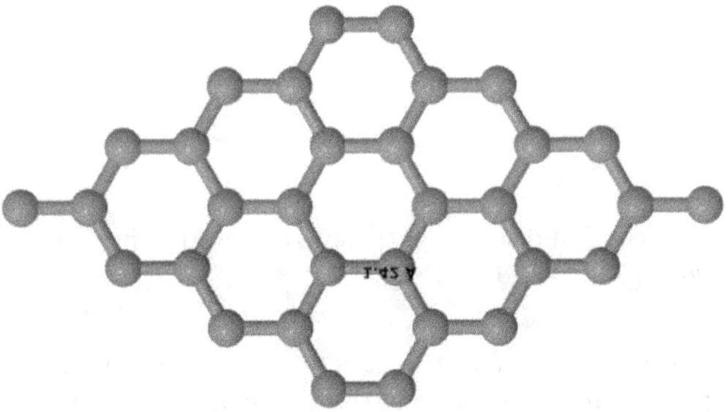

Figure 9.1 Optimized structure of IG sheet.

9.3.1 Adsorption of N_2O on IG

The interaction of N_2O with the graphene sheet was investigated in two modes based on the binding atom of N_2O, labelled as N-end and O-end. In N-end binding mode, the N atom of N_2O is close to the graphene sheet, whereas in O-end binding mode, the O atom of the gas molecule is close to the graphene sheet. The adsorption mechanism of N_2O on IG is discussed here. The nature of interactions of N_2O molecule with graphene is studied from the calculations of adsorption energies of N_2O molecule adsorbed on IG *via* N-end and O-end binding mode. After N_2O adsorption, the atomic geometry of IG remains unchanged (Figure 9.2). Figure 9.2 (a) and (b) show the optimized structures of N_2O adsorbed on IG through the N-end and O-end binding mode respectively. N_2O molecule either N-end (E_{ad} = -0.011 eV) or O-end (E_{ad} = -0.015 eV) was adsorbed on top of the carbon atom on IG (Figure 9.2), and the IG-N_2O binding distance is

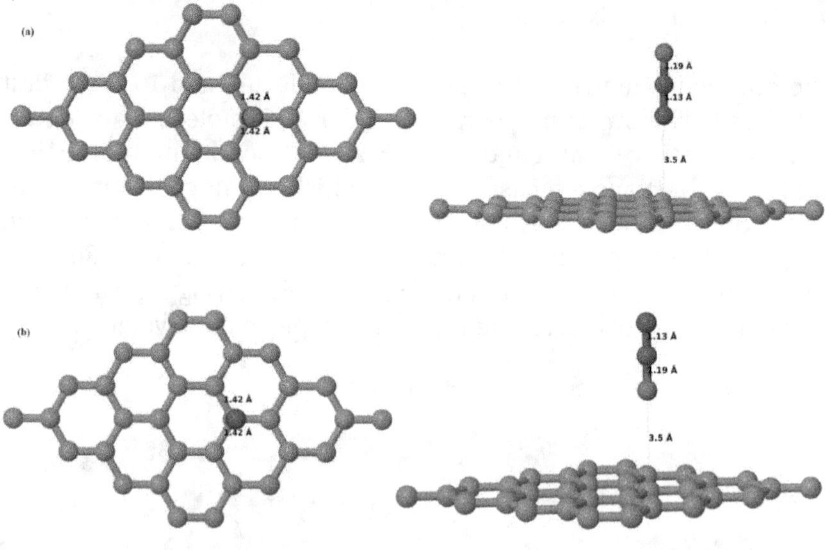

Figure 9.2 Top (left) and side (right) views of the adsorption structures of N_2O on IG *via* (a) N-end and (b) O-end.

3.5 Å in both cases, which indicates weak interaction between N_2O molecule and IG, in agreement with the results reported by Lv *et al.* [36].

9.3.2 Adsorption of N_2O on BG, NG, SiG, PG and GG

The adsorption of N_2O on graphene doped with B, N, Si, P and Ga is considered next. Initially, the doped graphene sheets are allowed to relax. After relaxation, both BG and NG retain the planar form of un-doped graphene as seen in Figure 9.3 (a) and (b). The carbon-dopant atom distance is found to be 1.49 Å (B-C bond) for BG and 1.41 Å (N-C bond) for NG. The substitution of a single carbon atom by other dopant atoms such as Si, P and Ga in SiG, PG and GG respectively results in a distorted geometry compared to IG, due to the stress introduced by the bigger sized dopant atoms compared to the host carbon atoms. Figure 9.3 (c-e) present the relaxed structures of SiG, PG and GG sheet. The dopant atom protrudes instead out of the plane, at a distance d from the plane of d = 0.989 Å for Si, d = 1.106 Å for P, d = 1.220 Å for Ga. The bond around the dopant atom expands to 1.74 Å (Si-C bond) in SiG and 1.76 Å (P-C bond) in PG and 1.83 Å (Ga-C bond) in GG from the ideal C–C bond length of 1.42 Å (Figure

9.3 (c-e)). These results are in good agreement with previous reported theoretical works [37-40].

Having determined the energetically most stable configuration of BG, NG, SiG, PG and GG sheets, the next step is to investigate the N_2O adsorption process. N_2O molecule is initially placed on the top site of the dopant atom on the above mentioned doped graphene sheets *via* two binding modes (N-end and O-end).

Figure 9.4 (a) and (b) present the relaxed structures of N_2O adsorbed on BG and NG system respectively. There is no significant change in the structure of the BG and NG after N_2O adsorption. For both N_2O adsorbed on BG and NG, the energetically favourable geometry is the one in which the O-end binding mode is tilted with the oxygen atom of N_2O facing towards the dopant atom (Figure 9.4 (a), (b)). The carbon-dopant atom distance remains unchanged in both BG and NG even after N_2O interaction. The low absorption energy (-0.027 eV for BG and -0.022 eV for NG) and long binding distance (3.72 Å for BG and 3.58 Å for NG) indicate weak forces and suggest that the interaction involved physisorption.

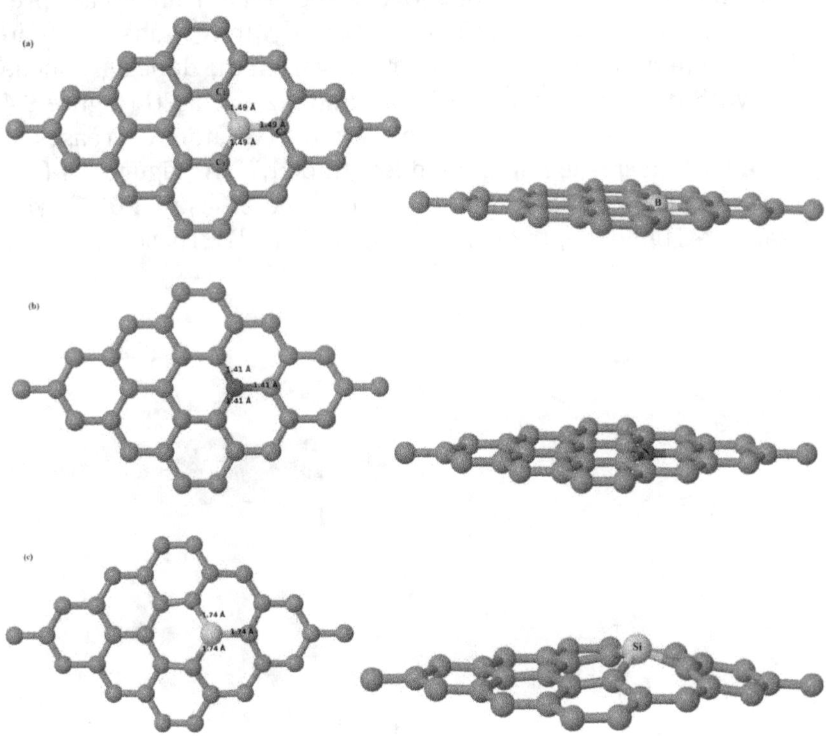

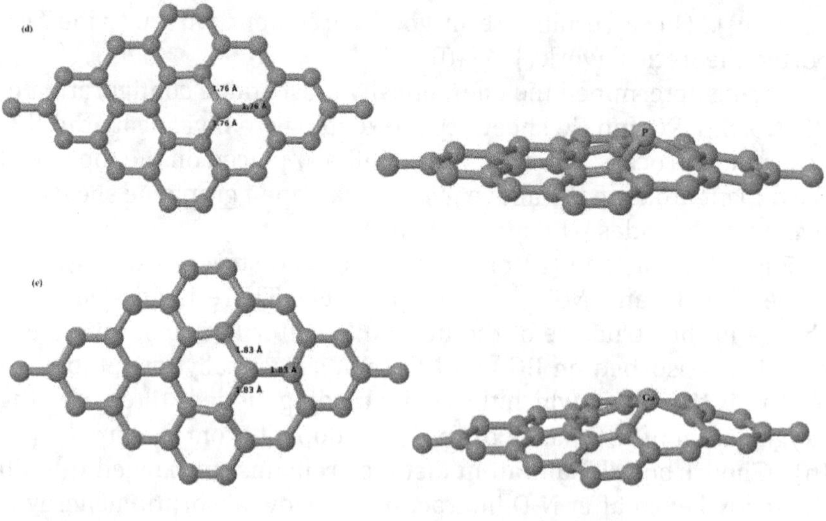

Figure 9.3 Top (left) and side (right) views of (a) BG, (b) NG, (c) SiG (d) PG, and (e) GG.

The optimized structures of N_2O adsorbed on SiG and PG are presented in Figures 9.4 (c) and (d) respectively. In N_2O adsorbed-SiG and -PG systems, N_2O molecule is far away from the doped graphene sheets with distances of 3.41 Å in SiG and 4.28 Å in PG (Figure 9.4 (c-d)). Upon N_2O adsorption, the elevation of Si atom decreases to 0.863 Å with decreased Si–C bond length of 1.73 Å (Figure 9.4 (c)). In the case of PG, the elevation of P atom decreases to 1.01 Å with the same P-C bond length of 1.76 Å (Figure 9.4 (d)). These structural

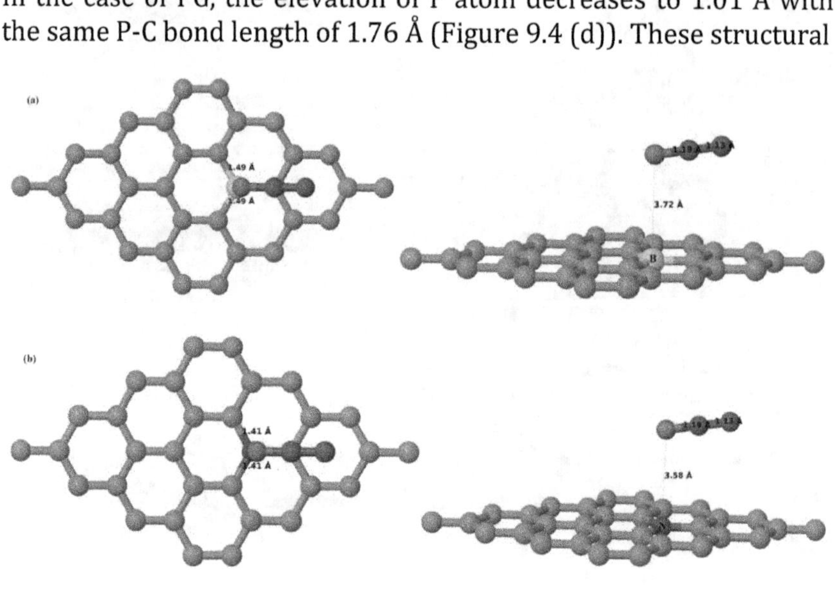

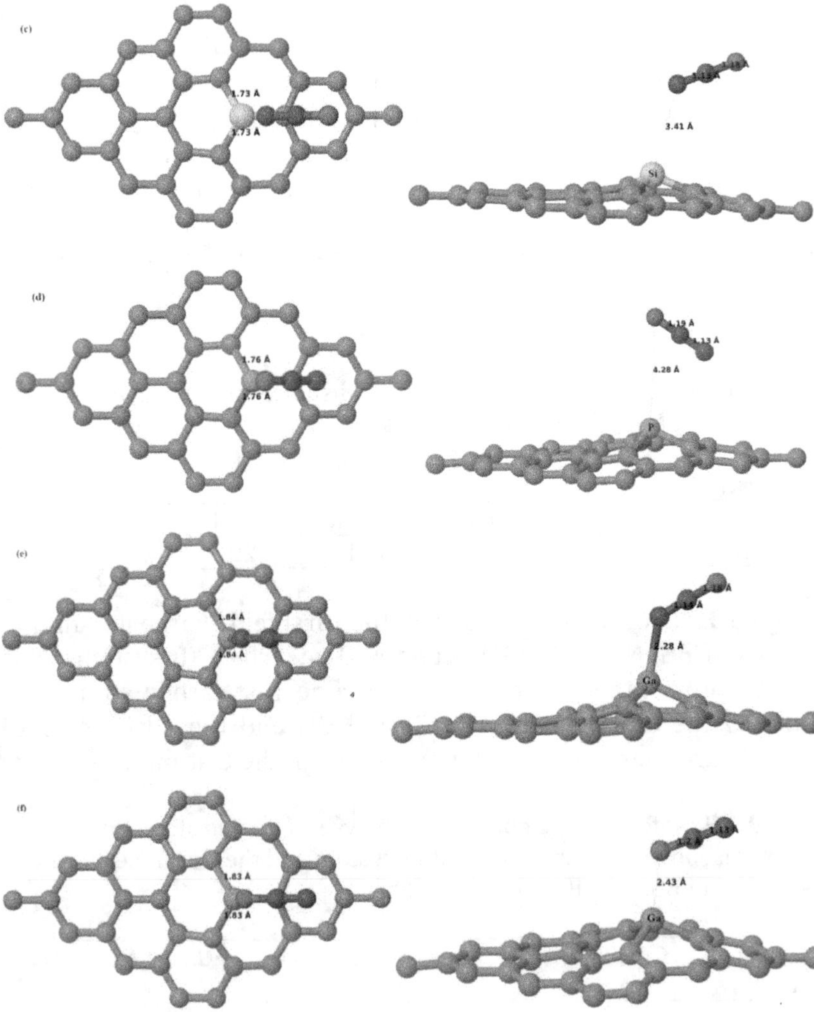

Figure 9.4 (contd.) Top (left) and side (right) views of optimized adsorption structures of N_2O molecule on (a) BG, (b) NG, (c) SiG, (d) PG, (e) GG *via* N-end and (f) GG *via* O-end.

parameters indicate the repulsive force between the N_2O molecule and the Si (or P) atom. The calculated adsorption energies for N_2O on SiG and PG are found to be almost the same (-0.030 eV), which is a clear demonstration of physisorption of N_2O on the surface of SiG and PG.

N_2O was adsorbed onto the top site of the Ga atom of GG via N-end (E_{ad} = -0.239 eV) and O-end (E_{ad} = -0.164 eV) and the corresponding dopant atom-gas molecule distance are 2.28 Å (Figure 9.4 (e)) and 2.43 Å (Figure 9.4 (f)) respectively, which shows that the interaction of Ga atom with N_2O is stronger for N-end adsorbed on GG. Figure 9.4 (e)) depicts the optimized structure of N_2O adsorbed on GG *via* the N-end with a strong chemical bond between Ga atom and the adsorbed N atom. The elevation of Ga atom and the Ga–C bond length increase to 1.245 Å and 1.84 Å on N_2O adsorption (Figure 9.4 (e)). These results are consistent with previous reported values [36]. The obtained adsorption energies are presented in Table 9.1.

Table 9.1 Values of adsorption energy and distance of N_2O above graphene surface for BG, NG, SiG, PG and GG sheets

System	BG	NG	SiG	PG	GG
E_{ad} in eV	-0.027	-0.022	-0.030	-0.030	-0.239
x in Å	3.72	3.58	3.41	4.28	2.28

Table 9.2 shows the results from the Hirshfeld charge population analysis of BG, NG, SiG, PG, GG and N_2O adsorbed-XG systems. The charge states of the dopant atoms, three nearest carbon atoms surrounding the dopant atom (C_1, C_2 and C_3) and the adsorbed N_2O molecule are presented in Table 9.2. In IG, all the C atoms are charge

Table 9.2 Hirshfeld charge population analysis of doping atom, three nearest carbon atoms surrounding doped atom and the N_2O molecule on the surface of BG, NG, SiG, PG and GG sheet[a]

System	C_1	C_2	C_3	B	N	Si	P	Ga	N	O
IG	0	0	0	-	-	-	-	-	-	-
BG	-0.061	-0.061	-0.061	0.023	-	-	-	-	-	-
NG	0.045	0.045	0.045	-	0.030	-	-	-	-	-
SiG	-0.084	-0.084	-0.084	-	-	0.205	-	-	-	-
PG	-0.046	-0.046	-0.046	-	-	-	0.172	-	-	-
GG	-	-	-	-	-	-	-	0.0	-	-

	0.028	0.028	0.028					20		
O-end N_2O-BG	-0.061	-0.061	-0.065	0.018	-	-	-	-	-	-0.0944
O-end N_2O-NG	0.044	0.044	0.041	-	0.028	-	-	-	-	-0.0999
N-end N_2O-SiG	-0.085	-0.085	-0.088	-	-	0.222	-	-	-0.086	-
O-end N_2O-PG	-0.046	-0.046	-0.046	-	-	-	0.170	-	-	-0.105
N-end N_2O-GG	-0.033	-0.033	-0.040	-	-	-	-	0.006	-0.030	-

[a]A negative sign indicates electrons gained, whereas positive sign implies electron lost by the atom (the unit of charge is electron)

neutral and have a charge state of zero. In BG, the three neighboring C atoms have a charge state of -0.061 and the B-dopant has a charge state of +0.023, since the neighboring C atoms attract electrons from the B atom. Similarly in SiG, the C neighbors attain a charge state of -0.084 and the Si-dopant lose electronic charge of +0.205. Similar behavior of the charge state of C neighbors and the dopant atom in BG, SiG, PG and GG is due to the fact that the electronegativity of C atom is higher than B, Si, P, and Ga atoms. As N atom is more electronegative than C atom, the neighboring C atoms possess a charge state of +0.045 and the N-dopant has a charge state of +0.030 in NG.

In BG, NG, SiG and PG, the charge state of the dopant atom and the C neighbors does not show any significant change after N_2O adsorption. These results indicate that there is negligible charge transfer between these doped graphene systems and N_2O. But when N_2O molecule is adsorbed on the top site of GG, the three carbon atoms neighboring the Ga atom attain electrons and the Ga atom loses valence electrons featuring 0.006 for N-end, which suggests that a chemical bond is formed between the Ga and N_2O.

9.3.3 Adsorption of N_2O on CG and MG

The adsorption of N_2O on CG and MG sheets are discussed in this section. The optimized atomic structures of CG and MG are shown in Figure 9.5. In CG and MG, the dopant atom is displaced outward forming a bump (Figure 9.5) with an elevation of 1.318 Å and 1.304 Å respectively. The carbon-dopant atom distances of 1.85 Å (Cr-C bond) and 1.82 Å (Mn-C bond), are in good agreement with the results reported in [41].

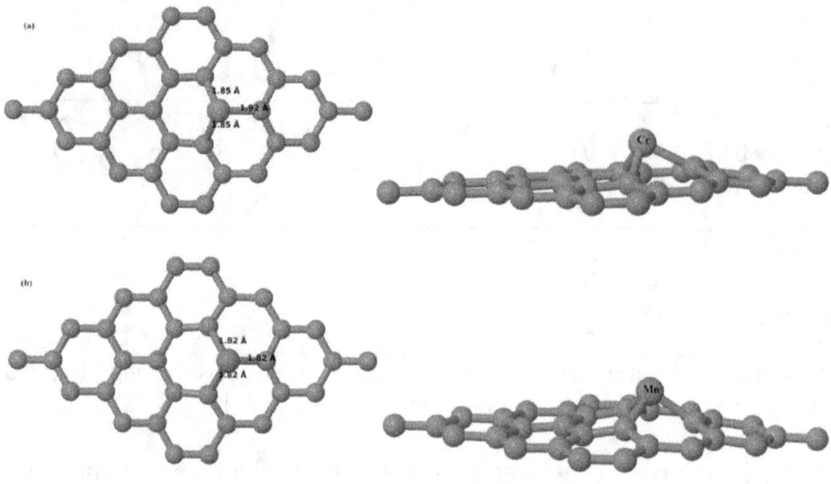

Figure 9.5 Top (left) and side (right) views of optimized structures of (a) CG and (b) MG sheets.

After N_2O adsorption, the structures are changed dramatically. The stable adsorption configurations of N_2O molecule adsorbed on CG and MG *via* N-end and O-end are presented in Figure 9.6. From the figure, it can be seen that N_2O molecule can strongly bond to CG and MG sheets by the O atom or by the two N atoms with dopant atom-gas molecule distance of around 2 Å. The calculated adsorption energies of these structures for CG and MG are listed in Table 9.3, suggesting that, for CG, the structure bonded by O and N atoms is the most stable one, whereas for MG, the structure bonded by N atoms is the most stable one, as these have the largest E_{ad}. Figure 9.6 (b) presents the most energetically stable adsorption configuration of N_2O on CG, where a partial dissociation of N_2O into N_2 and O-species on the Cr-dopant can be seen. The resulting NNCrO species remains bonded to the doped graphene sheet through the Cr dopant.

In both CG and MG, the carbon-dopant atom distance got extended after N_2O adsorption (Figure 9.6). The elevation of Cr and Mn above the graphene sheet has extended significantly to 1.550 Å and 1.396 Å, upon N_2O interaction. The above results show that N_2O is chemisorbed on both CG and MG with large adsorption energies (Table 9.3) and small binding distances.

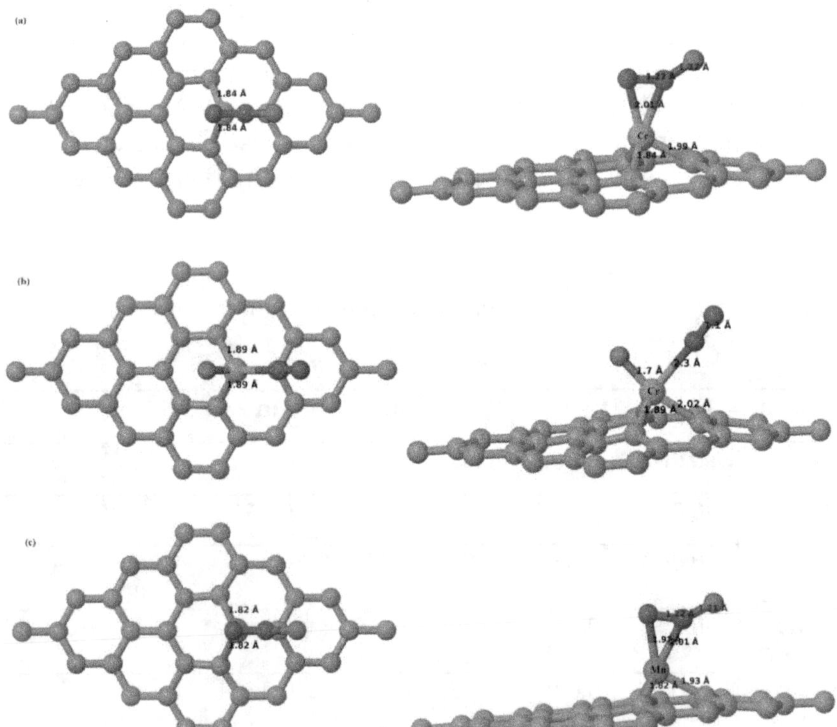

Figure 9.6 Top (left) and side (right) views of the adsorption structures of (a) N_2O on CG *via* N-end (b) N_2O on CG *via* O-end (c) N_2O on MG *via* N-end, and (b) N_2O on MG *via* O-end.

Table 9.3 Adsorption energies released during N_2O chemisorption on CG and MG sheets

System	CG *via* N-end	CG *via* O-end	MG *via* N-end	MG *via* O-end
E_{ad} in eV	-1.168	-2.406	-1.088	-0.478

Table 9.4 shows the results of Hirshfeld charge distribution analysis performed on CG, MG and N_2O adsorbed-CG and MG systems. As the C atoms around the dopant atom attract electrons from the dopant atom, the C neighbors attain a charge state of -0.067 and -0.037 in CG and MG respectively. The dopant atoms lose electronic charge and attain a charge state of +0.327 and +0.151in CG and MG respectively.

Table 9.4 Hirshfeld charge distribution analysis of the dopant atom, three C atoms around dopant atom and the N_2O molecule on the surface of CG and MG sheet

System	C_1	C_2	C_3	Cr	Mn	N	O
CG	-0.067	-0.067	-0.076	0.327	-	-	-
MG	-0.037	-0.037	-0.037	-	0.151	-	-
N-end N_2O-CG	-0.062	-0.062	-0.055	0.277	-	-0.160	-
O-end N_2O-CG	-0.052	-0.052	-0.100	0.303	-	-	-0.293
N-end N_2O-MG	-0.025	-0.025	-0.022	-	0.064	-0.116	-
O-end N_2O-MG	-0.035	-0.035	-0.040	-	0.084	-	-0.049

Upon N_2O adsorption on both CG and MG, the charge state of Cr and Mn reduces for both binding ends of the N_2O molecule. For N-end adsorbing on CG and MG, the carbon neighbors loses electrons. For the scheme of N_2O adsorbed on CG and MG through the O-end, both C_1 and C_2 neighbors around the Cr and Mn-dopant loses electronic charge, whereas the third C neighbor attain more electrons. The adsorption of N_2O on CG and MG result in significant change in the charge state of the dopant atoms, which could be attributed to the charge transfer between the Cr- or Mn-dopant and the N_2O molecule. The charge of O atom for O-end adsorbing on the CG is larger than the charge of N for N-end binding mode, whereas in the case of N_2O adsorbed on MG, the charge of N atom for N-end is larger. The binding atoms of N_2O get charge state of -0.160 for N-end and -0.293 for O-end on CG, and -0.116 for N-end and -0.049 for O-end on MG respectively, which implies that a chemical bond exists in the region between N_2O and the Cr- and Mn-dopants.

9.4 Summary

We have carried out first-principles studies to investigate the interaction of an isolated N_2O molecule on intrinsic and heteroatom-doped (B, N, Si, P, Ga, Cr and Mn) graphene. The stable adsorption geometries and adsorption energies are obtained on the basis of density functional theory calculations. Our calculations indicate that intrinsic graphene, B-, N-, Si- and P-doped graphene are not sensitive for N_2O molecule, due to weak adsorption. In contrast, Ga-doped graphene was found to be sensitive towards N_2O molecule, due to the strong interaction of N_2O with Ga-dopant. Cr- and Mn-doped graphene show chemisorption of N_2O on their surface with comparatively high adsorption energies and short binding distances. The charge population analysis shows that the electronic properties of Ga, Cr- and Mn- doped graphene are sensitive to the adsorption of a single N_2O molecule. The results also indicate that Cr- and Mn-doped graphene can be more suitable for the detection of N_2O gas.

References

1. Geim, A. K. and Novoselov, K. S. (2007) The rise of graphene. *Nature Materials,* **6**, 183-191.
2. Jang, H., Park, Y. J., Chen, X., Das, T., Kim, M.-S., and Ahn, J.-H. (2016) Graphene-based flexible and stretchable electronics. *Advanced Materials,* **28**, 4184-4202.
3. Tozzini, V., and Pellegrini, V. (2013) Prospects for hydrogen storage in graphene. *Physical Chemistry Chemical Physics,* **15**, 80-89.
4. Gadipelli, S., and Guo, Z. X. (2015) Graphene-based materials: Synthesis and gas sorption, storage and separation. *Progress in Materials Science,* **69**, 1-60.
5. Pumera, M. (2011) Graphene-based nanomaterials for energy storage. *Energy & Environmental Science,* **4**, 668-674.
6. Brownson, D. A. C., Kampouris, D. K., and Banks, C. E. (2011) An ov-

erview of graphene in energy production and storage applications. *Journal of Power Sources,* **196**, 4873-4885.

7. Goenka, S., Sant, V., and Sant, S. (2014) Graphene-based nanomaterials for drug delivery and tissue engineering. *Journal of Controlled Release,* **173**, 75-88.
8. Feng, L., and Liu, Z. (2011) Graphene in biomedicine: opportunities and challenges. *Nanomedicine,* **6**, 317-324.
9. Machado, B. F., and Serp, P. (2012) Graphene-based materials for catalysis. *Catalysis Science & Technology,* **2**, 54-75.
10. Zhang, Y., Tan, Y.-W., Stormer, H. L., and Kim, P. (2005) Experimental observation of the quantum Hall effect and Berry's phase in graphene. *Nature,* **438**, 201-204.
11. Novoselov, K. S., Geim, A. K., Morozov, S. V., Jiang, D., Katsnelson, M. I., Grigorieva, I. V., Dubonos, S. V., and Firsov, A. A. (2005) Two-dimensional gas of massless Dirac fermions in graphene. *Nature,* **438**, 197-200.
12. Danneau, R., Wu, F., Craciun, M. F., Russo, S., Tomi, M. Y., Salmilehto, J., Morpurgo, A. F., and Hakonen, P. J. (2008) Shot noise in ballistic graphene. *Physical Review Letters,* **100**, 196802.
13. Lee, C., Wei, X., Kysar, J. W., and Hone, J. (2008) Measurement of the elastic properties and intrinsic strength of monolayer graphene. *Science,* **321**, 385-388.
14. Schedin, F., Geim, A. K., Morozov, S. V., Hill, E. W., Blake, P., Katsnelson, M. I., and Novoselov, K. S. (2007) Detection of individual gas molecules adsorbed on graphene. *Nature Materials,* **6**, 652-655.
15. Yavari, F., and Koratkar, N. (2012) Graphene-based chemical sensors. *The Journal of Physical Chemistry Letters,* **3**, 1746-1753.
16. Wang, T., Huang, D., Yang, Z., Xu, S., He, G., Li, X., Hu, N., Yin, G., He, D., and Zhang, L. (2016) A review on graphene-based gas/vapor sensors with unique properties and potential applications. *Nano-Micro Letters,* **8**, 95-119.
17. Ko, G., Kim, H. Y., Ahn, J., Park, Y. M., Lee, K. Y., and Kim, J. (2010) Graphene-based nitrogen dioxide gas sensors. *Current Applied Physics,* **10**, 1002-1004.
18. Leenaerts, O., Partoens, B., and Peeters, F. M. (2008) Adsorption of H_2O, NH_3, CO, NO_2, and NO on graphene: A first-principles study. *Physical Review B,* **77**, 125416.
19. Giannozzi, P., Car, R., and Scoles, G. (2003) Oxygen adsorption on graphite and nanotubes. *The Journal of Chemical Physics,* **118**, 1003-1006.
20. Yong-Hui, Z., Ya-Bin, C., Kai-Ge, Z., Cai-Hong, L., Jing, Z., Hao-Li, Z., and Yong, P. (2009) Improving gas sensing properties of graphene by introducing dopants and defects: a first-principles study. *Nanotechnology,* **20**, 185504.
21. Shao, L., Chen, G., Ye, H., Wu, Y., Qiao, Z., Zhu, Y., and Niu, H. (2013)

Sulfur dioxide adsorbed on graphene and heteroatom-doped graphene: a first-principles study. *The European Physical Journal B,* **86**, 1-5.
22. Wuebbles, D. J. (2009) Nitrous oxide: No laughing matter. *Science,* **326**, 56-57.
23. Kramlich, J. C., and Linak, W. P. (1994) Nitrous oxide behavior in the atmosphere, and in combustion and industrial systems. *Progress in Energy and Combustion Science,* **20**, 149-202.
24. Forster, P., Ramaswamy, V., Artaxo, P., Berntsen, T., Betts, R., Fahey, D. W., Haywood, J., Lean, J., Lowe, D. C., Myhre, G., Nganga, J., Prinn, R., Raga, G., Schultz, M., and Van Dorland, R. (2007) *Changes in Atmospheric Constituents and In Radiative Forcing,* Cambridge University Press, United Kingdom.
25. Ravishankara, A. R., Daniel, J. S., and Portmann, R. W. (2009) Nitrous oxide (N_2O): The dominant ozone-depleting substance emitted in the 21st century. *Science,* **326**, 123-125.
26. Perez-Ramírez, J. (2007) Prospects of N_2O emission regulations in the European fertilizer industry. *Applied Catalysis B: Environmental,* **70**, 31-35.
27. Rapson, T. D., and Dacres, H. (2014) Analytical techniques for measuring nitrous oxide. *TrAC Trends in Analytical Chemistry,* **54**, 65-74.
28. Gonze, X., Amadon, B., Anglade, P. M., Beuken, J. M., Bottin, F., Boulanger, P., Bruneval, F., Caliste, D., Caracas, R., Cote, M., Deutsch, T., Genovese, L., Ghosez, P., Giantomassi, M., Goedecker, S., Hamann, D. R., Hermet, P., Jollet, F., Jomard, G., Leroux, S., Mancini, M., Mazevet, S., Oliveira, M. J. T., Onida, G., Pouillon, Y., Rangel, T., Rignanese, G. M., Sangalli, D., Shaltaf, R., Torrent, M., Verstraete, M. J., Zerah, G., and Zwanziger, J. W. (2009) ABINIT: First-principles approach to material and nanosystem properties. *Computer Physics Communications,* **180**, 2582-2615.
29. Perdew, J. P., Burke, K., and Ernzerhof, M. (1996) Generalized gradient approximation made simple. *Physical Review Letters,* **77**, 3865-3868.
30. Troullier, N., and Martins, J. L. (1991) Efficient pseudopotentials for plane-wave calculations. *Physical Review B,* **43**, 1993-2006.
31. Monkhorst, H. J., and Pack, J. D. (1976) Special points for Brillouin-zone integrations. *Physical Review B,* **13**, 5188-5192.
32. Schlegel, H. B. (1982) Optimization of equilibrium geometries and transition structures. *Journal of Computational Chemistry,* **3**, 214-218.
33. Hirshfeld, F. L. (1977) Bonded-atom fragments for describing molecular charge densities. *Theoretica Chimica Acta,* **44**, 129-138.
34. Dresselhaus, M. S., Dresselhaus, G., Saito, R., and Jorio, A. (2005) Raman spectroscopy of carbon nanotubes. *Physics Reports,* **409**,

47-99.

35. Castro Neto, A. H., Guinea, F., Peres, N. M. R., Novoselov, K. S., and Geim, A. K. (2009) The electronic properties of graphene. *Reviews of Modern Physics,* **81**, 109-162.
36. Lv, Y.-a., Zhuang, G.-l., Wang, J.-g., Jia, Y.-b., and Xie, Q. (2011) Enhanced role of Al or Ga-doped graphene on the adsorption and dissociation of N_2O under electric field. *Physical Chemistry Chemical Physics,* **13**, 12472-12477.
37. Dai, J., Yuan, J., and Giannozzi, P. (2009) Gas adsorption on graphene doped with B, N, Al, and S: A theoretical study. *Applied Physics Letters,* **95**, 232105.
38. Zou, Y., Li, F., Zhu, Z. H., Zhao, M. W., Xu, X. G., and Su, X. Y. (2011) An ab initio study on gas sensing properties of graphene and Si-doped graphene. *The European Physical Journal B,* **81**, 475-479.
39. Jiayu, D., and Jianmin, Y. (2010) Modulating the electronic and magnetic structures of P-doped graphene by molecule doping. *Journal of Physics: Condensed Matter,* **22**, 225501.
40. Sharma, S., and Verma, A. S. (2013) A theoretical study of H_2S adsorption on graphene doped with B, Al and Ga. *Physica B: Condensed Matter,* **427**, 12-16.
41. Dai, J., and Yuan, J. (2010) Adsorption of molecular oxygen on doped graphene: Atomic, electronic, and magnetic properties. *Physical Review B,* **81**, 165414.

10

Theoretical Insights into Gas Molecular Adsorption on Graphene: Implications for Gas Sensing Applications

10.1 Introduction

Graphene, a two-dimensional hexagonal sheet of carbon atoms, has grabbed immense research and industrial interest since its first isolation in 2004 [1] owing to its potentially tuneable and exotic structural, electronic, mechanical and optical properties [2-11]. Graphene has emerged as an attractive candidate for applications such as energy storage, electronics, field emission, photovoltaics, catalysis, biomedicine etc. [12-20]. It has a very large surface area of 2630 m^2/g (300 times higher than graphite and two times higher than single-walled carbon nanotubes), which provides large active area for sensing applications [21-27].

The unique two-dimensional nature which maximizes the interactions of gas molecules with graphene [28], metallic conductivity, low Johnson noise [2,3,8,28,29] and the limited crystal defects [2,3,8,28] contribute to the excellent gas sensing behavior of graphene. A little change of carrier concentration of graphene can cause a notable variation of electrical conductivity [28]. Apart from the above features, graphene monocrystals allow the fabrication of four-probe devices that can avoid the influence of the contact resistance and can extend the limit of sensitivity [30].

The linear energy-momentum dispersion near the Dirac point and the zero- energy band gap in the electronic spectrum of pure graphene [8] make graphene more accessible to doping with either electrons or holes by the gas molecules adsorbed on the surface of graphene [28,31]. Thus, the electronic properties of graphene are strongly influenced by the adsorption of the gas molecules in the surrounding environment, which makes this material attractive for gas sensing applications. Graphene can detect minute concentrations of gases present in the environment by measuring the change

Seba S. Varghese[a,b], Sundaram Swaminathan[c], Krishna K. Singh[b] and Vikas Mittal[a]
[a]Khalifa University of Science and Technology, Abu Dhabi, UAE; [b]Birla Institute of Technology and Science, Dubai, UAE; [c]DIT University, India

of graphene's electrical conductivity, caused by the change in the carrier concentration of graphene induced by the adsorbed gas molecules. The possibility of using graphene as a highly sensitive gas sensor has been experimentally demonstrated by Schedin *et al.* [28] with detection limit of the order of 1 parts per billion (ppb) for NO_2, H_2O, NH_3 and CO. The ultimate sensitivity of an individual molecule of NO_2 was also reported.

The excellent gas sensing properties of graphene have inspired theoreticians to investigate the mechanisms involved in the adsorption of gas molecules on graphene's surface. Numerous theoretical researches based on first-principles have been performed to analyze the interactions between gas molecules and graphene, so as to exploit the potential of graphene as gas sensors. In this chapter, the state of the art research on first-principles theoretical studies on gas molecular adsorption on graphene and chemically functionalized graphene has been reviewed. In addition, the effect of heteroatom dopants and defects on the adsorption potential of graphene in the light of reported simulation based studies has been discussed. Before discussing the latest computer simulation based works on adsorption properties of doped and defected graphene for various gas molecules, research studies on gas molecular adsorption on intrinsic graphene (IG), graphene nanoribbon (GNR) and graphene oxide (GO) are also described. Finally, the existing challenges and future perspectives of gas molecular adsorption on chemically functionalized graphene are also provided.

10.2 Theoretical Studies on Graphene-gas Molecular Interaction

Understanding the interactions between gas molecules and graphene are of great significance in developing graphene based gas sensors. Quantum mechanical simulations on the gas molecular adsorption of graphene initially decide on the most favorable adsorption configuration of the gas molecule on the surface of graphene and calculate the adsorption energy to understand the nature of interactions. The adsorption mechanisms are also discussed from charge transfers and density of states (DOS). The changes in the electronic structures of graphene caused by the adsorption of gas molecules are determined from charge transfer calculations. The effect of gas molecules caused by the physi- or chemi-sorption of gas molecules on the electronic conductivity of graphene are verified from the DOS of gas molecule-graphene adsorption systems. The properties such as adsorption energies of gas molecules on graphene surfaces and the charge transfers between gas molecules and graphene are studied from the simulated gas molecule-graphene adsorption systems. The results from the theoretical calculations are useful for application of graphene as gas sensors. The results can direct in designing effective gas sensors with optimized sensing performances.

10.2.1 Gas Molecules on Intrinsic Graphene

There have been many reports on the adsorption of gas molecules on IG. Wehling *et al.* [31], for the first time, presented a combined experimental and theoretical work on doping of graphene by gas adsorbates and its relevance to gas sensor applications. The relationship between the adsorbate-induced doping strength and the shell configuration of the adsorbate was demonstrated by considering the open- and closed-shell configurations of NO_2. The authors observed that single, open shell NO_2 molecule acts as a strong acceptor that results in strong doping of graphene, whereas its closed shell dimer N_2O_4 acts as a weak dopant. The effect of NO_2 and N_2O_4 on the electronic structure of graphene was well understood from the DOS shown in Figure 10.1, which reveals a strong acceptor level at 0.4 eV below the Dirac point in the NO_2 adsorption geometries (Figure 10.1 a,b). The authors pointed out that as the graphene's DOS spectrum has zero gap, a small chemical potential mismatch between the adsorbate and graphene is sufficient to induce a donor or acceptor level in the spectrum, as compared to conventional semiconductors. Hence, gas sensors made from graphene are expected to be more sensitive than those made from other semiconductors [31].

Later, Dan *et al.* [32] measured the intrinsic chemical responses of graphene devices by removing the resist residue from the lithography step and observed sharp decrease in sensing responses with the clean graphene samples (Figure 10.2). The electrical response of graphene devices was found to be small even upon exposure to strong analytes such as ammonia vapor. I-V characteristics and the device current showed minimal change after exposure to ammonia concentrations as high as 1000 ppm. The authors observed that the contamination layer from the conventional nanolithography process chemically dopes graphene and acts as an adsorbent layer which enhances the interactions between the gas molecules and graphene surface, thus, leading to the observed ultrahigh sensitivity of IG. This work confirmed the potential of graphene for vapor sensing applications due to the demonstrated desirable properties of graphene vapor sensors such as fast response and recovery (tens of seconds) and reversibility without heating or other refreshing methods. The authors also suggested the possibility of tuning the chemical and electronic properties of graphene through doping with impurities which could result in enhanced reactivity of graphene towards gas molecules [32].

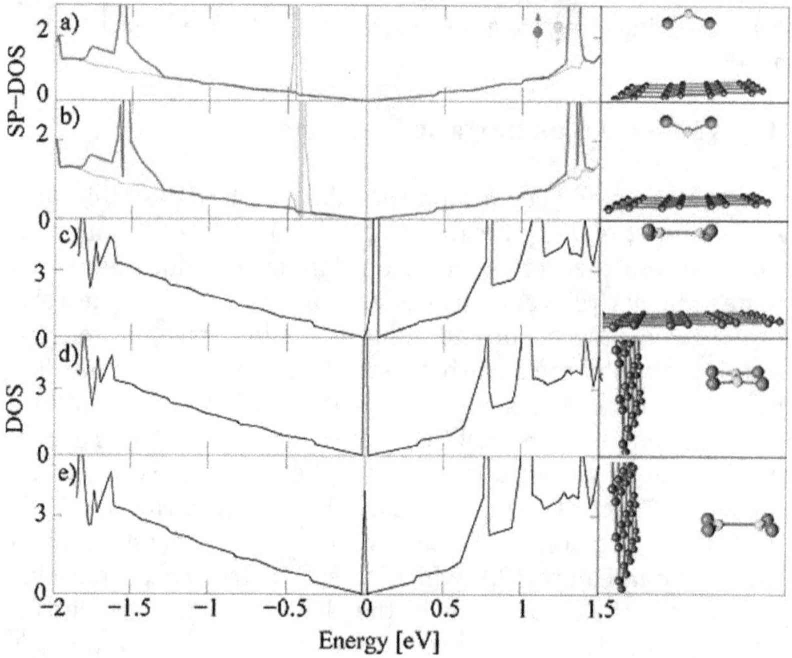

Figure 10.1 Left: Spin-polarized DOS of the graphene supercells with adsorbed NO_2 (a, b) and DOS of graphene with N_2O_4 (c-e), in various adsorption geometries. The energy of the Dirac points is defined as $E_D = 0$. In the case of NO_2, the Fermi level E_f of the supercell is below the Dirac point, directly at the energy of the spin down partially occupied molecular orbital, whereas for N_2O_4 E_f is directly at the Dirac points. Right: Adsorption geometries obtained with generalized gradient approximation (GGA). The carbon atoms are shown in blue, nitrogen is shown in green, and oxygen is shown in red. Reproduced from Reference 31 with permission from American Chemical Society.

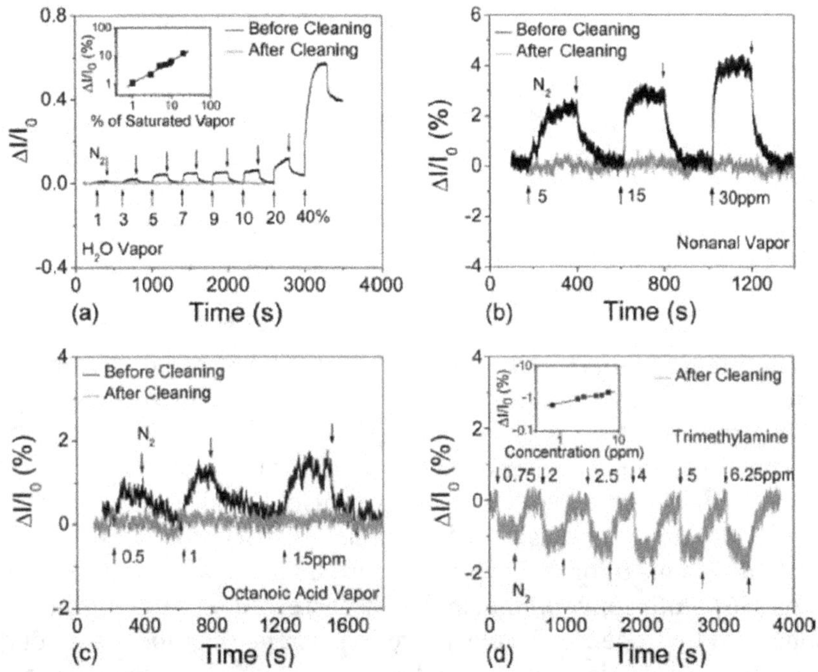

Figure 10.2 Measured sensor responses, before (black) and after (red) sample cleaning, to vapors of (a) water, (b) nonanal, (c) octanoic acid, and (d) trimethylamine. Reproduced from Reference 32 with permission from American Chemical Society.

The adsorption of H_2O, NH_3, CO, NO_2 and NO on IG studied from first-principles based on density functional theory (DFT) calculations have shown physisorption of these gas molecules on IG surface [33]. The signs of the calculated charge transfer between the gas adsorbates and IG were found to be in good agreement with the experimental results reported by Schedin *et al.* [28]. The authors compared the doping strength induced by the paramagnetic gas adsorbates, NO_2 and NO from which they observed that NO_2 induces strong doping, whereas NO induces only weak doping. The result for NO was found to be in contrary to the claim by Wehling *et al.* [33] that paramagnetic molecules induce strong doping.

The adsorption of NH_3 molecule onto IG has been investigated using molecular mechanics and DFT based calculations [34]. The most stable site for the adsorption of NH_3 on graphene surface was observed to be the top of the center of the carbon hexagon (H) from both DFT and molecular mechanics based studies. The calculated adsorption energy of ammonia molecule on the hydrogen terminated graphene flake was 0.01 eV, which indicates weak physisorption. The DOS plots of graphene flake with and without NH_3 molecule revealed that the interactions between NH_3 and graphene flake does not change the electronic structure of graphene flake significantly [34].

Lee and Kim [35] studied the chemisorption and physisorption of CO_2 on graphene based on both DFT and second-order Møller-Plesset (MP2) methods, since MP2 is more accurate than DFT for geometric relaxation of physisorbed molecules. The determined adsorption energies for both types of adsorption showed reasonably good agreement with experimental findings and previously reported theoretical works.

The adsorption mechanism of hydrogen onto various graphene flakes was investigated using first-principles calculations where non-local van der Waals (vdW) density functional (B3LYP-D3) method was used for structural relaxation and total energy estimations [36]. Both conventional and vdW corrected DFT based calculations for a hydrogen molecule above a coronene (a graphene model) surface were carried out to analyze the effect of using these approaches on the hydrogen molecule-graphene adsorbed system. The obtained adsorption energy (5.013 kJ mol^{-1}) for physisorbed hydrogen on coronene surface using B3LYP-D3 method was found to be in agreement with the experimental findings. It was found that non-local dispersion interactions are important in getting accurate results for hydrogen adsorption on graphene surface [36].

Theoretical works on gas molecular adsorption on IG showed very low adsorption energies and negligible charge transfers, as it has highly stable carbon atoms in the honeycomb lattice and has no dangling bonds on its surface. Also, the adsorption of most of the gaseous species on IG does not change its electronic properties, as the interactions of gas molecules on IG are very weak [37,38]. Due to the weak binding of gas molecules with IG, significant change in the electrical conductivity of the system does not occur. Thus, IG even being a 2D structure is not an ideal material as gas sensor.

10.2.2 Gas Molecules on Graphene Nanoribbons

Theoreticians have also explored the feasibility of employing GNR having free reactive edges, compared to graphene for the detection of NH_3 [38,39], NO_2 [38,39], CO [38-40], O_2 [38,39], N_2 [38], CO_2 [38-

41] and NO [38]. DFT investigation of the adsorption of various gas molecules around the sites of dangling bond defects on semiconducting GNRs with armchair-shaped edges (AGNRs) indicated energetically favorable adsorption of CO, NO, NO_2, O_2, CO_2, and NH_3 on AGNR, except that the N_2 adsorption is endothermic in nature [38]. Figure 10.3 also demonstrates the optimized structure of AGNRs with adsorbed CO, NO, NO_2, O_2, CO_2, and NH_3 [38]. The adsorption energies of CO, NO, NO_2, and O_2 on AGNR were larger than 1 eV, which indicates strong chemisorption. For CO_2 and NH_3 adsorption,

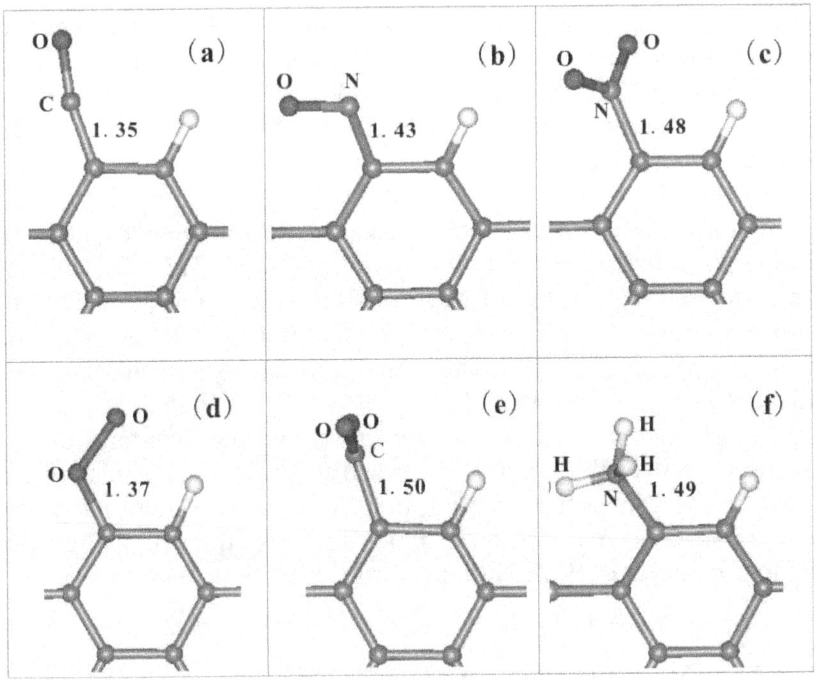

Figure 10.3 Optimized structure of AGNRs with gas molecule adsorption: (a) CO, (b) NO, (c) NO_2, (d) O_2, (e) CO_2, and (f) NH_3. Reproduced from Reference 38 with permission from American Chemical Society.

the adsorption energies were -0.31 eV and -0.18 eV respectively which indicate that the adsorptions lie between strong physisorption and weak chemisorption. The authors also exhibited that all gas molecules influence the electronic structure of AGNRs. Quantum transport calculations indicated that semiconducting AGNRs can be employed for selective detection of NH_3 out of the other gas molecules due to the modification of the conductance of AGNR by NH_3

adsorption, while all others have little effect on conductance. The charge transfer analysis on the adsorbed configurations showed that CO, NO, NO_2, O_2, and CO_2 act as electron acceptors whereas NH_3 act as electron donor. The adsorption energies and the charge transfers of gas molecules on GNR edges are found to be larger than that observed on the surface of graphene or GNR [33,37]. These results led to the conclusion that the interactions of the gas molecules with GNR edges are much stronger than that with graphene or GNR surface [38].

Paulla and Farajian [40] used MP2 calculations and DFT to examine the detection capability of AGNR having ~1 nm width for CO and CO2 gas molecules. These gas molecules undergo physisorption on AGNR with binding energies of -0.35 eV for CO2 and -0.252 eV for CO with low charge transfer values ranging from -0.005 e- to +0.005 e- respectively. Quantum conductance calculations on AGNR based nanosensor showed shift in the conductance characteristics on the order of few meV, compared to pristine AGNR for the adsorption of one and two molecules per two unit cells of AGNR. The effect of concentration of O_2 molecule adsorption on the electronic properties of semiconducting AGNRs with hydrogenated edges (HEAGNR) has been investigated via DFT. The adsorption energy calculations for different positions of gas molecule on HEAGNRs indicated that the carbon atoms at the edges are the most favorable centers for adsorption. The results showed decrease in the energy gap with increase in the concentration of O_2 molecule and the energy gap disappears at gas concentration of 0.02. This dependence of the electronic properties of HEAGNR on gas concentration could be used for sensing O_2 gas [42].

10.2.3 Gas Molecules on Graphene Oxide

Compared to graphene, GO serves as a suitable candidate for high performance gas molecular sensing under practical situations [43]. The enhanced interactions between the gas adsorbates and the reactive oxygen rich functional groups on GO improves the sensitivity of GO towards gas molecules. Several studies have focused on using GO as gas molecular sensors [43,44]. Peng *et al.* [43] investigated the adsorption of NH_3 on IG and GO using first-principles calculations. The results have shown that doping of graphene with -O and -OH species could increase the adsorption energy and the charge transfer of the NH_3 adsorbate. The calculated adsorption energy and the charge transfer for NH_3 on IG's surface were small, whereas for GO, the surface epoxy and hydroxyl groups promote the interactions between NH_3 and GO. The enhancement of the adsorption energy for the hydroxyl groups was found to be greater than that for the epoxy groups on the GO surface. Adsorption of NH_3 on epoxy or hydroxyl groups that exist on the basal plane of graphene does not show an appreciable increase in the adsorption energy [43].

Another DFT study on the interactions of nitrogen oxides NO_x (x = 1, 2, 3) and N_2O_4 with IG and GO also showed stronger adsorption of nitrogen oxides on GO than that on IG due to the presence of active defect sites such as hydroxyl, carbonyl functional groups and the carbon atom near these groups [44]. The sufficient active sites on GO increase the binding energies and enhance the charge transfer from nitrogen oxides to GO, which confirms the chemisorption of these gas molecules on the surface of GO. The DOS graphs of gas molecule-adsorbed GO showed strong hybridization of frontier orbitals of NO_2 and NO_3 with the electronic states around the Fermi level of GO, which resulted in strong acceptor doping of GO by these gas molecules [44].

Mattson *et al.* [45] reported *in-situ* infrared (IR) microspectroscopy investigation of NH_3 adsorption on reduced graphene oxide (RGO) to study the dynamics of adsorption under various realistic working conditions (i.e., ambient pressure) along with DFT results. The authors observed that the residual oxygen functional groups and structural defects in the RGO lattice act as highly active centers for NH_3 adsorption. The results suggested that the interactions of the epoxide groups and carbon vacancies of RGO with NH_3 will lead to the molecularly physisorbed NH_3 and also a variety of chemisorbed species arising from the dissociation of NH_3. The calculations have shown that the chemisorbed fragments such as NH_2, OH, and CH can produce small electron donor effect (Figure 10.4) [45].

10.2.4 Gas Molecules on Doped and Defected Graphene

Heteroatom doping has proved to be an efficient method for tailoring the physical, chemical and electronic properties of graphene [46]. Heteroatom doping of graphene can significantly alter the electronic structure of graphene [47] and make it appropriate to use as a highly sensitive gas sensor. As heteroatom doping allows tuning of the chemical reactivity of graphene, this approach has been verified for improving the sensing efficiency using quantum mechanical simulations. For instance, Zhang *et al.* [37] investigated the interactions of intrinsic, boron-, nitrogen-doped and defective graphenes (hereafter abbreviated as IG, BG, NG and DG respectively) with small gas molecules such as CO, NO, NO_2 and NH_3 using density functional computations. The authors observed enhancement in the interactions between gas molecules and doped or defective graphenes due

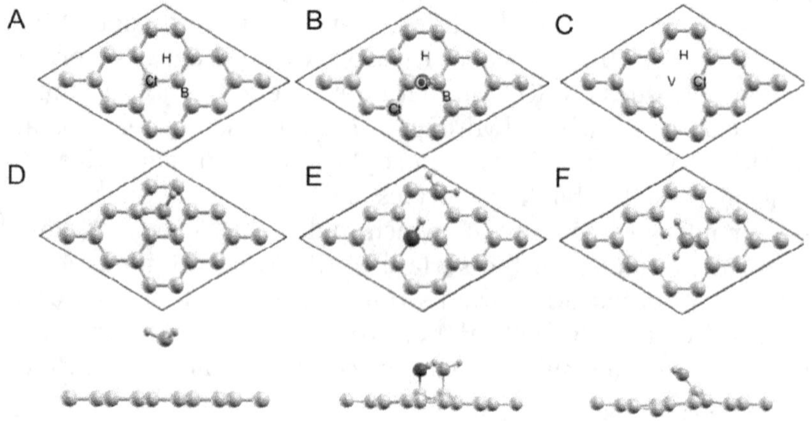

Figure 10.4 Structural models of RGO and NH_3 adsorption on graphene and RGO. (A-C) 3 × 3 supercells used to model bare RGO containing (A) pristine graphene, (B) graphene with a single epoxide group, and (C) graphene with a carbon vacancy. (D) NH_3 molecules adsorb molecularly on pristine graphene, (E) NH_3 molecules that adsorb near epoxide groups dissociate, resulting in NH_2 and OH groups, and (F) NH_3 molecules adsorbed at defects are dissociated into NH_2 and H bonded to next-nearest-neighbor carbon atoms. Reproduced from Reference 45 with permission from American Chemical Society.

to the much stronger adsorption of gas molecules on graphene modified with dopants and defects than that on IG. NG had weak interactions with CO, NO and NH_3, but had strong binding with NO_2. B-doping of graphene improved its adsorption properties for NO, NO_2 and NH_3, out of which NH_3 had tightest binding with BG. The adsorption energies, charge transfers and the electronic total charge density plots for systems of CO, NO and NO_2 adsorbed on DG showed that vacancy on DG acts as strong chemisorption binding site for CO, NO and NO_2 molecules. Therefore, the authors concluded that DG can be used as a suitable sensing platform for the detection of CO, NO and NO_2 as compared to other graphenes such as IG, BG and NG. The calculations also showed dramatic changes to graphene's electronic properties due to the strong interactions between adsorbed molecules and modified graphenes. It was concluded from the DOS plots of gas molecule-graphene adsorption systems that DG, BG and NG are appropriate for sensing CO, NO and NO_2, respectively. The current-voltage curve of gas sensor based on BG showed considerable increase of current on NO_2 adsorption and sensitivity of nearly two orders of magnitude higher than that of IG was observed at an optimum bias voltage of 1.0 V for NO_2. BG was found to be more sensitive to NO_2 than NH_3 as the observed sensitivity to NH_3 was only one order of magnitude higher than that of IG, when the bias voltage is greater than 1.0 V. The work revealed that the sensitivity of graphene based chemical gas sensors could be drastically improved by introducing appropriate dopant or defect [37].

Recently, Rad *et al.* [48] used DFT calculations to investigate the adsorption of SO_x ($x = 2, 3$) molecules on the surface of IG and NG. The study showed that the binding energy and the net charge transfer of the molecules on graphene got increased by doping with N. The DOS plots displayed major orbital hybridization between SO_x and NG, while there was no hybridization between SO_x and IG molecules. Physical and chemical adsorption of SO_2 and SO_3 molecules were observed on the surface of NG with binding energies of −27.5 and 65.2 kJmol^{-1}, respectively. From the results, the authors concluded that NG can be used as sensitive sensor for the detection of SO_2 and SO_3 [48]. Another work on the adsorption properties of O_3, SO_2 and SO_3 on BG reported strong chemisorption for O_3 with dissociation on the surface of BG [49]. The very low adsorption energies of SO_2 and SO_3 on BG indicated weak interactions of these molecules with BG. Rad *et al.* [49] pointed out that the n-type semiconducting property of BG is responsible for the improved sensitivity of BG to O_3 molecule. The observed high sensitivity of BG to O_3 can be exploited for highly selective O_3 sensors.

The enhanced sensitivity of graphene to CO through aluminum (Al) doping has been demonstrated theoretically [50]. DFT based calculations showed strong chemisorption of CO molecule on Al-doped graphene (hereafter abbreviated as AlG) through the formation of a strong Al-C bond with significant electron transfer from AlG to CO, whereas weak physisorption of CO was observed on IG. The charge difference caused by CO adsorption in AlG ($Q = 0.027\ e^-$) was almost one order larger than that in IG ($Q = 0.003\ e^-$), due to the modified electronic properties of graphene by Al doping. The noticeable increase in the DOS value of AlG near the Fermi level and the closure of energy band gap after CO adsorption indicated the introduction of extra number of shallow acceptor states in AlG through CO interaction. The large increase in the electrical conductivity of AlG upon chemisorption of CO molecule can be exploited for developing highly sensitive CO sensors [50].

Wang *et al.* [51] recently studied the adsorption of CO on IG, NG and AlG and observed very low adsorption energies of CO molecule on IG and NG. Similar to the findings in reported in Ref. [50], large adsorption energy and large charge transfer were observed for CO molecule adsorption on AlG. Although N-doping enhances the interactions of CO molecule with graphene, the adsorption still remained to be weak. The calculated band structures, DOS, and the charge density difference for the doped graphene sheets with and without adsorbed CO molecule showed significantly different adsorption on AlG compared to that on IG and NG. The study concluded that AlG is much more sensitive than IG and NG to the adsorption of CO [51].

AlG was also suggested to be a promising sensing material for detecting formaldehyde (H_2CO) from DFT based study [52]. The chemisorption of H_2CO molecule on the surface of AlG was confirmed by the calculated high adsorption energy value, short binding distance and the observed orbital hybridization between H_2CO and AlG, compared to that calculated for IG. The Mulliken charge analysis of H_2CO-adsorbed IG and AlG systems showed a charge transfer of 0.085 e^- from AlG to H_2CO molecule, which was about 4 times more than the charge transfer of 0.019 e^- from IG to H_2CO molecule [52].

High sensitivity of AlG to NO_2 and N_2O has also been reported [53]. The larger adsorption energy, significant net charge transfer and small binding distance of NO_2 and N_2O on the surface of AlG compared to that calculated for IG indicated strong chemical interactions of these gas molecules with AlG. The calculated DOS plots showed orbital hybridization between NO_2 as well as N_2O and AlG during adsorption, while no hybridization between the gas molecules and IG was observed [53]. Another work on the comparison of the adsorption behaviors of CO, CO_2 and H_2O on the surface of IG and AlG reported very high adsorption on AlG in contrast to relatively low adsorption for IG [54]. The weak physisorption of the gas molecules on IG and the strong chemisorption on AlG was confirmed from natural bond orbital analysis and DOS plots apart from the analysis of adsorption energies and charge transfers. The enhancement in the adsorption of the gases on graphene through Al-doping could be attributed to the modification in the electronic properties of graphene with increased electron density on carbon atoms surrounding the Al-dopant and decreased electron density on Al-dopant. This charge relocation makes the dopant atom as the reactive site for gas molecules [54].

In another study, Milowska and Majewski [55] reported the transport properties of sensors based on covalently modified graphene monolayer (GML) functionalized by $-CH_3$, $-CH_2$, $-NH_2$, -NH and $-OH-CH_3$, $-CH_2$, $-NH_2$, -NH and –OH Green's function (NEGF). For instance, Figure 10.5 demonstrates the current-voltage characteristics of pure and functionalized GML. The authors observed that carbene introduces significant changes in the transmission spectrum and current-voltage characteristics, whereas the methyl groups were observed to cause minimal change. These findings provide fundamental insights into the design of sensors using doped and functionalized graphene for various gases.

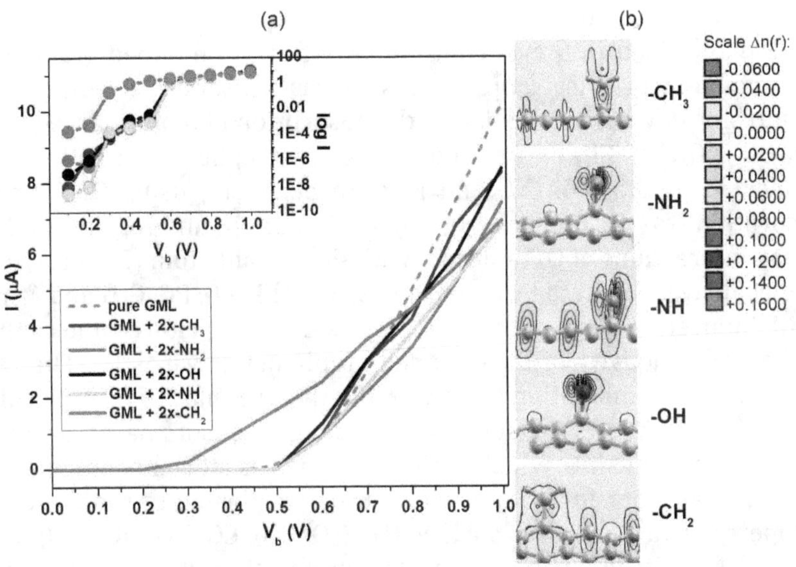

Figure 10.5 (a) The current-voltage characteristics of pure/functionalized GML; (b) the difference of the valence pseudocharge density and the superposition of the atomic valence pseudocharge densities. Reproduced from Reference 55 with permission from American Chemical Society.

A systematic theoretical investigation of the adsorption of common gas molecules such as H_2, H_2O, O_2, CO_2, CO, NO_2, NO, SO_2, NH_3, and N_2 on graphene doped with substitutional impurities such as boron, nitrogen, aluminium and sulfur has also been reported [56]. It was found that BG possesses the ability to chemically bind NO and NO_2, whereas S-doped graphene (hereafter abbreviated as SG) can only bind NO_2 molecule. The Lowdin charge analysis showed a charge transfer of about 0.35 e^- from BG to NO_2, and of about 0.76 e^- from SG to NO_2, which could produce large changes in the conductivity of these doped graphenes. Out of the considered doped graphenes, AlG was found to be highly reactive and binds all the considered gases except H_2 through strong Al–X (X = O, N, C) bonds, of length around 2 Å, which renders it as a highly sensitive gas sensing material. The large local curvature induced by the Al atom in the graphene lattice tends to make AlG more chemically reactive. The authors suggested that BG and SG could be used as good sensors for detecting polluting gases such as NO and NO_2. The results also predict the suitability of using SG as a selective sensor for NO_2 detection in the presence of other gas molecules such as NO, NH_3, CO, CO_2, H_2O, SO_2, O_2, H_2 and N_2 [56].

Dai and Yuan [57] evaluated the effect of O_2 on doped graphene for enabling practical applications of gas sensors, electronic and spintronic devices by simulating the adsorption of molecular oxygen on BG, NG, AlG, Si-, P-, Cr- and Mn-doped graphene (hereafter abbreviated as SiG, PG, CrG and MnG respectively) using DFT. Physisorption was observed for O_2 adsorption on BG and NG due to the weak interactions of O_2 molecule with the dopant atom (B and N). O_2 molecule was found to be chemisorbed on AlG, SiG, PG, CrG and MnG with induced obvious changes in the electronic structures due to the enhanced reactivity of these doped graphenes towards O_2. The results from the simulations indicate that the presence of O_2 molecule in air greatly affects the sensitivity of the gas sensors based on AlG, SiG, PG, CrG and MnG, which limit their usage as efficient gas sensors [57]. The authors, in another study, also investigated the adsorption property of various gases such as H_2, H_2O, CO_2, CO, NO, NO_2, SO_2, N_2, O_2 and NH_3 on PG. The calculations proved that PG can only physisorb gas molecules such as H_2, H_2O, CO_2, CO, N_2 and NH_3, whereas firm adsorption of NO, NO_2, SO_2 and O_2 was observed with large adsorption energies and short graphene-gas molecule distances through the formed P–X (X = O, N, S) bonds [58].

Zou *et al.* [59] investigated the adsorptions of small gas molecules such as CO, O_2, NO_2 and H_2O on IG and SiG using *ab initio* calculations. IG was found to be sensitive to the presence of O_2 and NO_2 molecules, as the electronic properties of IG got influenced by the adsorption of these gas molecules. However, there was no change in the electronic properties of IG upon the adsorption of CO and H_2O molecules. Doping of graphene with silicon makes it more reactive to the adsorption of CO, O_2, NO_2 and H_2O molecules and hence could be used as a good sensor for CO, O_2, NO_2 and H_2O. The strong interactions between SiG and the adsorbed molecules induced dramatic changes to the electronic properties of SiG [59]. Another DFT investigation of the adsorption of NO, N_2O and NO_2 on IG and SiG predicted the possibility of using SiG as an excellent candidate for the detection of NO and NO_2, as well as a metal-free catalyst for N_2O reduction [60]. The results indicated significant enhancement in the interactions of these gas molecules with graphene sheet upon Si doping. It was found that the adsorption of NO and NO_2 result in large changes in the electronic properties of SiG [60]. Zhang *et al.* [61] reported that H_2S adsorption was found to be much stronger on SiG through the formation of Si–S bond, compared to the weak adsorption of H_2S molecule observed on IG, BG and NG.

Lv *et al.* [62] carried out first-principles study of N_2O interaction with IG, AlG and Gallium (Ga) doped graphene (GaG). N_2O was found to be chemisorbed on the surface of AlG and GaG by the stronger covalent bond formed between the dopant atom and the N_2O molecule via the *p* orbitals coupling. The authors observed that the N_2O molecule adsorbed on the surface of AlG and GaG can be easily decomposed to N_2 and O_2 with the application of a perpendicular electric field. Therefore, both AlG and GaG can be used as excellent candidates for the detection and the dissociation of N_2O, thus protecting the ozone layer and controlling the global warming effects [62].

Sharma *et al.* [63] evaluated the H_2S adsorption potential of IG and group III (B, Al and Ga)-doped graphene using DFT modelling. It was found that H_2S weakly binds with IG and BG, on the other hand, chemisorption was observed for AlG and GaG with large adsorption energies and short binding distances. The authors also reported that the increased reactivity of AlG and GaG towards H_2S is due to the local curvature in these doped graphenes formed by the long X-C (X = Al or Ga) bonds compared to C-C bonds [63]. DFT study of phenol adsorption on IG and group III (B, Al and Ga)-doped graphene by Avila *et al.* [64] showed weak physical interactions of phenol molecule with IG and BG. The structural and electronic properties of IG and graphene doped with B atom remain unaffected by phenol adsorption. Much stronger interaction was seen between the phenol molecule and graphene doped with Al and GaG, corresponding to chemical adsorption [64]. Based on the report of Shao *et al.* [65] on the adsorption of SO_2 on IG and heteroatom (B, N, Al, Si, Cr, Mn, Ag, Au, and Pt)-doped graphene sheets using first-principles approach, AlG, SiG, CrG, MnG, Ag-, Au- and Pt-doped graphene (hereafter abbreviated as AgG, AuG and PtG respectively) can be used as good candidates as SO_2 gas sensors due to the strong chemisorption of SO_2 on these doped graphenes. SO_2 molecule had weak adsorption on IG, BG and NG. The authors observed that CrG and MnG are the best choices for SO_2 adsorption due to large adsorption energies and charge transfers of SO_2-adsorbed CrG and MnG systems, compared to that calculated for other doped graphene-SO_2 adsorbed systems [65]. Another study on the adsorption of H_2CO on BG, NG, SiG, AlG, CrG, MnG, and AuG suggested that BG and NG are not sensitive to H_2CO, whereas SiG, AlG, CrG, MnG, and AuG are sensitive to H_2CO as they can strongly adsorb H_2CO molecule [66]. Al and Mn were found to be the most appropriate dopants on graphene for the detection of H_2CO molecule because of their relatively large adsorption energies, large charge transfers and significant changes of DOS before and after adsorption.

In another study, Pramanik and Kang [67] reported the adsorption of paramagnetic O_2 and NO on pristine as well as N-doped and P-doped graphene (GrS) using DFT calculations. The authors observed that the van der Waals interaction significantly affects the physisorption energy along with the adsorption geometry of the gases in pristine and N-doped graphene. The authors also observed adsorption of the gases on P-doped graphene to be stronger than that on pristine or N-doped graphene. It was concluded that the O_2-adsorption is a normal chemisorption, whereas NO-adsorption was weak chemisorption. For instance, Figure 10.6 also demonstrates the optimized geometry of P-doped GrS-O_2.

In another study, Choudhuri *et al.* [68] performed DFT calculations to study the gas (CO, CO_2, NO, and NO_2) sensing mechanisms prevailing in pure and doped (B@, N@, and B–N@) graphene surfaces. Figure 10.7 represents the optimized structures of (a) CO, (b) CO_2, (c) NO, and (d) NO_2 molecules adsorbed on 6.25% B–N@graphene surface. The authors observed that the doping improves adsorption energy and selectivity. The electronic properties of the B–N@graphene surfaces were reported to change significantly in comparison with pure and B@ and N@graphene surfaces, when the gas molecules were adsorbed. It was also reported that the electronic properties changed maximum for NO and NO_2, but not for

CO and CO_2. The authors reported that the B–N@graphene is a promising semiconductor based gas sensor, which is a significant finding.

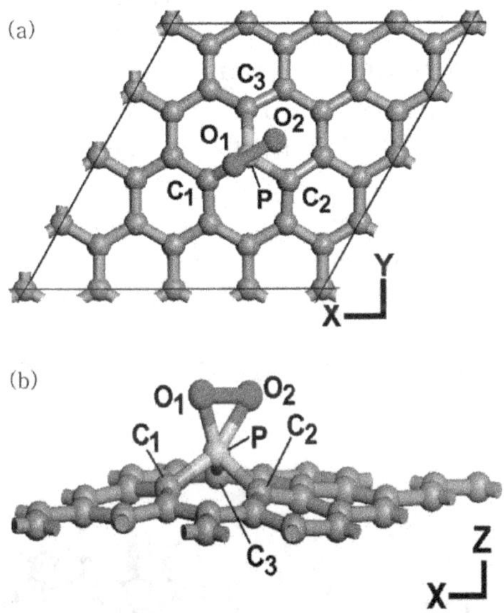

Figure 10.6 Optimized geometry of P-doped GrS-O_2 in XY (a) and XZ (b) planes. Reproduced from Reference 67 with permission from American Chemical Society.

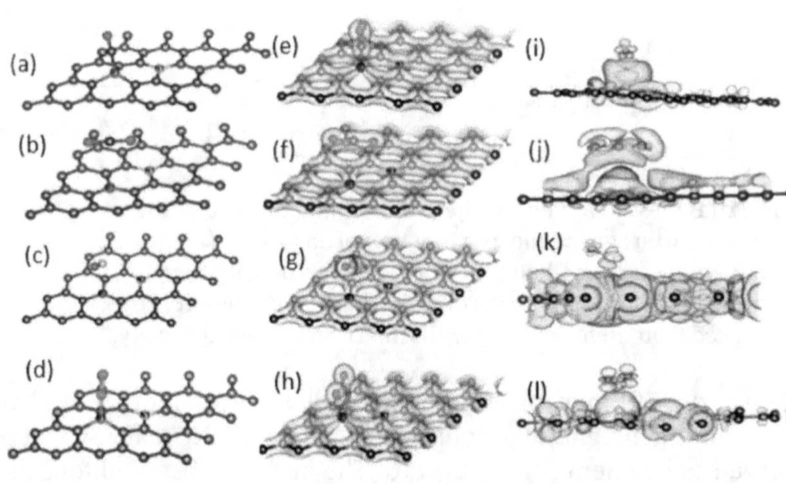

Figure 10.7 Optimized structures of (a) CO, (b) CO_2, (c) NO, and (d) NO_2 molecules adsorbed on 6.25% B–N@graphene surface. Reproduced from Reference 68 with permission from American Chemical Society.

Jiang *et al.* [69] also reported first principles density functional theory calculations to analyze the effect of designed sub nanometer pores on the permeability and selectivity of graphene sheets. High selectivity for H_2 was observed for H_2/CH_4 gas pair, along with high permeability for H_2 a nitrogen-functionalized pore. On the other hand, very high selectivity for H_2 was observed for an all-hydrogen passivated pore. Figure 10.8 also demonstrates the interaction energy between H_2 and the all-hydrogen passivated porous graphene as a function of adsorption height. Though the developed porous graphene was useful as a gas separation membrane, however, the use of designer nanopores can also find potential applications towards gas sensing.

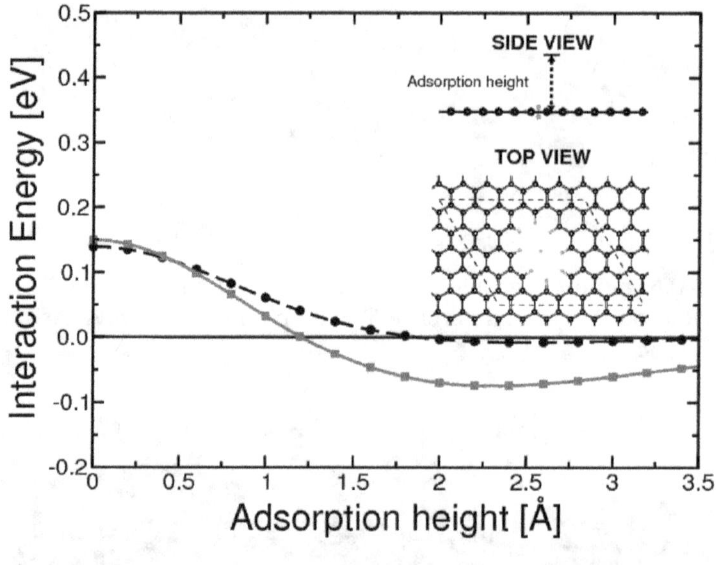

Figure 10.8 Interaction energy between H_2 and the all-hydrogen passivated porous graphene as a function of adsorption height. Squares and solid line: Rutgers-Chalmers van der Waals density functional (vdW-DF); circles and dashed line: Perdew, Burke, and Erzenhoff (PBE). Reproduced from Reference 69 with American Chemical Society.

Lee *et al.* [70] investigated the effect of adsorption of N_2, O_2, NO_2 and SO_2 on the magnetic properties of PtG. N_2, O_2, NO_2 and SO_2 were observed to be chemisorbed on PtG. The authors observed four different types of magnetic properties of PtG based on the chemisorbed molecules on the surface of PtG. Even though there was no spin polarization for N_2 adsorption, other gas molecules changed the magnetic properties of PtG. O_2 adsorption on PtG led to local polarization on the gas molecule, whereas the adsorption of NO_2 and SO_2 on PtG introduced two types of complete polarization on PtG, which differs in the spin direction of PtG and gas adsorbate. These results suggest that the differences in magnetic properties of the PtG based on the type of adsorbed gas molecules could be exploited for developing graphene-based gas sensors [70].

The H_2S sensing behavior of graphene sheet has been found to be significantly enhanced by the substitution of graphene with transition metal (TM) atoms such as Ca, Co, Fe and modification with single vacancy defect from DFT calculations and non-equilibrium Green's function formalism [61]. The observed dramatic changes of the electronic and magnetic properties of TM-doped graphene upon H_2S adsorption suggested strong interactions of H_2S molecule with TM-doped graphene. The electron transport properties of gas sensors using Fe-doped graphene (abbreviated as FeG) sheets as sensing materials exhibited much higher sensitivity than that of the devices made with IG. Similar study on the effect of doped Fe atom on the H_2S sensing behavior of graphene also showed significantly improved interactions between H_2S and FeG sheet from the results of adsorption energy, electron density difference, and DOS plots [71]. Different H_2S adsorbed FeG systems with distinct binding distances (2.5-7.0 Å) were simulated to investigate the influence of the distance between H_2S and the FeG sheet on their interactions. The adsorption energies approach to zero for binding distances larger than 5.0 Å, which is similar to that of H_2S on IG. The DFT calculations proved that the adsorbed H_2S dissociates into S and H_2 [71].

The analyses of the adsorption energies of molecular halogens, CH_3OH, CH_3SH, H_2O, and H_2S on graphene, SG and 2SG showed that the SG is the best adsorbent, fluorine is the best adsorbate, and all molecular halogens adsorb on graphene better than other molecules. The enhancement in the electrical conductivity of SG after the interactions with molecular halogens was predicted from the molecular orbital results and DOS graphs [72].

Wanno *et al.* [73] used DFT calculations to investigate the use of VIIIB TM (Fe, Ru, Os, Co, Rh, Ir, Ni, Pd, and Pt)-doped graphene sheets for CO sensing and observed that CO molecule forms a strong chemisorption bond to VIIIB TM-doped graphene. Among the considered VIIIB TM-doped graphene sheets, Os- and Fe-doped graphene displayed strong chemical interactions with C and O atoms of CO molecule respectively [73]. Later Nasehnia and Seifi [74] considered the effect of O_2 molecule on VIIIB TM-doped graphene structures to evaluate the suitability of these TM-doped graphenes as toxic gas sensing materials in air. O_2 molecule was found to be chemisorbed on these doped graphene sheets with large adsorption energies (−0.653 eV to −1.851 eV) and appreciable amounts of charge transfers (0.375 e^- to 0.650 e^-) from doped graphene sheets to O_2 molecule. The enhanced reactivity of VIIIB TM-doped graphene structures to O_2 could be attributed to the metallic doping and the pyramidalization of the doped graphene sheets. The adsorption energies for both O_2 and CO molecules on VIIIB TM-doped graphene sheets have been found to be similar. Hence, these materials may not be considered as a good choice for toxic gas sensing application [74].

Ma *et al.* [75] analyzed the sensitivity of IG and Pd-doped graphene (abbreviated as PdG) toward small gas molecules such as CO, NH_3, O_2 and NO_2 using first-principles. The authors observed weak adsorption of the gases on IG and significant increase in the interactions of graphene with gas molecules by the introduction of Pd dopants. The authors proposed that PdG is more appropriate for the detection of CO, NH_3, O_2 and NO_2 compared to IG owing to the dramatically increased adsorption energy and elevated charge transfers of gas molecule-Pd doped graphene systems [75].

Cr-doped zigzag graphene nanoribbons (Cr-ZGNRs) have shown high sensitivity to SO_2 molecule than pure ZGNRs from the investigation of the effect of SO_2 molecule on the structural, electronic and transport properties of pure and Cr-ZGNR using DFT and non-equilibrium Green's functional formalism [76]. The adsorption of SO_2 did not induce any change in the I-V curves of pure ZGNRs, which shows the insensitivity of pure ZGNRs towards SO_2. The observed significant increase of molecular current through Cr-ZGNR after SO_2 molecule adsorption revealed the potential of using Cr-ZGNR as a good SO_2 gas sensor [76].

DFT studies have been carried out by Aguila *et al.* [77] to elucidate the interactions between three different graphene structures (intrinsic, divacancy and the inclusion of N_2 in the graphene) and four H_2S concentrations (1 H_2S, 2 H_2S, 3 H_2S and 4 H_2S). H_2S binds to IG layer, graphene layer modified with defects and impurities with adsorption energies in the range of 43-433 meV, that correspond to the physisorption of H_2S on the considered graphene structures. Largest binding energies were observed for all the three graphene structures upon the adsorption of 3 H_2S molecules, which indicates the concentration of 3 H_2S adsorbed molecules saturate both intrinsic and modified graphene sheets [77].

DFT based study of the adsorption of H_2S molecules onto defective graphene as a function of vacancy concentration showed stronger interactions of the molecule with the carbon atoms surrounding the vacancy than with the carbon atoms in a perfect arrangement, which corresponds to chemisorption process followed by the release of H_2 molecule. The DOS of IG, DG and DG with chemisorbed sulfur demonstrated changes in the electrical conductivity. The most favorable adsorption configuration was found to be the one in which the sulfur atom of the H_2S molecule is at the center of the vacancy defect, which facilitates covalent binding with three carbon atoms through unsaturated bonds [78].

Using first-principles calculations, Ma *et al.* [79] showed chemisorption of CO, NO and O_2 gas molecules on vacancy-defected graphene (VG) and hence VG cannot selectively detect CO in air, as the chemisorbed NO and O_2 produce charge transfer which strongly disturb the sensing signals of VG for CO. From the investigation of the adsorption of NO, CO, N_2 and O_2 molecules on pyridinic-like N-doped graphene (NG), it was found that NG has high selectivity for CO in the presence of N_2, O_2 and NO due to chemisorption of CO along with large charge transfer compared to the physical adsorption of N_2, O_2 and NO on the surface of NG [79].

The adsorption studies of H_2CO on graphene modified with a combination of vacancy and dopants such as B, N and S suggested that VG is more sensitive to the presence of H_2CO molecule that IG [80]. However, H_2CO molecule had physisorption on VG with small adsorption energy. The calculations have shown chemisorption of H_2CO on B-doped, N-doped and S-doped vacancy-defected graphene (abbreviated as BVG, NVG and SVG respectively) for which the adsorption energies and net charge transfer are larger than that of VG without dopants. The enhanced adsorption of H_2CO molecule on the dopant-defect combination due to the strong orbital interaction between the *p* orbitals of dopant atoms and carbon atoms around the vacancy was also evident from the results of partial DOS [80].

TM-doped Stone-Wales (SW) defected graphene has also been verified as a promising sensing material for H_2CO detection due to the enhanced adsorption on its surface [81]. H_2CO molecule undergoes chemisorption on TM atoms (Cr, Mn and Co) doped perfect graphene (abbreviated as Cr-PG, Mn-PG and Co-PG respectively) and SW defective graphene structures. The larger binding energy and shorter binding distance of H_2CO molecule on TM-doped defected graphene (Cr-SWG, Mn-SWG and Co-SWG) than that on TM-doped perfect graphene structures showed that the presence of SW-defect improved the adsorption of H_2CO. It was found that the adsorption of H_2CO affect the electronic conductance of the Cr-doped and Mn-doped defective graphene, and hence by measuring the change in electronic conductance, the presence of H_2CO molecule can be detected sensitively. These works theoretically revealed the combined effect of defect and doping on the sensitivity of graphene towards H_2CO gas, for developing future H_2CO sensing devices [81].

Hussain *et al.* [82] investigated the adsorption capability of hydrogenated graphene (graphane) sheet for H_2S and NH_3 gases using DFT. Both pristine graphane sheet and defected graphane sheet (obtained by removing a few surface H atoms from the graphane sheet) have been found to have low affinity for both H_2S and NH_3 gas molecules. The sensing affinity of graphane sheet to these gas molecules was found to be stronger on doping the graphane sheet with Li adatoms. Li-doped graphane sheet showed higher affinity for NH_3 gas compared to H_2S gas due to the stronger hybridization of *s* orbitals of Li adatom with *p* orbitals of nitrogen atom of NH_3 compared to that of with the *p* orbitals of sulfur atom of H_2S. The adsorption energy calculations indicated that Li-doped graphane sheet is suitable for the detection of H_2S and NH_3 gases [82].

In another work by Tanveer *et al.* [83], pristine graphane sheet showed less affinity towards CO, H_2O and NO_2 gas molecules. Defected graphane sheet had strong affinity towards these gas molecules. However, CO and H_2O were found to be weakly physisorbed to the defected graphane sheet. Defected graphane sheet chemically binds NO_2 molecule due to the strong hybridization of the N (*p*) and O (*p*) states with the most active C (*p*) states which lie at the defected site of the graphane sheet. The observed increase in the trend of the work function of the defected graphane sheet on the adsorption of CO, H_2O and NO_2 gas molecules opens up possibilities for next generation gas sensors [83].

Tawfik *et al.* [84] examined the sensitivity of ZGNR with titanium or tin (Sn) atom adsorbed in a double carbon vacancy (DV ZGNR) to H_2S and SO_2 gas molecules. Both Ti- and Sn-doped DV ZGNRs (abbreviated as Ti/DV ZGNR and Sn/DV ZGNR respectively) have shown high sensitivity to H_2S gas. The observed weak sensitivity of Sn/DV ZGNR and the mild sensitivity of Ti/DV ZGNR to SO_2 indicate that both these sensor structures are poor sensors of SO_2. The strong interaction of O_2 molecule with Ti/DV ZGNR through chemisorption process with associated dissociation resulted in reduction of the adsorption energy of both H_2S and SO_2 on Ti/DV ZGNR. In the case of Sn/DV ZGNR, there was no significant reduction in the adsorption energy of subsequent H_2S, while the adsorption energy of SO_2 was found to be significantly reduced after the adsorption of O_2. These results propose the reusability of oxidized Ti/DV ZGNR for both gases, whereas Sn/DV ZGNR exhibited reusability only with respect to SO_2. The authors observed that oxidation does not affect the sensitivity of Ti/DV ZGNR to H_2S but prevents the dissociation of the H–S bond during the adsorption of H_2S on Ti/DV ZGNR, which makes Ti/DV ZGNRs reusable for this gas [84].

Recently, Xie *et al.* [85] explored the gas sensing behavior of molecular devices constructed with an azulene-like dipole molecule sandwiched between two GNR electrodes using DFT and nonequilibrium Green's function method. The authors showed that toxic gas molecules such as CO, NO and NO_2 would adsorb on the doped B atoms of the AGNR. Different gas molecule adsorption on molecular devices with symmetric B-doped AGNR electrodes resulted in different change in the current–voltage profile and hence these devices can be employed for the detection of these gases. The adsorption of these gas molecules induced changes in the sub-bands of B-doped AGNR near the Fermi energy, which depends on the type of the adsorbed molecule. These changes led to obvious variation in the current profile and were employed for specific detection of gas molecules [85].

As shown in Figure 10.9 [86], a transient voltage across the rGO was applied to generate defects in the rGO. The authors observed that the voltage activation process led to an increase in the number of epoxide and ether functional groups in the rGO. The activated rGO was observed to be sensitive to NO_2. The sensitivity of the activated rGO was 500% higher than that of the original rGO, with the detection limit to very low concentration of 50 ppb. The authors observed through density functional theory (DFT) calculations that the high sensitivity to NO_2 and sensing was due to charge transfer from ether groups to NO_2.

Kumar *et al.* [87] also studied the effect of external defects in sensing the performance of graphene field-effect transistors (chemFETs). The authors observed that the sensitivity of graphene chemFETs was not solely due to graphene and can be generated by external defects in the insulating substrate (Figure 10.10). These external defects in the insulating substrate can help to tune the electronic properties of graphene.

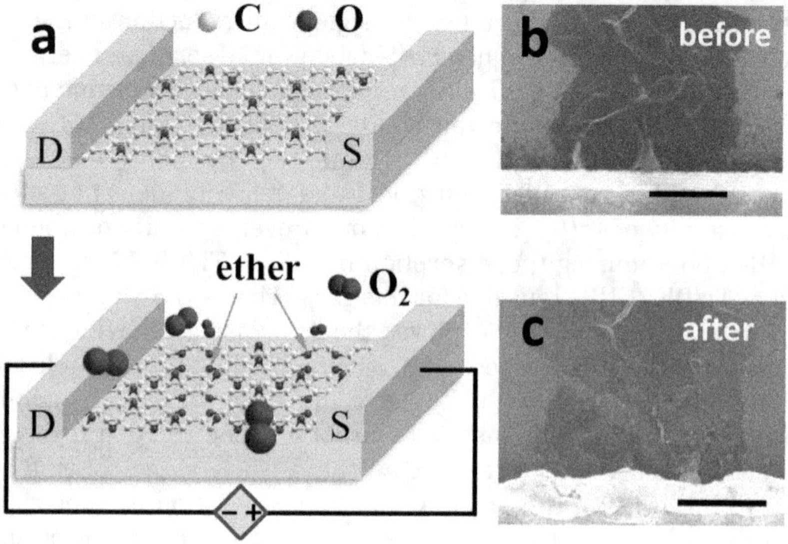

Figure 10.9 Schematic illustration of the voltage activation process for rGO in the air. Reproduced from Reference 86 with permission from American Chemical Society.

To verify the effect of different dopant atoms such as N, Al, Zn, Ti, Zr on the interactions between hydrogen molecules and graphene sheets, Zhang *et al.* [88] performed DFT based calculations on the hydrogen molecule-intrinsic and doped graphene adsorbed systems. The interaction energy with the hydrogen molecule was found to be the largest for Ti-doped graphene (abbreviated as TiG), which is followed by the Zn-doped graphene (abbreviated as ZnG) and then AlG. N-doping did not enhance the adsorption characteristics of graphene sheet with H_2 molecule [88]. The authors, in another report, studied the effect of doped N and Ti on the interactions of CO, NO, SO_2, and HCHO with graphene and observed that Ti-doping can greatly improve the interactions of these gas molecules with graphene with the order of interactions as SO_2 > NO > HCHO > CO from the analyses of adsorption energy, electron density difference and DOS [89]. TiG also enabled selective gas adsorption, whereas NG did not exhibit selective adsorption behavior. Thus, TiG sheet can be used as an effective sensor material than IG and NG sheets for the detection and removal of these harmful gases [89]. TiG was also found to be very promising for HF detection [88]. The calculated charge transfers and the band structure plots of TiG with and without HF molecule indicated that the adsorption of HF on TiG will lead to large increase in the electrical conductivity of TiG sheet [90].

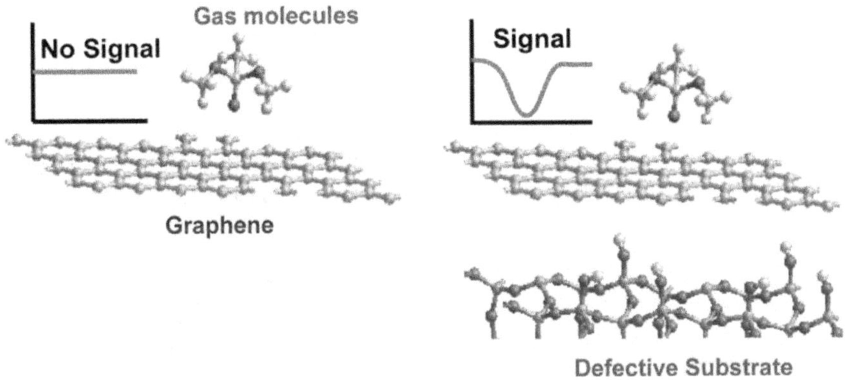

Figure 10.10 Representation of the sensing performance in the presence of extremal defects. Reproduced from Reference 87 with permission from American Chemical Society.

Table 10.1 also summarizes the results of the first-principles simulation studies on different graphene-gas systems.

Table 10.1 Adsorption energy E_a (eV), equilibrium graphene-molecule distance (d) and charge transfer (Q) for gas molecules adsorbed on intrinsic, doped and defected graphene

Type of graphene	Gas molecule	E_a	d (Å)	Q (e)	Ref.
	CO	-0.12	3.02	-0.01	
	NO	-0.30	2.43	0.04	[37]
	NO_2	-0.48	2.73	-0.19	
	NH_3	-0.11	2.85	0.02	
IG	CO	-0.0102	3.8561	-0.0011	[51]
	H_2CO	-0.083	3.127	0.085	[52]
	CO	-0.05	2.80	-0.01	
	O_2	-0.18	2.34	-0.15	[59]
	NO_2	-0.38	2.80	-0.24	
	H_2O	-0.14	2.31	-0.14	
	H_2S	-0.16	2.64	0.01	[61]
	SO_2	0.012	3.279	-0.077	[65]
BG	CO	-0.14	2.97	-0.02	
	NO	-1.07	1.99	0.15	

	NO_2	-1.37	1.67	-0.34	[37]
	NH_3	-0.50	1.66	0.40	
	O_2	-0.0232	3.5099	-0.0863	[57]
	H_2S	-0.11	3.26	0.01	[61]
	SO_2	0.205	3.162	-0.110	[65]
	H_2CO	-0.018	3.845	0.002	[66]
NG	CO	-0.14	3.15	0	[37]
	NO	-0.40	2.32	0.01	
	NO_2	-0.98	2.87	-0.55	
	NH_3	-0.12	2.86	0.04	
	CO	-0.0261	3.5540	-0.0027	[51]
	O_2	-0.1228	3.3196	0.0235	[57]
	H_2S	-0.14	3.22	0.01	[61]
	SO_2	0.172	3.478	-0.263	[65]
	H_2CO	-0.046	3.193	-0.073	[66]
DG	CO	-2.33	1.33	0.26	[37]
	NO	-3.04	1.34	-0.29	
	NO_2	-3.04	1.42	-0.38	
	NH_3	-0.24	2.61	0.02	
AlG	H_2S	-0.91	1.77	0.74	[61]
	CO	-2.6928	2.0623	0.2346	[51]
	H_2CO	-3.721	1.876	0.019	[52]
	O_2	-1.5589	1.8770	0.4367	[57]
	SO_2	1.262	1.825	-0.744	[65]
	H_2CO	-1.262	1.858	0.502	[66]
SiG	O_2	-1.3132	1.7109	0.9184	[57]
	CO	-0.58	2.08	0.24	[59]
	O_2	-3.99	1.71	-1.10	
	NO_2	-3.89	1.76	0.12	
	H_2O	-0.59	2.01	0.24	
	H_2S	-0.94	2.53	0.26	[61]
	SO_2	0.902	1.737	-0.959	[65]
	H_2CO	-0.769	1.722	0.772	[66]
PG	O_2	-1.0359	1.6275	0.8181	[57]
	O_2	-1.0895	1.6279	0.8181	[58]
	H_2	-0.0143	3.3625	-	
	H_2O	-0.0469	3.4650	-	
	N_2	-0.0091	4.2252	-	
	CO_2	-0.0107	4.0019	-	
	CO	-0.0685	4.3139	-	
	NH_3	-0.0146	4.5685	-	
	NO_2	-1.8875	1.5364	0.6198	
	NO	-0.5093	1.8074	0.4995	
	SO_2	-0.3216	1.6580	0.5007	

	O_2	-2.6098	1.7935	0.2166	[57]
CrG	SO_2	1.675	1.927	-0.672	[65]
	H_2CO	-1.215	1.960	0.275	[66]
	O_2	-2.0918	1.8641	0.1865	[57]
MnG	SO_2	1.729	1.905	-0.599	[65]
	H_2CO	-1.025	1.859	0.357	[66]
AgG	SO_2	0.968	2.173	-0.454	[65]
AuG	SO_2	1.284	2.167	-0.479	[65]
	H_2CO	-0.748	2.111	0.211	[66]
	SO_2	1.018	2.229	-0.550	[65]
	N_2	0.414	2.084	-	
PtG	O_2	1.487	2.038	-	
	NO_2	2.210	2.234	-	[70]
	SO_2	1.060	2.250	-	
	CO_2	0.086	2.197	-	
	H_2O	0.095	2.914	-	
CaG	H_2S	-0.66	4.81	0.04	
CoG	H_2S	-1.80	3.77	1.16	[61]
FeG	H_2S	-1.92	3.70	1.23	

10.2.5 Gas Molecules on Other Modified Graphene

Omidvar and Mohajeri [91] investigated the interactions between small gas molecules (O_2, N_2, CO and NO) and different armchair graphene nanoflakes (GNFs) such as intrinsic, B- or N-doped as well as functionalized GNFs with COOH, CN, and NO_2. In this study, pristine, B-, N-doped and functionalized GNF were found to be not sensitive to N_2, O_2 and CO. Pristine GNFs exhibited high sensitivity and reactivity towards NO molecule. The edge functionalization of GNFs with electron withdrawing groups such as COOH-groups significantly enhanced the adsorption of NO molecule which enables highly sensitive and selective detection of NO. The authors observed that the stronger interactions between NO and edge COOH-functionalized GNFs result in large variation of the energy band gap of about 555% after NO adsorption [91].

The adsorption energies and the charge transfer calculations of H_2S adsorption on Cu_2O modified graphene (Cu_2O-GS) showed strong adsorption of H_2S on the surface of Cu_2O-GS. DOS of Cu_2O-GS exhibited considerable variations after H_2S adsorption and also the electrical conductance of the nanosensor at room temperature got significantly enhanced by H_2S adsorption. From the obtained results, it was predicted that Cu_2O-GS nanosensor can be exploited as a nanosensor for H_2S detection. The obtained results were found to be in excellent agreement with previous experimental results [92].

The adsorption of various gas molecules on graphene based materials have been implemented in quantum mechanical simulation software packages such as Atomistix Toolkit [93], Materials Studio DMol^3 [94], CASTEP [95], Gaussian [96], VASP [97], Quantum Espresso [98], SIESTA [99] and ABINIT [100].

10.3 Conclusion and Future Perspectives

Graphene has been considered as one of the most promising candidates for gas sensing applications owing to its excellent structural and electronic properties. Graphene based gas sensors have already demonstrated to have the highest sensitivities and lowest detection limits compared to sensors based on one-dimensional carbon nanotubes and semiconductor nanowires due to the lower electrical noise of graphene arising from its two-dimensional crystal structure.

Theoretical studies based on quantum mechanical calculations enable us to gain a deep understanding of the interactions of gas molecules with graphene, before going for expensive device fabrication. Modeling of graphene-gas molecule adsorbed systems has verified weak physisorption of gas molecules on intrinsic graphene, which poses a major limitation for future applications of graphene based gas sensors. Graphene oxide and reduced graphene oxide have shown highly sensitive sensing response compared to intrinsic graphene, by virtue of the surface rich oxygen-containing functional groups. Theoretical studies have revealed that the interactions of graphene with gas molecules could be improved by heteroatom doping, introducing defects or by the combination of dopants and defects, thereby enhancing the sensitivity of graphene gas sensors. The predictions from quantum mechanical simulations can stimulate interest in experimental works on suitable modification of graphene and future development of novel micro devices based on modified graphene for enabling high sensitivity detection of gas molecules.

In spite of the great advancements in the field of graphene based gas sensors during the past 10 years, there are some major challenges that need to overcome to meet the increasing demands of the industry for high performance gas sensors. Stronger binding of gas molecules required for gas sensing applications, achieved by oxidized, doped or defected graphene would result in noticeable changes in electrical conductivity of oxidized, doped and defected graphene upon exposure to gases. The main drawback with these modified graphene based gas sensors is the long recovery period, as the thermal energy at room temperature is not sufficient to overcome the activation energy needed for releasing the strongly adsorbed gas molecules from the surface. The adsorption and desorption processes on intrinsic graphene are energetically viable as compared to the chemisorption on modified graphene. Strong adsorption on the surface of modified graphene makes gas desorption extremely difficult without the assistance of ultra-violet light illumination or annealing at high temperatures or by application of an electric field. These treatments make the adsorption process reversible in practical time periods. Thus, a proper balance should be maintained between sensitivity and recovery of the modified graphene based gas sensors. Modified graphene based gas sensors are best choices for ultrahigh sensitivity detection of gas molecules, but as just "on off" sensors. Hence, future work needs to focus on functionalization schemes that could control the binding energy of the gas molecules on graphene which could enable fast regeneration of the sensor at room temperature. The realization of highly sensitive and reversible gas sensors based on graphene can ultimately lead to single molecule detection of polluting gases in practical situations.

As graphene gets strongly influenced by a range of different gas species and mixtures, specific identification of gases by selective adsorption is challenging. The adsorption of electron withdrawing gases such as O_2 and NO_2 on intrinsic graphene will lead to an increase in electrical conductivity of graphene for both gases, whereas under controlled environment, which makes graphene difficult to distinguish between these oxidizing gases. Similarly, the adsorption of electron donating gases will result in decrease in the electrical conductivity of graphene and thus specific identification of electron donating gases is also difficult. Apart from this, a mixture containing equal amounts of oxidizing gases (for example, O_2 and NO_2) and reducing gases (for example, NH_3 and CO), will not create any change in the conductivity of graphene as the p-type and n-type doping induced by oxidizing and reducing gases cancel out each other. Thus, future work needs to concentrate on functionalization of graphene with appropriate capture agents that specifically binds the target gas to graphene's surface.

Modified graphene has been widely studied as highly sensitive gas sensors. The inherent sensitivity of modified graphene to most of the gases makes it uninteresting for developing selective gas sensing devices. For example, Al-doped graphene has shown high chemical reactivity to almost all gases and hence Al-doped graphene based sensor is not a suitable choice for the detection of a particular gas of interest in atmospheric air, as the other constituent gases of air strongly affect the sensing signal of Al-doped graphene for the desired gas. Similar is the case with other modified graphene materials, barring the few exceptions. Several modified graphene materials have demonstrated high selectivity. Among these, S-doped graphene is highly selective to NO_2, B-doped graphene is selective to NO and NO_2 in the presence of other common gas molecules, pyridinic-like N-doped graphene has shown high selectivity for CO in the presence of N_2, O_2 and NO. Until now, most of the previous studies have centered on different modifications of graphene for improving the interactions between gas molecules and graphene. In addition to the focus on increased reactivity, theoretical works need to focus on selectivity aspects of modified graphene based gas sensors and also on the suitability of using them under practical environments. Overall, the potential of the use of graphene for functional and efficient gas sensors is immense and is expected to be realized in greater details in the coming years.

References

1. Novoselov, K. S., Geim, A. K., Morozov, S. V., Jiang, D., Zhang, Y., Dubonos, S. V., Grigorieva, I. V., and Firsov, A. A. (2004) Electric field effect in atomically thin carbon films. *Science,* **306**, 666-669.
2. Novoselov, K. S., Geim, A. K., Morozov, S. V., Jiang, D., Katsnelson, M. I., Grigorieva, I. V., Dubonos, S. V., and Firsov, A. A. (2005) Two-dimensional gas of massless Dirac fermions in graphene. *Nature,* **438**, 197-200.
3. Zhang, Y., Tan, Y.-W., Stormer, H. L., and Kim, P. (2005) Experimental observation of the quantum Hall effect and Berry's phase in graphene. *Nature,* **438**, 201-204.
4. Lee, C., Wei, X., Kysar, J. W., and Hone, J. (2008) Measurement of the elastic properties and intrinsic strength of monolayer graphene. *Science,* **321**, 385-388.
5. Nair, R. R., Blake, P., Grigorenko, A. N., Novoselov, K. S., Booth, T. J., Stauber, T., Peres, N. M., and Geim, A. K. (2008) Fine structure constant defines visual transparency of graphene. *Science,* **320**, 1308.

6. Katsnelson, M. I. (2007) Graphene: carbon in two dimensions. *Materials Today,* **10**, 20-27.
7. Geim, A. K., and Novoselov, K. S. (2007) The rise of graphene. *Nature Materials,* **6**, 183-191.
8. Castro Neto, A. H., Guinea, F., Peres, N. M. R., Novoselov, K. S., and Geim, A. K. (2009) The electronic properties of graphene. *Reviews of Modern Physics,* **81**, 109-162.
9. Geim, A. K. (2009) Graphene: Status and prospects. *Science,* **324**, 1530-1534.
10. Craciun, M. F., Russo, S., Yamamoto, M., and Tarucha, S. (2011) Tuneable electronic properties in graphene. *Nano Today,* **6**, 42-60.
11. Novoselov, K. S., Falko, V. I., Colombo, L., Gellert, P. R., Schwab, M. G., and Kim, K. (2012) A roadmap for graphene. *Nature,* **490**, 192-200.
12. Singh, V., Joung, D., Zhai, L., Das, S., Khondaker, S. I., and Seal, S. (2011) Graphene based materials: Past, present and future. *Progress in Materials Science,* **56**, 1178-1271.
13. Xiang, Q., Yu, J., and Jaroniec, M. (2012) Graphene-based semiconductor photocatalysts. *Chemical Society Reviews,* **41**, 782-796.
14. Zhao, G., Wen, T., Chen, C., and Wang, X. (2012) Synthesis of graphene-based nanomaterials and their application in energy-related and environmental-related areas. *RSC Advances,* **2**, 9286-9303.
15. Zhu, J., Yang, D., Yin, Z., Yan, Q., and Zhang, H. (2014) Graphene and graphene-based materials for energy storage applications. *Small,* **10**, 3480-3498.
16. Weiss, N. O., Zhou, H., Liao, L., Liu, Y., Jiang, S., Huang, Y., and Duan, X. (2012) Graphene: An emerging electronic material. *Advanced Materials,* **24**, 5782-5825.
17. Tao, L., Wang, D., Jiang, S., Liu, Y., Xie, Q., Tian, H., Deng, N., Wang, X., Yang, Y., and Ren, T.-L. (2016) Fabrication techniques and applications of flexible graphene-based electronic devices. *Journal of Semiconductors,* **37**, 041001.
18. Lahiri, I., Verma, V. P., and Choi, W. (2011) An all-graphene based transparent and flexible field emission device. *Carbon,* **49**, 1614-1619.
19. Bonaccorso, F., Sun, Z., Hasan, T., and Ferrari, A. C. (2010) Graphene photonics and optoelectronics. *Nature Photonics,* **4**, 611-622.
20. Yang, Y., Asiri, A. M., Tang, Z., Du, D., and Lin, Y. (2013) Graphene based materials for biomedical applications. *Materials Today,* **16**, 365-373.
21. Shao, Y., Wang, J., Wu, H., Liu, J., Aksay, I. A., and Lin, Y. (2010) Graphene based electrochemical sensors and biosensors: A review. *Electroanalysis,* **22**, 1027-1036.

22. Basu, S., and Bhattacharyya, P. (2012) Recent developments on graphene and graphene oxide based solid state gas sensors. *Sensors and Actuators B: Chemical,* **173**, 1-21.
23. Liu, Y., Dong, X., and Chen, P. (2012) Biological and chemical sensors based on graphene materials. *Chemical Society Reviews,* **41**, 2283-2307.
24. Yuan, W., and Shi, G. (2013) Graphene-based gas sensors. *Journal of Materials Chemistry A,* **1**, 10078-10091.
25. Varghese, S. S., Lonkar, S., Singh, K. K., Swaminathan, S., and Abdala, A. (2015) Recent advances in graphene based gas sensors. *Sensors and Actuators B: Chemical,* **218**, 160-183.
26. Wang, T., Huang, D., Yang, Z., Xu, S., He, G., Li, X., Hu, N., Yin, G., He, D., and Zhang, L. (2016) A review on graphene-based gas/vapor sensors with unique properties and potential applications. *Nano-Micro Letters,* **8**, 95-119.
27. Varghese, S., Varghese, S., Swaminathan, S., Singh, K., and Mittal, V. (2015) Two-dimensional materials for sensing: Graphene and beyond. *Electronics,* **4**, 651-687.
28. Schedin, F., Geim, A. K., Morozov, S. V., Hill, E. W., Blake, P., Katsnelson, M. I., and Novoselov, K. S. (2007) Detection of individual gas molecules adsorbed on graphene. *Nature Materials,* **6**, 652-655.
29. Danneau, R., Wu, F., Craciun, M. F., Russo, S., Tomi, M. Y., Salmilehto, J., Morpurgo, A. F., and Hakonen, P. J. (2008) Shot noise in ballistic graphene. *Physical Review Letters,* **100**, 196802.
30. Hill, E. W., Vijayaragahvan, A., and Novoselov, K. (2011) Graphene sensors. *IEEE Sensors Journal,* **11**, 3161-3170.
31. Wehling, T. O., Novoselov, K. S., Morozov, S. V., Vdovin, E. E., Katsnelson, M. I., Geim, A. K., and Lichtenstein, A. I. (2008) Molecular doping of graphene. *Nano Letters,* **8**, 173-177.
32. Dan, Y., Lu, Y., Kybert, N. J., Luo, Z., and Johnson, A. T. C. (2009) Intrinsic response of graphene vapor sensors. *Nano Letters,* **9**, 1472-1475.
33. Leenaerts, O., Partoens, B., and Peeters, F. M. (2008) Adsorption of H_2O, NH_3, CO, NO_2, and NO on graphene: A first-principles study. *Physical Review B,* **77**, 125416.
34. Seyed-Talebi, S. M., and Beheshtian, J. (2013) Computational study of ammonia adsorption on the perfect and rippled graphene sheet. *Physica B: Condensed Matter,* **429**, 52-56.
35. Lee, K.-J., and Kim, S.-J. (2013) Theoretical investigation of CO_2 adsorption on graphene. *Bulletin of the Korean Chemical Society,* **34**, 3022-3026.
36. Ganji, M. D., Hosseini-Khah, S., and Amini-Tabar, Z. (2015) Theoretical insight into hydrogen adsorption onto graphene: a first-principles B3LYP-D3 study. *Physical Chemistry Chemical Physics,* **17**, 2504-2511.

37. Zhang, Y.-H., Chen, Y.-B., Zhou, K.-G., Liu, C.-H., Zeng, J., Zhang, H.-L., and Peng, Y. (2009) Improving gas sensing properties of graphene by introducing dopants and defects: a first-principles study. *Nanotechnology,* **20**, 185504.
38. Huang, B., Li, Z., Liu, Z., Zhou, G., Hao, S., Wu, J., Gu, B.-L., and Duan, W. (2008) Adsorption of gas molecules on graphene nanoribbons and its implication for nanoscale molecule sensor. *The Journal of Physical Chemistry C,* **112**, 13442-13446.
39. Rostam, M., Yawar, M., and Nader, G. (2008) Investigation of gas sensing properties of armchair graphene nanoribbons. *Journal of Physics: Condensed Matter,* **20**, 425211.
40. Paulla, K. K., and Farajian, A. A. (2013) Concentration effects of carbon oxides on sensing by graphene nanoribbons: Ab initio modeling. *The Journal of Physical Chemistry C,* **117**, 12815-12825.
41. Akbari, E., Yousof, R., Ahmadi, M. T., Kiani, M. J., Rahmani, M., Feiz Abadi, H. K., and Saeidmanesh, M. (2014) The effect of concentration on gas sensor model based on graphene nanoribbon. *Neural Computing and Applications,* **24**, 143-146.
42. Fathalian, A., Jalilian, J., and Shahidi, S. (2013) Effects of gas adsorption on the electronic properties of graphene nanoribbons. *Physica B: Condensed Matter,* **417**, 75-78.
43. Peng, Y., and Li, J. (2013) Ammonia adsorption on graphene and graphene oxide: a first-principles study. *Frontiers of Environmental Science & Engineering,* **7**, 403-411.
44. Tang, S., and Cao, Z. (2011) Adsorption of nitrogen oxides on graphene and graphene oxides: Insights from density functional calculations. *The Journal of Chemical Physics,* **134**, 044710.
45. Mattson, E. C., Pande, K., Unger, M., Cui, S., Lu, G., Gajdardziska-Josifovska, M., Weinert, M., Chen, J., and Hirschmugl, C. J. (2013) Exploring adsorption and reactivity of NH_3 on reduced graphene oxide. *The Journal of Physical Chemistry C,* **117**, 10698-10707.
46. Wang, X., Sun, G., Routh, P., Kim, D.-H., Huang, W., and Chen, P. (2014) Heteroatom-doped graphene materials: syntheses, properties and applications. *Chemical Society Reviews,* **43**, 7067-7098.
47. Denis, P. A. (2010) Band gap opening of monolayer and bilayer graphene doped with aluminium, silicon, phosphorus, and sulfur. *Chemical Physics Letters,* **492**, 251-257.
48. Shokuhi Rad, A., Esfahanian, M., Maleki, S., and Gharati, G. (2016) Application of carbon nanostructures toward SO_2 and SO_3 adsorption: a comparison between pristine graphene and N-doped graphene by DFT calculations. *Journal of Sulfur Chemistry,* **37**, 176-188.
49. Rad, A. S., Shabestari, S. S., Mohseni, S., and Aghouzi, S. A. (2016) Study on the adsorption properties of O_3, SO_2, and SO_3 on B-doped graphene using DFT calculations. *Journal of Solid State Chemistry,*

237, 204-210.

50. Ao, Z. M., Yang, J., Li, S., and Jiang, Q. (2008) Enhancement of CO detection in Al doped graphene. *Chemical Physics Letters,* **461**, 276-279.
51. Wang, W., Zhang, Y., Shen, C., and Chai, Y. (2016) Adsorption of CO molecules on doped graphene: A first-principles study. *AIP Advances,* **6**, 025317.
52. Chi, M., and Zhao, Y.-P. (2009) Adsorption of formaldehyde molecule on the intrinsic and Al-doped graphene: A first principle study. *Computational Materials Science,* **46**, 1085-1090.
53. Rad, A. S. (2015) First principles study of Al-doped graphene as nanostructure adsorbent for NO_2 and N_2O: DFT calculations. *Applied Surface Science,* **357**, Part A, 1217-1224.
54. Rad, A. S., and Foukolaei, V. P. (2015) Density functional study of Al-doped graphene nanostructure towards adsorption of CO, CO_2 and H_2O. *Synthetic Metals,* **210**, Part B, 171-178.
55. Milowska, K. Z., and Majewski, J. A. (2014) Graphene based sensors: Theoretical study. *Journal of Physical Chemistry C,* **118**(31), 17395-17401.
56. Dai, J., Yuan, J., and Giannozzi, P. (2009) Gas adsorption on graphene doped with B, N, Al, and S: A theoretical study. *Applied Physics Letters,* **95**, 232105.
57. Dai, J., and Yuan, J. (2010) Adsorption of molecular oxygen on doped graphene: Atomic, electronic, and magnetic properties. *Physical Review B,* **81**, 165414.
58. Dai, J., and Yuan, J. (2010) Modulating the electronic and magnetic structures of P-doped graphene by molecule doping. *Journal of Physics: Condensed Matter,* **22**, 225501.
59. Zou, Y., Li, F., Zhu, Z. H., Zhao, M. W., Xu, X. G., and Su, X. Y. (2011) An ab initio study on gas sensing properties of graphene and Si-doped graphene. *The European Physical Journal B,* **81**, 475-479.
60. Chen, Y., Gao, B., Zhao, J.-X., Cai, Q.-H., and Fu, H.-G. (2012) Si-doped graphene: an ideal sensor for NO- or NO_2-detection and metal-free catalyst for N_2O-reduction. *Journal of Molecular Modeling,* **18**, 2043-2054.
61. Zhang, Y.-H., Han, L.-F., Xiao, Y.-H., Jia, D.-Z., Guo, Z.-H., and Li, F. (2013) Understanding dopant and defect effect on H_2S sensing performances of graphene: A first-principles study. *Computational Materials Science,* **69**, 222-228.
62. Lv, Y.-a., Zhuang, G.-l., Wang, J.-g., Jia, Y.-b., and Xie, Q. (2011) Enhanced role of Al or Ga-doped graphene on the adsorption and dissociation of N_2O under electric field. *Physical Chemistry Chemical Physics,* **13**, 12472-12477.
63. Sharma, S., and Verma, A. S. (2013) A theoretical study of H_2S adsorption on graphene doped with B, Al and Ga. *Physica B: Condens-*

ed Matter, **427**, 12-16.

64. Avila, Y., Cocoletzi, G. H., and Romero, M. T. (2014) First principles calculations of phenol adsorption on pristine and group III (B, Al, Ga) doped graphene layers. *Journal of Molecular Modeling,* **20**, 1-9.
65. Shao, L., Chen, G., Ye, H., Wu, Y., Qiao, Z., Zhu, Y., and Niu, H. (2013) Sulfur dioxide adsorbed on graphene and heteroatom-doped graphene: a first-principles study. *The European Physical Journal B,* **86**, 1-5.
66. Liu, X.-Y., and Zhang, J.-M. (2014) Formaldehyde molecule adsorbed on doped graphene: A first-principles study. *Applied Surface Science,* **293**, 216-219.
67. Pramanik, A., and Kang, H. S. (2011) Density functional theory study of O_2 and NO adsorption on heteroatom-doped graphenes including the van der Waals interaction. *Journal of Physical Chemistry C,* **115**, 10971-10978.
68. Choudhuri, I., Patra, N., Mahanta, A., Ahuja, R., Pathak, B. (2015) B-N@Graphene: Highly sensitive and selective gas sensor. *Journal of Physical Chemistry C,* **119**(44), 24827-24836.
69. Jiang, D.-e., Cooper, V. R., and Dai, S. (2009) Porous graphene as the ultimate membrane for gas separation. *Nano Letters,* **9**(12), 4019-4024.
70. Lee, Y., Lee, S., Hwang, Y., and Chung, Y.-C. (2014) Modulating magnetic characteristics of Pt embedded graphene by gas adsorption (N_2, O_2, NO_2, SO_2). *Applied Surface Science,* **289**, 445-449.
71. Zhang, H.-p., Luo, X.-g., Song, H.-t., Lin, X.-y., Lu, X., and Tang, Y. (2014) DFT study of adsorption and dissociation behavior of H_2S on Fe-doped graphene. *Applied Surface Science,* **317**, 511-516.
72. Tavakol, H., and Mollaei-Renani, A. (2014) DFT, AIM, and NBO study of the interaction of simple and sulfur-doped graphenes with molecular halogens, CH_3OH, CH_3SH, H_2O, and H_2S. *Structural Chemistry,* **25**, 1659-1667.
73. Wanno, B., and Tabtimsai, C. (2014) A DFT investigation of CO adsorption on VIIIB transition metal-doped graphene sheets. *Superlattices and Microstructures,* **67**, 110-117.
74. Nasehnia, F. and Seifi, M. (2014) Adsorption of molecular oxygen on VIIIB transition metal-doped graphene: A DFT study. *Modern Physics Letters B,* **28**, 1450237.
75. Ma, L., Zhang, J.-M., Xu, K.-W., and Ji, V. (2015) A first-principles study on gas sensing properties of graphene and Pd-doped graphene. *Applied Surface Science,* **343**, 121-127.
76. Shao, L., Chen, G., Ye, H., Niu, H., Wu, Y., Zhu, Y., and Ding, B. (2014) Sulfur dioxide molecule sensors based on zigzag graphene nanoribbons with and without Cr dopant. *Physics Letters A,* **378**, 667-671.
77. Castellanos Aguila J. E., Cocoletzi, H. H., and Cocoletzi, G. H. (2013)

A theoretical analysis of the role of defects in the adsorption of hydrogen sulfide on graphene. *AIP Advances,* **3**, 032118.

78. Borisova, D., Antonov, V., and Proykova, A. (2013) Hydrogen sulfide adsorption on a defective graphene. *International Journal of Quantum Chemistry,* **113**, 786-791.
79. Ma, C., Shao, X., and Cao, D. (2014) Nitrogen-doped graphene as an excellent candidate for selective gas sensing. *Science China Chemistry,* **57**, 911-917.
80. Zhou, Q., Yuan, L., Yang, X., Fu, Z., Tang, Y., Wang, C., and Zhang, H. (2014) DFT study of formaldehyde adsorption on vacancy defected graphene doped with B, N, and S. *Chemical Physics,* **440**, 80-86.
81. Zhou, Q., Wang, C., Fu, Z., Tang, Y., and Zhang, H. (2014) Adsorption of formaldehyde molecule on Stone–Wales defected graphene doped with Cr, Mn, and Co: A theoretical study. *Computational Materials Science,* **83**, 398-402.
82. Hussain, T., Panigrahi, P., and Ahuja, R. (2014) Enriching physisorption of H_2S and NH_3 gases on a graphane sheet by doping with Li adatoms. *Physical Chemistry Chemical Physics,* 16, 8100-8105.
83. Hussain, T., Panigrahi, P., and Ahuja, R. (2014) Sensing propensity of a defected graphane sheet towards CO, H_2O and NO_2. *Nanotechnology,* **25**, 325501.
84. Tawfik, S. A., Cui, X., Carter, D., Ringer, S., and Stampfl, C. (2015) Sensing sulfur-containing gases using titanium and tin decorated zigzag graphene nanoribbons from first-principles. *Physical Chemistry Chemical Physics,* **17**, 6925-6932.
85. Xie, Z., Zuo, X., Zhang, G.-P., Li, Z.-L., and Wang, C.-K. (2016) Detecting CO, NO and NO_2 gases by Boron-doped graphene nanoribbon molecular devices. *Chemical Physics Letters,* **657**, 18-25.
86. Cui, S., Pu, H., Mattson, E. C., Wen, Z., Chang, J., Hou, Y., Hirschmugl, C. J., and Chen, J. (2014) Ultrasensitive chemical sensing through facile tuning defects and functional groups in reduced graphene oxide. *Analytical Chemistry,* **86**, 7516-7522.
87. Kumar, B., Min, K., Bashirzadeh, M., Farimani, A. B., Bae, M.-H., Estrada, D., Kim, Y. D., Yasaei, P., Park, Y. D., Pop, E., Aluru, N. R., and Salehi-Khojin, A. (2013) The role of external defects in chemical sensing of graphene field-effect transistors. *Nano Letters,* **13**, 1962-1968.
88. Zhang, H.-p., Luo, X.-g., Lin, X.-y., Lu, X., and Leng, Y. (2013) Density functional theory calculations of hydrogen adsorption on Ti-, Zn-, Zr-, Al-, and N-doped and intrinsic graphene sheets. *International Journal of Hydrogen Energy,* **38**, 14269-14275.
89. Zhang, H.-p., Luo, X.-g., Lin, X.-y., Lu, X., Leng, Y., and Song, H.-t. (2013) Density functional theory calculations on the adsorption of formaldehyde and other harmful gases on pure, Ti-doped, or N-

doped graphene sheets. *Applied Surface Science,* **283**, 559-565.

90. Song, E. H., Zhu, Y. F., and Jiang, Q. (2013) Density functional theory calculations of adsorption of hydrogen fluoride on titanium embedded graphene. *Thin Solid Films,* **546**, 124-127.
91. Omidvar, A., and Mohajeri, A. (2014) Edge-functionalized graphene nanoflakes as selective gas sensors. *Sensors and Actuators B: Chemical,* **202**, 622-630.
92. Mohammadi-Manesh, E., Vaezzadeh, M., and Saeidi, M. (2015) Theoretical study on electronic structure, and electrical conductance at room temperature of Cu_2O–GS nanosensors and detection of H_2S gas. *Computational Materials Science,* **97**, 181-185.
93. Brandbyge, M., Mozos, J.-L., Ordejon, P., Taylor, J., and Stokbro, K. (2002) Density-functional method for nonequilibrium electron transport. *Physical Review B,* **65**, 165401.
94. Delley, B. (2000) From molecules to solids with the DMol3 approach. *The Journal of Chemical Physics,* **113**, 7756-7764.
95. Milman, V., Winkler, B., White, J. A., Pickard, C. J., Payne, M. C., Akhmatskaya, E. V., and Nobes, R. H. (2000) Electronic structure, properties, and phase stability of inorganic crystals: A pseudopotential plane-wave study. *International Journal of Quantum Chemistry,* **77**, 895-910.
96. Frisch, M. J., Trucks, G. W., Schlegel, H. B., Scuseria, G. E., Robb, M. A., Cheeseman, J. R., Scalmani, G., Barone, V., Mennucci, B., Petersson, G. A., Nakatsuji, H., Caricato, M., Li, X., Hratchian, H. P., Izmaylov, A. F., Bloino, J., Zheng, G., Sonnenberg, J. L., Hada, M., Ehara, M., Toyota, K., Fukuda, R., Hasegawa, J., Ishida, M., Nakajima, T., Honda, Y., Kitao, O., Nakai, H., Vreven, T., Montgomery, Jr., J. A., Peralta, J. E., Ogliaro, F., Bearpark, M., Heyd, J. J., Brothers, E., Kudin, K. N., Staroverov, V. N., Kobayashi, R., Normand, J., Raghavachari, K., Rendell, A., Burant, J. C., Iyengar, S. S., Tomasi, J., Cossi, M., Rega, N., Millam, J. M., Klene, M., Knox, J. E., Cross, J. B., Bakken, V., Adamo, C., Jaramillo, J., Gomperts, R., Stratmann, R. E., Yazyev, O., Austin, A. J., Cammi, R., Pomelli, C., Ochterski, J. W., Martin, R. L., Morokuma, K., Zakrzewski, V. G., Voth, G. A., Salvador, P., Dannenberg, J. J., Dapprich, S., Daniels, A. D., Farkas, O., Foresman, J. B., Ortiz, J. V., Cioslowski, J., Fox, D. J. (2009) *Gaussian 09*, Gaussian, Inc., USA.
97. Kresse, G., and Furthmuller, J. (1996) Efficient iterative schemes for *ab initio* total-energy calculations using a plane-wave basis set. *Physical Review B,* **54**, 11169-11186.
98. Giannozzi, P., Baroni, S., Bonini, N., Calandra, M., Car, R., Cavazzoni, C., Ceresoli, D., Chiarotti, G. L., Cococcioni, M., Dabo, I., Corso, A. D., de Gironcoli, S., Fabris, S., Fratesi, G., Gebauer, R., Gerstmann, U., Gougoussis, C., Kokalj, A., Lazzeri, M., Martin-Samos, L., Marzari, N., Mauri, F., Mazzarello, R., Paolini, S., Pasquarello, A., Paulatto, L., Sbraccia, C., Scandolo, S., Sclauzero, G., Seitsonen. A. P., Smogunov,

A., Umari, P., and Wentzcovitch, R. M. (2009) QUANTUM ESPRESSO: a modular and open-source software project for quantum simulations of materials. *Journal of Physics: Condensed Matter,* **21**, 395502.

99. Soler, J. M., Artacho, E., Gale, J. D., Garcia, A., Junquera, J., Ordejon, P., and Sanchez-Portal, D. (2002) The SIESTA method for ab initio order- N materials simulation. *Journal of Physics: Condensed Matter,* **14**, 2745.

100. Gonze, X., Amadon, B., Anglade, P. M., Beuken, J. M., Bottin, F., Boulanger, P., Bruneval, F., Caliste, D., Caracas, R., Cote, M., Deutsch, T., Genovese, L., Ghosez, P., Giantomassi, M., Goedecker, S., Hamann, D. R., Hermet, P., Jollet, F., Jomard, G., Leroux, S., Mancini, M., Mazevet, S., Oliveira, M. J. T., Onida, G., Pouillon, Y., Rangel, T., Rignanese, G. M., Sangalli, D., Shaltaf, R., Torrent, M., Verstraete, M. J., Zerah, G., and Zwanziger, J. W. (2009) ABINIT: First-principles approach to material and nanosystem properties. *Computer Physics Communications,* **180**, 2582-2615.

11

Ternary Cu/ZnO/Graphene Nanocomposites with Photocatalytic Activity for Methylene Blue Dye Degradation under Visible Light

11.1 Introduction

Aqueous waste from the textile industries is known to contaminate water resources, thus, creating significant environmental pollution. Specifically, water contamination occurring from various kinds of industrial organic dyes poses a hazard to human health and environment. Significant research efforts have focused on dye degradation in aqueous waste using photocatalysts [1-6]. Specifically, nanostructured semiconducting metal oxides such as titanium dioxide (TiO_2), zinc oxide (ZnO), copper oxide (CuO), tungsten trioxide (WO_3) and molybdenum trioxide (MoO_3) have gained importance in the photocatalytic applications due to their ability to degrade these organic dyes with high efficiency [7]. These semiconducting metal oxide nanomaterials have wide energy band gap, high stability and low toxicity as well as possess good optical and electrical properties [8-10]. Among the different metal oxides, ZnO is a well-accepted photocatalyst with a wide band gap of 3.37 eV and a large exciton binding energy of 60 meV for achieving an effective liquid phase degradation [11-14].

As compared to individual metal oxide nanomaterials, hybrid metal oxides have been reported to possess better efficiency for various applications, including photocatalytic applications [15]. The preparation of hybrid nanostructured materials has been reported *via* hybridization of different metal oxides, doping of metal nanoparticles into nanostructure of another metal oxide, decoration of metals and metal oxides onto carbon materials, etc. [16]. Among these, metal and metal oxide doped carbon nanomaterials have received more attention in the field of photocatalysis [17-21]. Graphene is an allotrope of carbon, which is an individual sheet of graphite with sp^2-hybridized two-dimensional structure with atomic level thickness. ZnO on graphene sheets has been reported to be an efficient photocatalyst

Liyamol Jacob, M. R. Vengatesan and Vikas Mittal, Khalifa University of Science and Technology, Abu Dhabi, UAE

where the fast hole-electron pair recombination of ZnO could be significantly reduced by the presence of graphene sheets [22-24]. The graphene sheets act as electron acceptor, which prevents the electron-hole recombination in ZnO [25,26]. However, the ZnO-graphene hybrid is activated only in the UV light medium, thus, in order to improve the photocatalytic efficiency of ZnO, the hybridization of ZnO with other metal oxides such as CuO [27] and TiO_2 [28] as well as metals like Cu [29,30], Ag [31,32] and Au [33,34] has also been reported.

Recently, the focus of research has shifted towards the photocatalysis reactions in the visible light medium [35,36]. Cu nanoparticles possess a localized surface plasmon resonance (LSPR) effect and exhibit LSPR absorption in the visible light range [37-40]. LSPR effect can be explained as the collective oscillation of conduction electrons in metallic nanoparticles, which resonate with the electromagnetic field of incident light in the visible light range [41]. The bare Cu nanoparticles have the limitation to be used as a photocatalyst due to heir low chemical stability under catalytic reaction conditions, as the Cu nanoparticles are easily oxidized to Cu_2O or CuO in air or in the presence of traces of molecular oxygen. Recently, Guo *et al.* [42] reported the stability of Cu nanoparticles in the presence of graphene sheets and Cu coated graphene was demonstrated as a photocatalyst for the degradation of aromatic nitro compounds under visible light medium. It was also recently reported that due to the LSPR effect, the conducting Cu nanoparticles convert the energy of incident photons and transfer the plasmonic energy to the semiconducting ZnO *via* resonant energy transfer (RET) [43]. RET generates electron-hole pairs in the semiconductor ZnO by the dipole-dipole interaction between the plasmonic metal (donor) and semiconductor (acceptor), which greatly enhances the visible-light photocatalytic activity as compared to the semiconductor alone.

In the present work, hybridization of both Cu and ZnO on the graphene sheets has been demonstrated. The photocatalytic efficiency of the hybrids was studied using the photodegradation of methylene blue (MB) solution under visible light medium. Subsequently, the photocatalytic activity of hybrid nanomaterials was compared with ZnO/graphene (ZG) and Cu/graphene (CG) hybrids. The morphological features of hybrid nanomaterials were confirmed through physio-chemical analysis techniques. The possible mechanism of the Cu/ZnO/graphene (CZG) hybrid for the synergetic photocatalysis reaction has also been proposed. In another study, these Cu/ZnO hybrid nanoparticle-decorated graphene nanocomposite were also reported as functional anode materials for the high-performance Li-ion batteries [44]. X-ray photoelectron spectroscopy (XPS) analysis of the hybrid materials revealed that the Cu nanoparticles existed in +2 oxidation state on the graphene nanostructure. In the electrochemical analysis of the hybrid materials, one of the hybrids was observed to exhibit a stable and specific capacity of 630 mAh.g^{-1} at a current rate of 100 mA.g^{-1} with almost 95% capacity retention for up to 100 cycles (Figure 11.1).

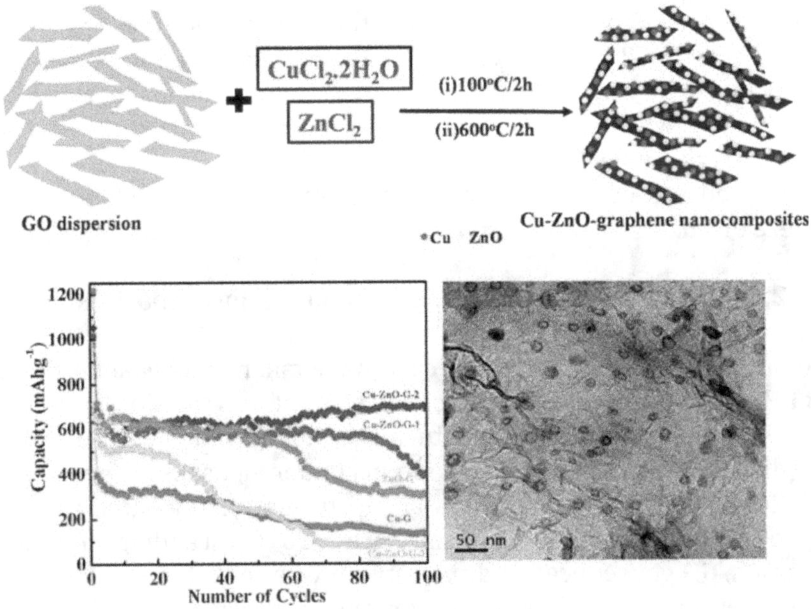

Figure 11.1 Representation of the hybrid nanomaterial synthesis and performance as anode materials for the high-performance Li-ion batteries. Reproduced from Reference 44 with permission from Elsevier.

11.2 Experimental

11.2.1 Materials

Polyvinylpyrrolidone (PVP), copper chloride dihydrate, zinc chloride, ortho-phosphoric acid, potassium permanganate ($KMnO_4$), concentrated sulfuric acid (H_2SO_4) and ammonia (20%) were received from Sigma Aldrich and were used without purification. Graphite powder was purchased from Sigma Aldrich with an average particle size of 10 µm.

11.2.2 Synthesis of Graphite Oxide (GO)

GO was synthesized using Tour's method [45]. Briefly, 300 mL of conc. H_2SO_4 and 75 mL of ortho-phosphoric acid were mixed together. Subsequently 10 g of graphite powder was added and the contents were stirred for 45 min at 30 °C. $KMnO_4$ (58 g) was added slowly to the mixture and the mixture was allowed to stir for 72 h at 30 °C. Subsequently, 1500 mL of 5% H_2O_2 solution was added to the mixture and stirred for 2 h at 30 °C. The content was allowed to settle and centrifuged, followed by removal of the superannuated solution. The residue was washed with 1 M HCl, followed by de-ionized water. The residue was further purified using cellulose membrane for 1 week in de-ionized water. Finally, the purified graphite oxide solution was freeze dried.

11.2.3 Preparation of Cu/ZnO/Graphene Nanocomposites

50 mL of 0.1 M $ZnCl_2$ and $CuCl_2$ each were taken in a beaker and 1 g of PVP was added to the mixture. The content was stirred for 15 min at 30 °C. Subsequently, the mixture was slowly added to 250 mL of GO dispersion (2 mg/mL solid content) under vigorous stirring at 30 °C. 20% ammonia solution was added to the mixture and pH was raised to 9, still maintaining the mixture at 30 °C. On adding ammonia solution, GO got reduced and the content was centrifuged to remove the superannuated solution. The content was washed with water (50 mL) followed by ethanol (50 mL). Subsequently, the sample was dried at 50 °C for 6 h. The as-synthesized dried hybrid sample was thermally treated at 600 °C for 2 h in nitrogen medium. The resultant compound was termed as CZG-1. The other two hybrids were prepared by the same procedure using 0.1 M $ZnCl_2$/0.05 M $CuCl_2$ and 0.1 M $ZnCl_2$/0.025 M $CuCl_2$ respectively. The corresponding hybrids were termed as CZG-2 and CZG-3. In addition, Cu/G (CG) and ZnO/G (ZG) hybrid materials were also generated using 0.1 M $CuCl_2.2H_2O$ and 0.1 M $ZnCl_2$.

11.2.4 Characterization

Wide-angle X-ray diffraction (WAXRD) analysis of the materials was performed on an analytical powder diffractometer (X'Pert PRO) using CuKα radiation (λ = 1.5406 Å). A zero-background holder was used to minimize the noise. The samples were step-scanned from 2θ = 5-60° using a step size of 2θ = 0.02° and a step time of 10 s. The morphology of hybrid samples was examined by transmission electron microscope (TEM) (FEI TECNAI) at 200 kV at room temperature without staining. The composition of elements was analyzed using energy dispersive X-ray analysis (EDX) FEI TECNAI, which was coupled with TEM instrument. The sample was dispersed in ethanol and the suspension was transferred to 400 mesh nickel electron microscopy grids (coated with holey carbon film). TEM image processing was performed using Digital Micrograph software (Gatan GIF 963, USA). The Bruner-Emmett-Teller (BET) surface area of the hybrids was measured using an ASAP-2010 surface analyzer. The samples were degassed at approximately 200 °C for 4 h before analysis. The morphology of the nanomaterials was also characterized in Hitachi field emission in-lens S-900 high resolution scanning electron microscope (SEM) at accelerating voltages of 10-20 kV. In addition, the optical properties of the Cu/ZnO/graphene nanocomposites were also studied using UV-vis diffused reflectance spectroscopy.

11.2.5 Photocatalytic Studies

The photocatalytic activity of the nanocomposites was investigated using a self-assembled apparatus with a metal halogen lamp (HQIBT, 400 W/D, OSRAM, Germany) as the visible light medium. 0.05 g sample of the hybrids was added to 100 ppm of MB dye solution in distilled water. The solution was stirred in dark for adsorption - desorption equilibrium for 30 min prior to the visible light irradiation (>430 nm). The concentration of MB prior to the irradiation was taken as the initial concentration. At specific time intervals, 2 mL sample was withdrawn by syringe from the irradiated suspension and the catalyst was removed using an in-built syringe filter. The λ_{max} of MB solution at 665 nm was determined with a calibration curve of MB (concentration vs. absorbance) using a Cary 500 UV-vis spectrophotometer (Varian, Palo Alto, CA).

11.3 Results and Discussion

Nanoparticle precursor solutions such as $ZnCl_2$ and $CuCl_2.2H_2O$ underwent hydrolysis in the presence of ammonia, thus, resulting in decorated nanoparticles in hydroxyl form on GO sheets and the dark brown colored colloidal GO solution turned simultaneously into black precipitate indicating partial reduction of GO. The hybrid metal/metal oxide/graphene were formed by annealing the hydroxyl form of the metal and metal oxide nanoparticles decorated on partially reduced GO at 600 °C. The powder X-ray diffraction patterns of CG, ZG, and CZG hybrid nanostructures are presented in Figure 11.2. The Cu nanoparticles decorated on graphene sheets displayed characteristic peaks at 2θ = 43.7° and 50.5°, which indicated the formation of a face-centered cubic (FCC) copper phase without significant oxides or other impurities [46]. The ZG hybrid exhibited characteristic peak of ZnO at 2θ = 31.8°, 34.4°, 36.2°, 47.7° and 56.5° corresponding to the characteristic ZnO crystal planes of (100), (002), (101), (102) and (110). This indicated that the ZnO nanoparticles had hexagonal wurtizite structure on the graphene plane which was in good agreement with the JCPDS card 36-1451 [47]. In the case of the CZG nanocomposites, characteristic peaks of both ZnO and Cu were observed. This indicated that the Cu and ZnO were distributed on the graphene surface with FCC copper phase as well as ZnO hexagonal wurtizite structure. In each hybrid, graphene exhibited a weak broad peak at 2θ = 26.5° corresponding to (002) plane.

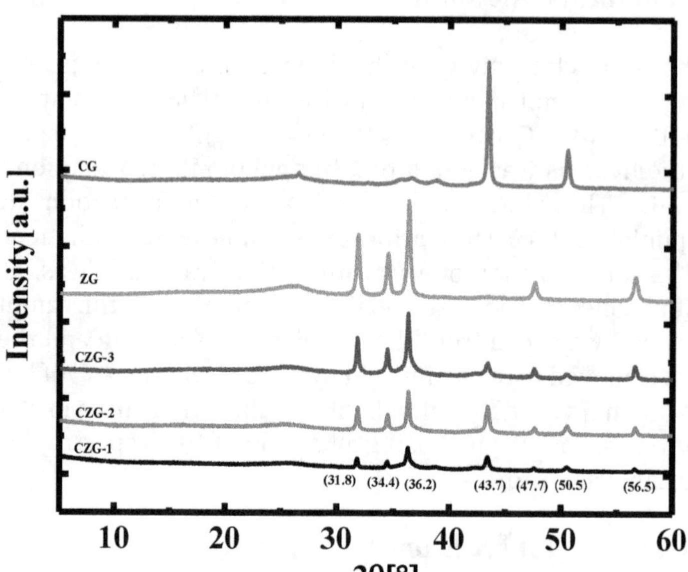

Figure 11.2 XRD spectra of CG, ZG and CZG nanocomposites.

The surface area of the nanomaterials is a key factor driving their adsorption as well as catalytic applications. Table 11.1 represents the textural properties of the hybrid materials in comparison with thermally reduced graphene (TRG). TRG was prepared from GO at 600 °C in an inert atmosphere and exhibited a surface area of 446 m^2/g with pore volume and pore radius of 5.2 cc/g and 18.8 Å respectively. After nanoparticle decoration on graphene, the surface area and pore volume of graphene were significantly reduced for most of the hybrids, however, there was no change in the pore radius. In addition, the surface area of hybrid CZG-2 exhibited almost same value as TRG with reduced pore volume of 2.2 cc/g. This indicated that the nanoparticles in CZG-2 were very small in size, which only filled the pores of the graphene without altering the surface area of graphene. These finding are in good agreement with the previously reported literature studies [27].

Table 11.1 Textural properties of hybrid nanocomposites

Sample	Surface area (m^2/g) (BET method)	Pore volume (BJH method) (cc/g)	Pore radius (BJH method) (Å)
CZG-1	140.3	1.3	19.1
CZG-2	403.7	2.2	19.2
CZG-3	160.6	1.2	19.1
ZG	101.3	0.4	19.1
CG	113.0	0.4	18.9
TRG	446.0	5.2	18.8

The morphology and composition of the hybrid samples were analyzed using TEM and EDX analysis. Figures 11.3(a) and (b) show the TEM images of the CZG-1 and CZG-2 hybrids respectively. It was evident that the graphene sheets had numerous folds and wrinkles which contributed to a higher surface area. The ZnO and Cu particles were well separated and uniformly distributed throughout the graphene sheets and had a uniform morphology. In addition, Cu and ZnO in hybrid nanomaterials exhibited hollow spherical structures with average size of 10-20 nm. However, in the CG and ZG hybrids, the nanoparticles exhibited only a solid spherical morphology with size of 10-20 nm (Figure 11.3(c) and (d)). Figure 11.4a shows the EDX spectrum of the CZG-2. Key elements such as Zn, Cu and C were observed to be present in the hybrid without the trace of any impurities. The presence of Ni could be attributed to the grids used for the analysis.

The SEAD pattern of CZG-2 also exhibited the line ring pattern which indicated the polycrystalline nature of Cu/ZnO in the graphene sheets (Figure 11.4b).

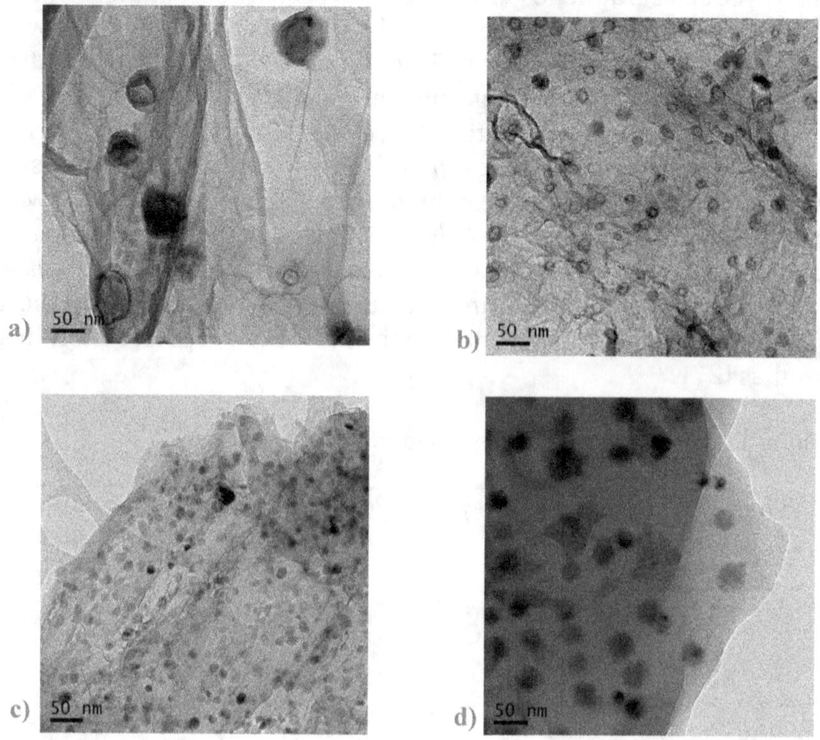

Figure 11.3 TEM images of a) CZG-1, b) CZG-2, c) CG and d) ZG.

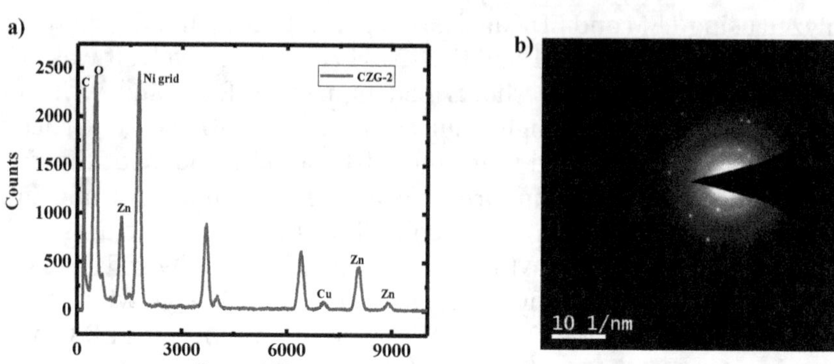

Figure 11.4 a) EDX spectrum of CZG-2 and b) SAED pattern of CZG-2.

Further, Figure 11.5 shows the SEM images of CG. ZG and CZG hybrids. The highly dispersed and random arrangement of the ZnO and Cu particles on the graphene sheets was confirmed from the images. The graphene platelets were also observed to be well exfoliated into single or few layer sheets.

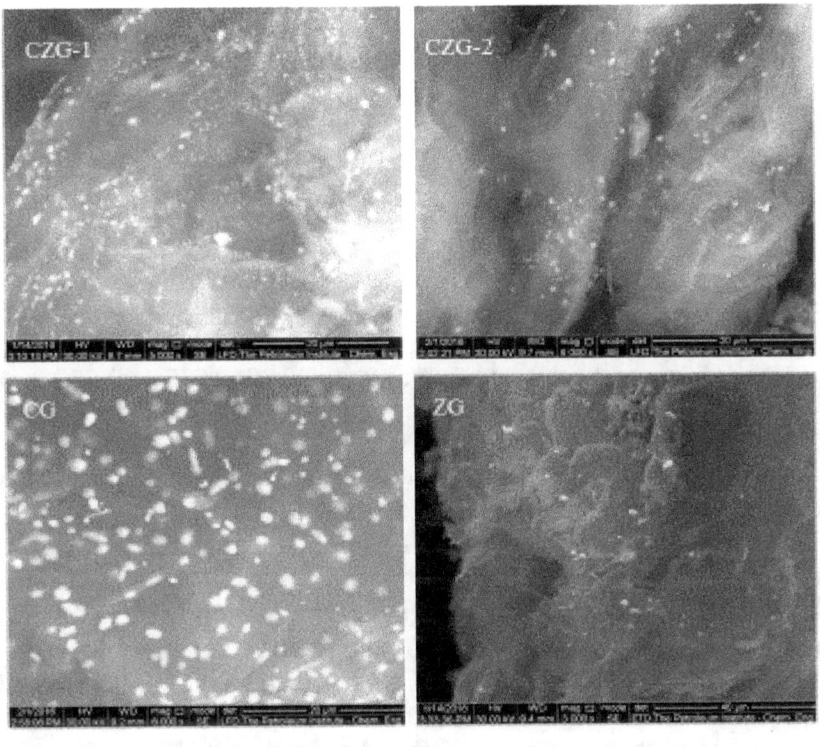

Figure 11.5 SEM images of CZG-1, CZG-2, CG and ZG.

The optical properties of the Cu/ZnO/graphene nanocomposites, studied using UV-vis diffused reflectance spectroscopy, are presented in Figure 11.6a. The optical band gap energy was calculated from the absorption spectra using Tauc's relation [48]:

$$(\alpha h\nu)^n = A\,(h\nu - E_g)$$

where α is the absorption coefficient, h is the Planck constant, ν is the light frequency, E_g is the band gap, n is either ½ for an indirect transition or 2 for a direct transition and A is a constant. The absorption band at around 200 nm was attributed to the π- π^* transition of C=C

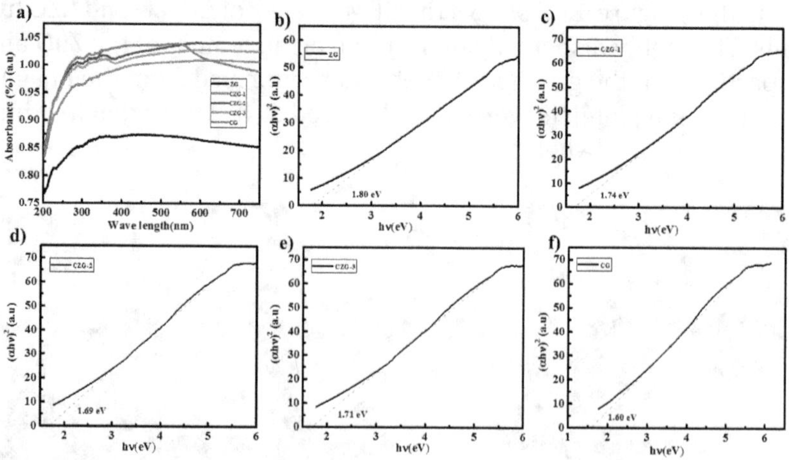

Figure 11.6 a) UV–vis spectra of (a) CZG, CG and ZG nanocomposites (curves from top to bottom in the plot correspond to CG, CZG-2, CZG-3, CZG-1 and ZG respectively), and Tauc's plots of b) ZG, c) CZG-1, d) CZG-2, e) CZG-3 and f) CG.

in the graphitic structure and the appearance of a shoulder peak around 300 nm was due to the n- π* transition of C=O [49]. Among the ternary nanocomposites, CZG-2 exhibited higher absorbance, indicating that the optimum content of metallic Cu increased the surface electric charge of ZnO [50]. Figures 11.6 b-f show the Tauc's plots of ZG, CZG-1, CZG-2, CZG-3 and CG. The observed energy band gaps were ZG = 1.8 eV, CZG-1 = 1.74 eV, CZG-2 = 1.69 eV, CZG-3 = 1.71 eV and CG = 1.60 eV respectively. From these values for the hybrids, it was observed that graphene exhibited a significant influence on the electronic energy level in the nanomaterials (for instance, ZG exhibited a narrow energy band gap of 1.8 eV compared to pure ZnO (3.04 eV)) [51]. The doping of Cu on the ZnO/graphene resulted in a decrease in band gap energy. The change in band gap energy might be due to quantum confinement effect and spin exchange interaction between the sp electrons of host and d electrons of dopant. The exchange of electrons led to negative correction in conduction band (CB) and positive correction in valance band, which resulted in a narrow band gap correction in the nanocomposites [52,53]. Among the ternary composites, CZG-2 exhibited a lower energy band gap, probably due to its small particle size [54].

The photocatalytic efficiency of hybrid materials was studied by the degradation of MB dye solution under visible light irradiation in

UV-vis spectroscopy. The strong absorbance peak at 665 nm was attributed to the monomeric hetero-polyaromatic linkage of MB dye which diminished due to dye degradation with time. The shoulder peak in the UV-vis spectrum at 292 nm, due to benzene ring of MB, has also been reported to exhibit reduction with degradation over time [10]. Figure 11.7 shows the UV spectra of degradation behavior

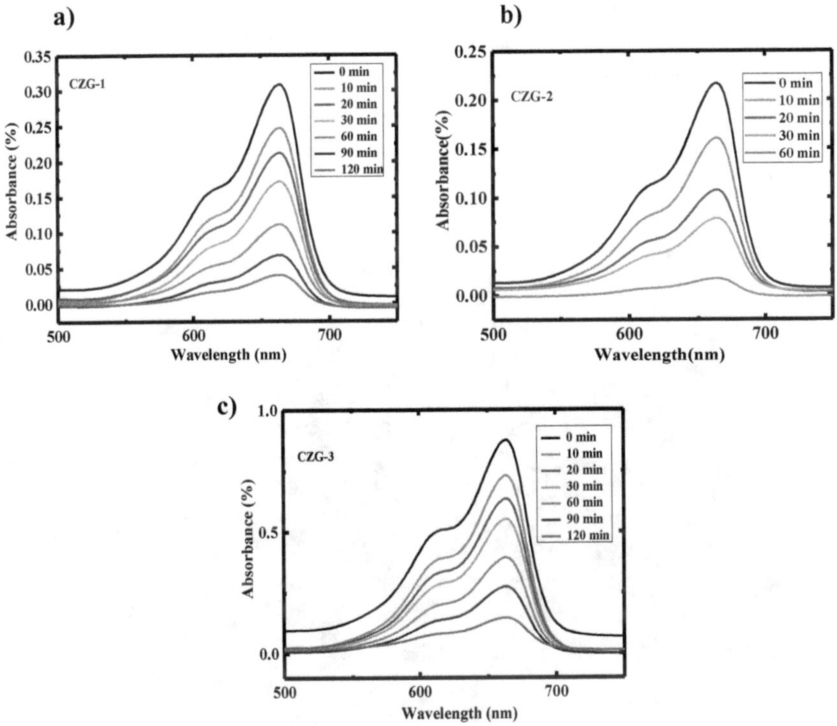

Figure 11.7 UV spectra of degradation behavior of MB dye solution: a) CZG-1, b) CZG-2 and c) CZG-3 catalysts at different time intervals. The curves in the figures correspond to the curves in the legends in the same sequence from top to bottom.

of MB dye solution with the three different hybrids (CZG-1, CZG-2, and CZG-3) as a function of time. The disappearance of the band at 665 nm confirmed the degradation of MB at specific time intervals. CZG-2 degraded the MB within 60 min of irradiation, while CZG-1 and CZG-3 took 2 h for the complete degradation of MB. This indicated that 0.05 M Cu loading provided optimum level of photocatalytic efficiency in the CZG-2 hybrid. The higher (CZG-1) and lower (CZG-3) loading of Cu nanoparticles in the hybrids resulted in lower photocatalytic efficiency as compared to CZG-2. As the surface area is an important factor contributing to the photocatalytic activity of the nanomaterials, higher surface area of 403 m^2/g for CZG-2 would have contributed to the observed accelerated MB degradation. Feng *et al.* [55] also studied the morphology dependent photocatalytic properties of hollow octahedral and spherical Cu_2O nanocrystals and reported that the hollow morphology exhibited higher photocatalytic efficiency as compared to the solid morphology [55]. Thus, hollow morphology achieved for CZG hybrids in the study would have helped to achieve higher efficiency for degradation of MB. Figure 11.8 also

a)

b)

c)

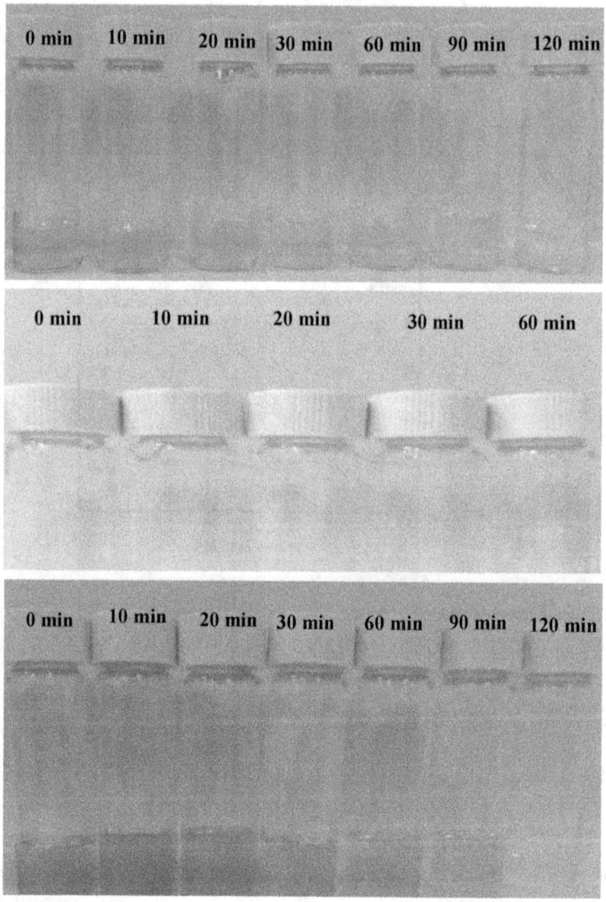

Figure 11.8 Visual observation of degradation behavior of MB dye solution: a) CZG-1, b) CZG-2 and c) CZG-3 catalysts at different time intervals.

shows the visual observation of change in color of the MB dye solution for different photocatalysts under visible light irradiation as a function of time. Increase in irradiation time indicated the disappearance of the color of MB solution linearly. The de-colorization efficiency of photocatalyst was inversely proportional to the concentration of dye solution. Thus, the order of decolourization efficiency of catalyst was as follows CZG-2 > CZG-3 > CZG-1 > ZG > CG. Figure 11.9

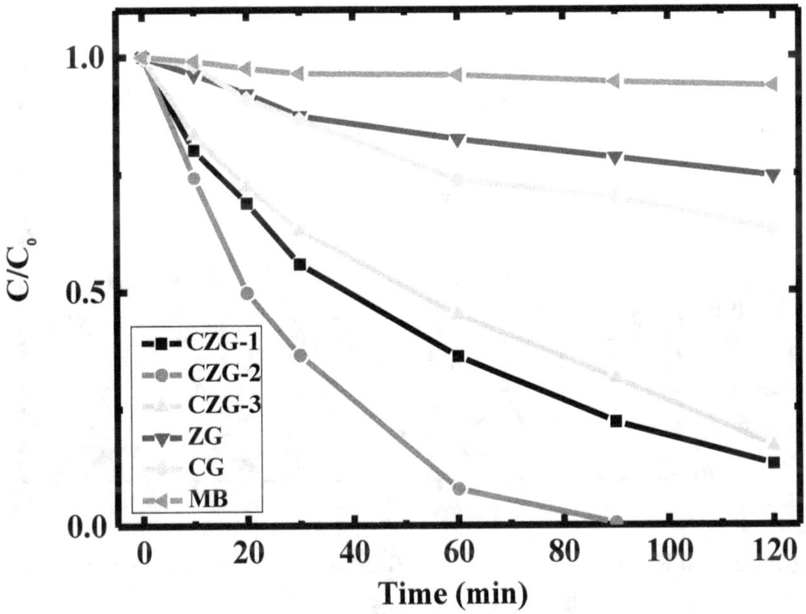

Figure 11.9 Change in concentration of MB dye solution with time.

also quantitatively shows the change in the concentration of MB dye solution with respect to time. In the absence of the catalyst, MB remained un-degraded even after the irradiation of visible light for more than 2 h. The whole process was fitted into a pseudo-first order reaction to gain more insights about the kinetics of the reaction. The rate constant was taken as the slope of the linear fitting curve, using the equation:

$$\ln (C_0/C) = K_{app}t$$

where C_0 is the initial methylene blue concentration, C is the concentration of MB at time t and K_{app} is the rate constant. As mentioned above, the maximum catalytic activity for the CZG-2 could be correlated to its highest surface area among the hybrids. High surface area led to higher number of adsorption sites and reaction centers, as well as faster flow rate of the molecules. This also correlated with the literature report that the reaction kinetics of organic pollutants can be described with Langmuir-Hinshelwood model [56]. Figure 11.10 also

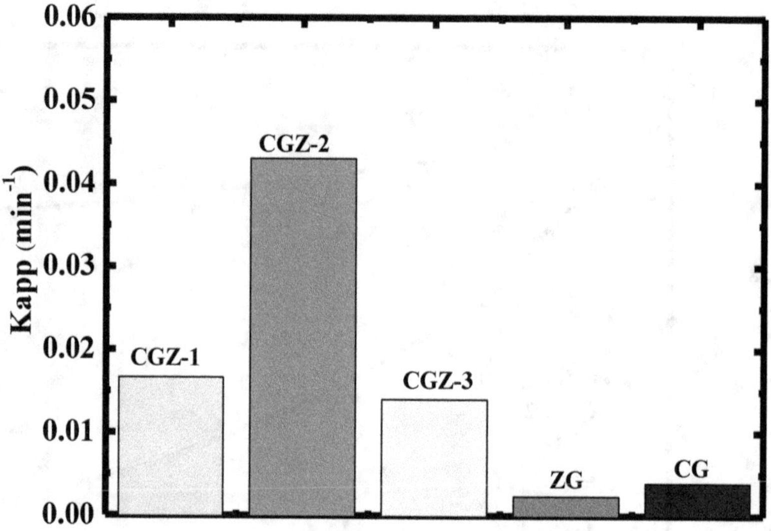

Figure 11.10 K_{app} values for CZG-1, CZG-2, CZG-3, CG and ZG.

compares the magnitude of K_{app} for CZG-1, CZG-2, CZG-3, CG and ZG. It has been reported that the average concentration of the dyes before and after irradiation (C/C_0) is proportional to the average change in the maximum absorbance values of dyes before and after irradiation (A/A_0) at different time intervals [7]. Based on this, the photocatalytic efficiency (η) of all samples was calculated by using the following equation:

$$\eta = (1 - C/C_0) \times 100$$

Photocatalytic efficiency of the samples at different time intervals is depicted in Table 11.2. CZG-2 degraded 93% of MB within 60 min of irradiation. It was also observed that the photocatalytic efficiency of CZG-2 was almost 5 times higher than the rate constant of ZG and 3 times higher than that of CG. Thus, the observed results indicated

Table 11.2 Photocatalytic efficiency of catalyst with time

Time (min)	Photocatalytic efficiency of hybrids (%)				
	CZG-1	CZG-2	CZG-3	ZG	CG
10	19.8	25.8	16.7	3.8	1.8
20	31.3	50.3	27.6	7.8	9.0
30	44.3	63.3	37.2	12.6	13.4
60	64.0	92.3	55.0	17.6	26.3
90	78.0		68.7	21.5	30.2
120	87.0		82.3	25.5	37.0

the superior photocatalytic efficiency of the CZG hybrid nanomaterials compared to the individual metal and metal oxides such as Cu and ZnO doped graphene (CG and ZG) hybrids. Due to the LSPR effect, the Cu nanoparticles in the ternary hybrid acted as light scavenger which enhanced the light energy of the absorption and transferred the photonic energy to semiconducting ZnO-graphene *via* RET mechanism. ZnO-graphene received the photonic energy due to the dipole-dipole interaction between the Cu and ZnO-graphene. In addition, the LSPR effect decreased the recombination effect of photo-generated electron hole pair which contributed to the higher photocatalytic efficiency of the hybrid CZG [57]. The photo-excited electrons in the conduction band of ZnO readily transferred to Cu nanoparticles which acted as electron sink due to the Schottky barrier effect at the Cu/ZnO/graphene interface [58]. In absence of Cu, the fast recombination of electron hole pairs of ZnO may result in minimized photocatalytic activity of the hybrid. Similarly, in absence of graphene, the ZnO nanoparticles may possibly convert the Cu nanoparticles into Cu_2O [59]. Thus, use of graphene as a stabilizer as well as electron hole acceptor helped to synergistically improve the photocatalytic performance. Thus, as also shown in Figure 11.11, the charge separation in the metal-semiconductor interface and negatively charged graphene prevented the recombination of electron hole pairs and, thus, enhanced the photocatalytic efficiency. At the same time, the adsorption on semiconducting ZnO-graphene hybrid could be enhanced by coupling with plasmonic Cu nanostructures *via* a local field enhancement effect, leading to an enhanced photocatalytic property of the Cu/ZnO/graphene hybrids. During the photoreaction, the holes in the valence bands of ZnO could oxidize water or hydroperoxide radicals to generate the hydroxyl radicals as well as super-oxides which degraded the dye into non-toxic CO_2 and water (see Figure 11.11 for

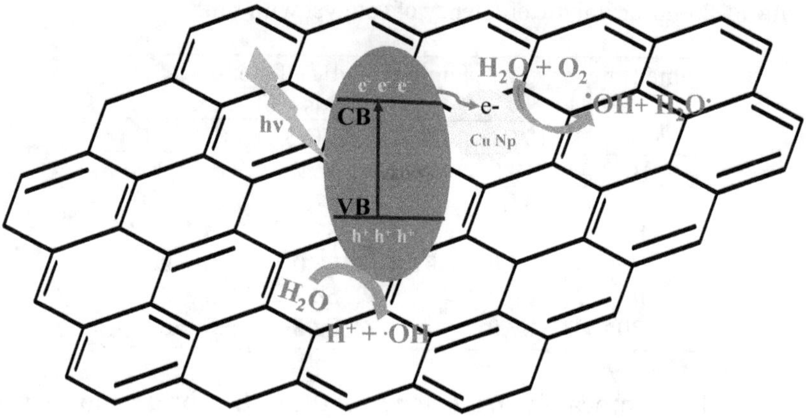

Figure 11.11 Schematic representation of photocatalytic mechanism of Cu/ZnO/graphene nano-hybrid.

the mechanism). This was in good agreement with the earlier reported photocatalytic study of Au/ZnO-rGO nanocomposites [60]. In addition, as mentioned above, the hollow sphere morphology and higher surface area of the ternary hybrids synergistically contributed to the higher photocatalytic efficiency. The photocatalytic efficiency of the CZG hybrids generated in the current study was also compared with the previously reported metal and metal oxide doped graphene nanocomposites, as shown in Table 11.3. It was obvious that the photocatalytic efficiency varied in different nanocomposites due to the experimental conditions, nature of the metal and metal oxides, reaction time and light medium, however, overall, Cu/ZnO/graphene hybrid catalyst reported in the present work exhibited better photocatalytic efficiency in faster degradation of methylene blue solution under visible light.

The practical applications of the developed photocatalysts rely on their effective reusability, thus, also optimizing the cost. To demonstrate the cyclic stability and reusability of the generated photocatalysts, recycling experiments were carried out with CZG-2 under visible light irradiation up to five cycles. The nanomaterial exhibited significant recyclability, as depicted in Figure 11.12. The decrease in absorption capacity after each run might be due to the difficulty in washing away the dye molecules from the active sites. The reduction in efficiency might also be due to the loss of some photocatalyst material after every run. However, overall, the efficiency did not change significantly even after five cycles, thus, confirming the effectiveness

Table 11.3 Comparison of photocatalytic efficiency of Cu/ZnO/graphene hybrids with other catalysts reported in literature

Photocatalyst	Photocatalytic efficiency of the catalyst in degrading methylene blue solution		Ref.
	Time (min)	Efficiency (%)	
Graphene/ZnO (UV light)	120	68	7
ZnO-graphene (UV-visible near IR range)	140	82	61
ZnO-rGO (UV light)	260	88	23
Copper doped ZnO	120	97.5	62
Mn doped ZnO/graphene (visible light)	180	91	63
Au doped ZnO-graphene (periodic black light)	150	80.5	60
Graphene-Ag/ZnO (sun light)	420	90	64
Cu/ZnO/graphene (visible light)	60	93	Present work

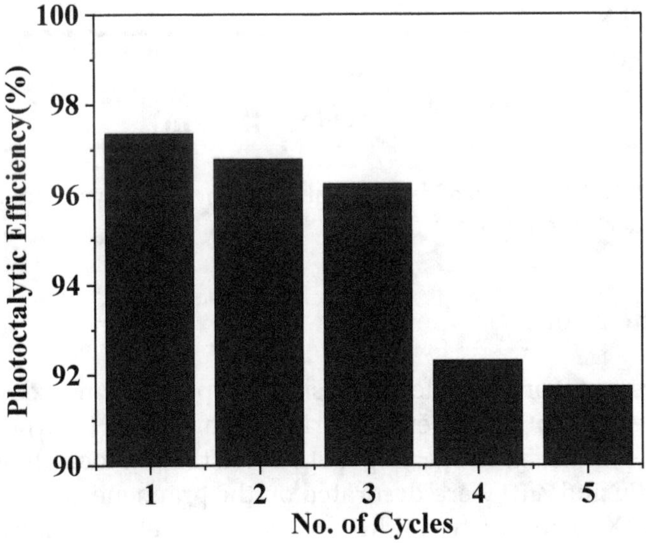

Figure 11.12 Photocatalytic efficiency of the catalyst vs. number of cycles.

of the developed CZG nanocomposites for the removal of aquatic contaminants. Also, the photocatalytic efficiency was confirmed by reacting the CZG-2 nanocomposite in dark for 2 h. From Figure 11.13, it can be seen that even after 2 h of continuous stirring in dark, the color of the dye remained almost unchanged. Therefore, the nanocomposite exhibited an effective methylene blue degradation under visible light medium.

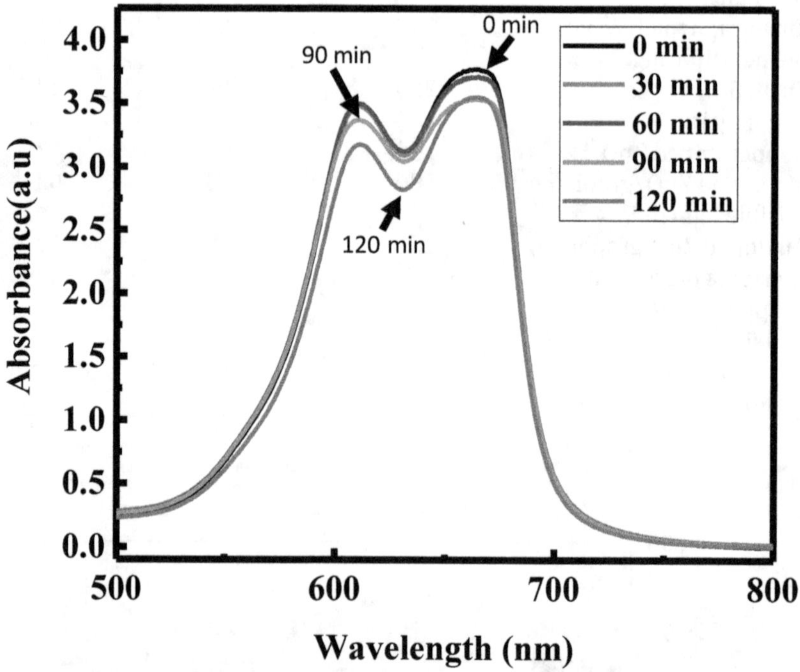

Figure 11.13 UV spectra of degradation behavior of MB dye solution using CZG-2 under dark medium.

11.4 Conclusions

In the present work, a facile synthesis of copper and zinc oxide doped graphene (Cu/ZnO/graphene) hybrid nanomaterials (CZG) has been reported *via* thermal annealing method for the photocatalytic applications. Cu and ZnO were decorated on the graphene sheets as confirmed by X-ray diffraction (XRD), transmission electron microscopy (TEM) and energy dispersive X-ray (EDX) analysis. The nanoparticles in the ternary hybrids exhibited hollow nano-sphere morphology, whereas the nanoparticles in the Cu/graphene (CG) and ZnO/graphene (ZG) hybrids had a solid spherical morphology on the graphene sheets. The photocatalytic activity of the CZG nanocomposites was confirmed to be superior as compared to CG and ZG hybrids. The photocatalysis experiments demonstrated that the hybrid graphene nanocomposite with composition of 0.5 M Cu/1 M ZnO (CZG-2) had 93% photocatalytic efficiency under visible light after 60 min of visible light exposure. Also, the CZG-2 nanocomposite exhibited retention of the photocatalytic efficiency to 92% after 5 cycles. In comparison, the photocatalytic activity of ZnO/graphene and Cu/graphene composites exhibited 26% and 37% photocatalytic efficiency respectively after 2 h of irradiation. The higher photocatalytic efficiency of hybrid CZG can be attributed to the synergetic effect of localized surface plasmon resonance (LSPR) effect and semiconducting mechanism as well as higher surface area and porosity.

Acknowledgement

The authors thank S. Stephen and A. Tharalekshmy for experimental support. L.J. and M.R.V. contributed equally to the work.

References

1. Wang, L., Shang, J., Hao, W., Jiang, S., Huang, S., Wang, T., Sun, Z., Du, Y., Dou, S., Xie, T., Wang, D., and Wang, J. (2014) A dye-sensitized visible light photocatalyst-Bi24O31Cl10. *Scientific Reports*, **4**, 7384.
2. Tayade, R. J., Natarajan, T. S., and Bajaj, H. C. (2009) Photocatalytic degradation of methylene blue dye using ultraviolet light emitting diodes. *Industrial and Engineering Chemistry Research*, **48**, 10262-10267.
3. Madhusudan, P., Zhang, A. J., Cheng, B., and Liu, G. (2013) Photocatalytic degradation of organic dyes with hierarchical Bi2O2CO3 microstructures under visible-light. *CrystEngComm*, **15**(2), 231-240.
4. Thirumalairajan, S., Girija, K., Mastelaro, V. R., and Ponpandian, N. (2014) Photocatalytic degradation of organic dyes under visible light irradiation by floral-like $LaFeO_3$ nanostructures comprised of nanosheet petals. *New Journal of Chemistry*, **38**(11), 5480-5490.
5. Tanaka, K., Padermpole, K., Hisanaga, T., (2000) Photocatalytic degradation of commercial azo dyes. *Water Research*, **34**, 327-333.
6. Vautie, M., Guillard, C., and Herrmann, J. M. (2001) Photocatalytic degradation of dyes in water: case study of indigo and of indigo carmine. *Journal of Catalysis*, **201**(1), 46-59.

7. Thangavel, S., Krishnamoorthy, K., Krishnaswamy, V., Raju, N., Kim, S. J., and Venugopal, G. (2015) Graphdiyne–ZnO nanohybrids as an advanced photocatalytic material. *Journal of Physics Chemistry C*, **119**, 22057-22065.
8. Ajmal, A., Majeed, I., Malik, R. N., Idriss, H., and Nadeem, M. A. (2014) Principles and mechanisms of photocatalytic dye degradation on TiO_2 based photocatalysts: a comparative overview. *RSC Advances*, **4**(70), 37003-37026.
9. Zou, Z., Ye, J., Sayama, K., and Arakaw, H. (2001) Direct splitting of water under visible light irradiation with an oxide semiconductor photocatalyst. *Nature*, **414**, 625-627.
10. Djurisic, A. B., Leung, Y. H., and Ng, A. M. C. (2014) Strategies for improving the efficiency of semiconductor metal oxide photocatalysis. *Materials Horizons*, **1**, 400-410.
11. Yu, J., and Yu, X. (2008) Hydrothermal synthesis and photocatalytic activity of zinc oxide hollow spheres. *Environmental and Science Technology*, **42**(13), 4902-4907.
12. Velayutham, T. S., Majid, W. H. A., Gan, W. C., Zak, A. K., and Gan, S. N. (2012) Theoretical and experimental approach on dielectric properties of ZnO nanoparticles and polyurethane/ZnO nanocomposites. *Journal of Applied Physics*, **112**(5), 054106.
13. Chen, K. J., Fang, T. H., Hung, F. Y., Ji, L. W., Chang, S. J., Young, S. J., and Hsiao, Y. J. (2008) The crystallization and physical properties of Al-doped ZnO nanoparticles. *Applied Surface Science*, **254**(18), 5791-5795.
14. Wang, Z. H., Geng, D. Y., Han, Z., and Zhang, Z. D. (2009) Characterization and optical properties of ZnO nanoparticles obtained by oxidation of Zn nanoparticles. *Material Letters*, **63**(29), 2533-2535.
15. Chen, J. W., Shi, J. W., Wang, X., Ai, H. Y., Cui, H. J., and Fu, M. L. (2013) Hybrid metal oxides quantum dots/TiO_2 block composites: Facile synthesis and photocatalysis application. *Powder Technology*, **246**, 108-116.
16. Khan, M. M., Adil, S. F., and Mayouf, A. A. (2015) Metal oxides as photocatalysts. *Journal of Saudi Chemical Society*, **19**(5), 462-464.
17. Zhang, H., Fan, X., Quan, X., Chen, S., and Yu, H. (2011) Graphene sheets grafted Ag@ AgCl hybrid with enhanced plasmonic photocatalytic activity under visible light. *Environmental Science and Technology*, **45**(13), 5731-5736.
18. Xiong, Z., Zhang, L. L., and Zhao, X. S. (2011) Visible-light-induced dye degradation over copper-modified reduced graphene oxide. *Chemistry - A European Journal*, **17**(8), 2428-2434.
19. Bhunia, S. K., and Jana, N. R. (2014) Reduced graphene oxide-silver nanoparticle composite as visible light photocatalyst for degradation of colorless endocrine disruptors. *ACS Applied Materials and Interfaces*, **6**(22), 20085-20092.

20. Khalid, N. R., Ahmed, E., Hong, Z., Ahmad, M., Zhang, Y., and Khalid, S. (2013) Cu-doped TiO_2 nanoparticles/graphene composites for efficient visible-light photocatalysis. *Ceramics International*, **39**(6), 7107-7113.
21. Luo, G., Jiang, X., Li, M., Shen, Q., Zhang, L., and Yu, H. (2013) Dye-sensitization-induced visible-light reduction of graphene oxide for the enhanced TiO_2 photocatalytic performance. *ACS Applied Materials and Interfaces*, **5**(8), 2924-2929.
22. Bai, X., Wang, L., Zong, R., Lv, Y., Sun, Y., and Zhu, Y. (2013) Performance enhancement of ZnO photocatalyst via synergic effect of surface oxygen defect and graphene hybridization. *Langmuir*, **29**(9), 3097-3105.
23. Lv, T., Pan, L., Liu, X., and Sun, Z. (2012) Enhanced photocatalytic degradation of methylene blue by ZnO-reduced graphene oxide-carbon nanotube composites synthesized via microwave-assisted reaction. *Catalysis Science and Technology*, **2**(11), 2297-2301.
24. Yin, D., Le, Z., Liu, B., and Wu, M. (2012) Preparation and characterization of ZnO-graphene composite photocatalyst. *Journal of Nanoscience and Nanotechnology*, **12**(2), 937-942.
25. Herring, N. P., Almahoudi, S. H., Olson, C. R., and Shall, M. S. E., (2012) Enhanced photocatalytic activity of ZnO-graphene nanocomposites prepared by microwave synthesis. *Journal of Nanoparticle Research*, **14**, 1277-1289.
26. Lv, R., Wang, X., Lv, W., Xu, Y., Ge, Y., He, H., Li, G., Wu, X., Li, X., and Li, Q. (2015) Facile synthesis of ZnO nanorods grown on graphene sheets and its enhanced photocatalytic efficiency. *Journal of Chemical Technology and Biotechnology*, **90**(3), 550-558.
27. Saravanan, R., Karthikeyan, S., Gupta, V. K., Sekaran, G., Narayanan, V., and Stephen, A. (2013) Enhanced photocatalytic activity of ZnO/CuO nanocomposite for the degradation of textile dye on visible light illumination. *Materials Science and Engineering C*, **33**(1), 91-98.
28. Karunakaran, C., Abiramasundari, G., Gomathisankar, P., Manikandan, G., and Anandi, V. (2011) Preparation and characterization of ZnO-TiO_2 nanocomposite for photocatalytic disinfection of bacteria and detoxification of cyanide under visible light. *Materials Research Bulletin*, **46**(10), 1586-1592.
29. Bai, H., Liu, Z., and Sun, D. D. (2011) Hierarchical ZnO/Cu "corn-like" materials with high photodegradation and antibacterial capability under visible light. *Physical Chemistry Chemical Physics*, **13**, 6205-6210.
30. Xu, C., Cao, L., Su, G., Liu, W., Liu, H., Yu, Y., and Qu, X. (2010) Preparation of ZnO/Cu_2O compound photocatalyst and application in treating organic dyes. *Journal of Hazardous Materials*, **176**(1-3), 807-813.

31. Ren, C., Yang, B., Wu, M., Xu, J., Fu, Z., Lv, Y., Guo, T., Zhao, Y., and Zhu, C. (2010) Synthesis of Ag/ZnO nanorods array with enhanced photocatalytic performance. *Journal of Hazardous Materials,* **182**(1-3), 123-129.
32. Lin, Y. G., Hsu, Y. K., Chen, Y. C., Wang, S. B., Miller, J. T., Chen, L. C., and Chen, K. H. (2012) Plasmonic $Ag@Ag_3(PO_4)_{1-x}$ nanoparticle photosensitized ZnO nanorod-array photoanodes for water oxidation. *Energy and Environmental Sci*ence, **5**, 8917-8922.
33. Wu, J. -J., and Tseng, C. H., (2006) Photocatalytic properties of nc-Au/ZnO nanorod composites. *Applied Catalysis B: Environmental,* **66**(1-2), 51-57.
34. Zhang, P., Shao, C., Li, X., Zhang, M., Zhang, X., Sun, Y., and Liu, Y. (2012) In situ assembly of well-dispersed Au nanoparticles on TiO2/ZnO nanofibers: A three-way synergistic heterostructure with enhanced photocatalytic activity. *Journal of Hazardous Materials,* **237-238**, 331-338.
35. Pelaez, M., Nolan, N. T., Pillai, S. C., Seery, M. K., Falaras, P., Kontos, A. G., Dunlop, P. S. M., Hamilton, J. W. J., Byrne, J. A., O'Shea, K., Entezari, M. H., and Dionysiou, D. D. (2012) A review on the visible light active titanium dioxide photocatalysts for environmental applications. *Applied Catalysis B: Environmental,* **125**, 331-349.
36. Moniz, S. J. A., Shevlin, S. A., Martin, D. J., Guob, Z. X., Tang, J. (2015) Visible-light driven heterojunction photocatalysts for water splitting–a critical review. *Energy and Environmental Sci*ence, **8**(3), 731-759.
37. Chan, G. H., Zhao, J., Hicks, E. M., Schatz, G. C., and Van Duyne, R. P. (2007) Plasmonic properties of copper nanoparticles fabricated by nanosphere lithography. *Nano Letters,* **7**(7), 1947-1952.
38. Marimuthu, A., Zhang, J. W., and Linic, S. (2013) Tuning selectivity in propylene epoxidation by plasmon mediated photo-switching of Cu oxidation state. *Science,* **339**(6), 1590-1593.
39. Ghodselahi, T., Zahrabi, H., Saani, M. H., and Vesaghi, M. A. (2011) CO gas sensor properties of Cu@ CuO core–shell nanoparticles based on localized surface plasmon resonance. *Journal of Physical Chemistry C,* **115**(45), 22126-22130.
40. Pedersen, D. B., and Wang, S. (2007) Surface plasmon resonance spectra of 2.8±0.5 nm diameter copper nanoparticles in both near and far fields. *Journal of Physical Chemistry C,* **111**(47), 17493-17499.
41. Chou. H.-T., Huang, W.-H., Wu, T.-M., Yu, Y.-K., and Hsu, H.-C. (2017) LSPR effects of Au nanoparticles/ZnO nano-composite films. *Sensing and Bio-Sensing Research,* **14**, 17-20.
42. Guo, X., Hao, C., Jin, G., Zhu, H.-Y., and Guo, X.-Y. (2014) Copper nanoparticles on graphene support: an efficient photocatalyst for coupling of nitroaromatics in visible light. *Angewandte Chemie Interna-*

tional Edition, **53**(7), 1973-1977.

43. Cushing, S. K., Li, J., Meng, F., Senty, T. R., Suri, S., Zhi, M., Li, M., Bristow, A. D., and Wu, N. (2012) Photocatalytic activity enhanced by plasmonic resonant energy transfer from metal to semiconductor. *Journal of American Chemical Society*, **134**(36), 15033-15041.
44. Jacob, L., Prasanna, K., Vengatesan, M. R., Santhoshkumar, P., Lee, C. W., and Mittal, V. (2018) Binary Cu/ZnO decorated graphene nanocomposites as an efficient anode for lithium ion batteries. *Journal of Industrial and Engineering Chemistry*, **59**, 108-114.
45. Marcano, D. C., Kosynkin, D. V., Berlin, J. M., Sinitskii, A., Sun, Z., Slesarev, A., Alemany, L. B., Lu, W., and Tour, J. M. (2010) Improved Synthesis of Graphene Oxide. *ACS Nano*, **4**(8), 4806-4814.
46. Kumar, A., Saxena, A., De, A., Shankar, R., and Mozumdar, S., (2013) Facile synthesis of size-tunable copper and copper oxide nanoparticles using reverse microemulsions. *RSC Advances*, **3**, 5015-5021.
47. Gayathri, S., Jayabal, P., Kottaisamy, M., and Ramakrishnan, V. (2014) Synthesis of ZnO decorated graphene nanocomposite for enhanced photocatalytic properties. *Journal of Applied Physics*, **115**, 173504.
48. Saranya, M., Ramachandran, R., and Wang, F. (2016) Graphene-zinc oxide (G-ZnO) nanocomposite for electrochemical supercapacitor applications. *Journal of Science: Advanced Materials and Devices*, **1**(4), 454-460.
49. Shown, I., Hsu, H. -C., Chang, Y. -C., Lin, C. -H., Roy, P. K., Ganguly, A., Wang, C. -H., Chang, J. -K., Wu, C. -I, Chen, L -C., and Chen, K. -H. (2014) Highly efficient visible light photocatalytic reduction of CO_2 to hydrocarbon fuels by cu-nanoparticle decorated graphene oxide. *Nano Letters*, **14**, 6097-6103.
50. Ravichandran, K., Chidhambaram, N., and Gobalakrishnan, S. (2016) Copper and graphene activated ZnO nanopowders for enhanced photocatalytic and antibacterial activities. *Journal of Physics and Chemistry of Solids*, **93**, 82-90.
51. Xu, J., Cui, Y., Han, Y., Hao, M., and Zhang, X. (2016) ZnO-graphene composites with high photocatalytic activities under visible light. *RSC Advances*, **6**, 96778-96784.
52. Zahirullah, S. S., Prince, J. J., and Inbaraj, P. F. H. (2017) Structural and optical properties of Cu-doped ZnO nanorods by silar method. *Materials Technology*, **32**(12), 755-763.
53. Robak, E., Coy, E., Kotkowiak, M, Jurga, S., Załęski, K., and Drozdowski, H. (2016) The effect of Cu doping on the mechanical and optical properties of zinc oxide nanowires synthesized by hydrothermal route. *Nanotechnology*, **27**, 175706.
54. Salaken, S. M., Farzana, and E., Podde, J. (2013) Effect of Fe-doping on the structural and optical properties of ZnO thin films prepared by spray pyrolysis. *Journal of Semiconductors*, **34**, 073003.

55. Feng, L., Zhang, C., Gao, G., and Cui, D. (2012) Facile synthesis of hollow Cu_2O octahedral and spherical nanocrystals and their morphology-dependent photocatalytic properties. *Nanoscale Research Letters*, **7**, 276.
56. Zhang, X., Qin, J., Hao, R., Wang, L., Shen, X., Yu, R., Limpanart, S., Ma, M., and Liu, R. (2015) Carbon-doped ZnO nanostructures: Facile synthesis and visible light photocatalytic applications. *Journal of Physical Chemistry C*, **119**(35), 20544-20554.
57. Ma, X., Dai, Y., Yu, L., and Huang, B. (2014) Noble-metal-free plasmonic photocatalyst: hydrogen doped semiconductors. *Scientific Reports*, **4**, 3986.
58. Pan, Y., Deng, S., Polavarapu, L., Gao, N., Yuan, P., Sow, C. H., and Xu, Q. -H., (2012) Plasmon-enhanced photocatalytic properties of Cu_2O nanowire–Au nanoparticle assemblies. *Langmuir*, **28**(33), 12304-12310.
59. Cui, Y. -L., Guo, X. -N., Wang, Y. -Y., and Guo, X. -Y. (2015) Visible-light-driven photocatalytic N-arylation of imidazole derivatives and arylboronic acids on Cu/graphene catalyst. *Scientific Reports*, **5**, 12005.
60. Khoa, N. T., Kim, S. W., Thuan, D. V., Tien, H. N., Hur, S. H., Kim, E. J., and Hahn, S. H. (2015) Fast and effective electron transport in a Au-graphene-ZnO hybrid for enhanced photocurrent and photocatalysis. *RSC Advances*, **5**, 63964-63969.
61. Chen, D., Wang, D., Ge, Q., Ping, G., Fan, M., Qin, L., Bai, L., Lv, C., and Shu, K. (2015) Graphene-wrapped ZnO nanospheres as a photocatalyst for high performance photocatalysis. *Thin Solid Films*, **574**, 1-9.
62. Bhuyan, T., Khanuja, M., Sharma, R., Patel, S., Reddy, M. R., Anand, S., and Varma A., (2015) A comparative study of pure and copper (Cu)-doped ZnO nanorods for antibacterial and photocatalytic applications with their mechanism of action. *Journal of Nanoparticle Research*, **17**, 288.
63. Ahmad, M., Ahmed, E., Ahmed, W., Elhissi, A., Hong, Z. L., and Khalid, N. R., (2014) Enhancing visible light responsive photocatalytic activity by decorating Mn-doped ZnO nanoparticles on graphene. *Ceramics International*, **40**(7), 10085-10097.
64. Ahmad, M., Ahmed, E., Hong, Z. L., Khalid, N. R., Ahmed, W., and Elhiss, A. (2013) Graphene-Ag/ZnO nanocomposites as high performance photocatalysts under visible light irradiation. *Journal of Alloys and Compounds*, **577**, 717-727.

12

Performance of Various Adsorbents towards Diverse Gases: A Comparative Study

12.1 Introduction

The ever-increasing consumption of fossil fuels and the resulting combustion products create disastrous impacts on the environments along with the depletion of energy sources. As the combustion of fossil fuels emerges many pollutant gases, namely carbon dioxide (CO_2), methane (CH_4), nitrogen (N_2), nitrous oxide (N_2O), sulfur dioxide (SO_2), etc., into the atmosphere, it directly enhances the accumulation of greenhouse gases responsible for universal climatic alterations and, thereby, causes global warming [1-4]. These growing issues can be rectified using different routes such as enhancement in energy competence, development of other alternative energy sources and making use of pollutant gas capture [5-7].

Among the various techniques for capture and storage of pollutant gases, methods based on adsorption technologies have received considerable attention in terms of efficiency and cost for reducing the concentration of pollutants form the environment. Adsorption is considered as a surface phenomenon caused by the existence of cohesive forces such as van der Waals forces and hydrogen bonding at the interfaces of adsorbent and gas molecules (adsorbate) [8,9]. This phenomenon can be generally categorized into physical adsorption (physisorption) and chemical adsorption (chemisorption) based on the interactions between adsorbent and adsorbate. Physisorption is characterized by the presence of weak interactions such as van der Waals or electrostatic forces while chemisorption is caused by the occupancy of strong chemical bonds [10,11]. In addition, the capturing of exhaust gases by adsorption can be executed using many routes in the process plants. For example, capturing of CO_2, a major constituent of the exhaust gases generated during the combustion of fossil fuels or biomass, can be achieved through any of the following routes such as post-combustion, pre-combustion and oxyfuel combustion. The post-combustion technique is commonly employed for

Anish M. Varghese and Vikas Mittal, Khalifa University of Science and Technology, Abu Dhabi, UAE

the capture of exhaust CO_2 from gaseous fuels, chemicals, fertilizers and power production, whereas pre-combustion is applicable for mainly power stations, etc. Oxyfuel (oxy-firing) combustion is generally implemented in coal fired power plants to yield economic CO_2 capturing [6,7,12-15]

The selection of a suitable adsorbent is the critical parameter which decides the overall performance of gas capture and storage processes. It should follow some requirements including great extent of capability to adsorb gases, superior gas selectivity, swift kinetics, easy regeneration, consistency for substantial adsorption-desorption stages, stability towards moisture and impurities, economical operation and so on [13,16]. Basically, the adsorbent is required to retain its effectiveness during periodic operations consisting of alternative steps of adsorption and desorption, where the desorption can be brought about by means of temperature, pressure or vacuum swing [17]. On the basis of chemical constituents, structural features and operation mechanisms, gas adsorbents may be divided into different categories. Specifically, various kinds of adsorbents have been introduced for the purpose of adsorption of diverse pollutant gas molecules including silica [18], clay [19], activated carbon [20], carbon fibers [21], graphene [22]/graphite oxide [23], nanotubes [24,25], zeolites [26], ion-exchange resins [27], polymers [28], alumina [29], metal oxides [30], metal organic frameworks [31], etc. For instance, Figure 12.1 demonstrates the adsorption isotherms of CH_4 and CO_2 on activated carbon [20]. The aim of this work is to overview the performance of various adsorbents other than metal oxides and metal organic frameworks towards the adsorption of gas molecules in the respect of environment and energy applications.

12.2 Silica Based Adsorbents

Silica based materials have received considerable attention as adsorbents for effective capture or removal of greenhouse and other pollutant gases because of their high specific surface area, highly arranged pore structures, ordered distribution of pore diameter, presence of amorphous pore walls, etc. [32].

12.2.1 Adsorption of CO_2

Belmabkhout *et al.* [33] reported the superior volumetric capacity of

MCM-41 silica adsorbent for CO_2 storage at ambient temperature and elevated pressure. Also, the adsorption of CO_2 on MCM-41silica was fast and absolutely reversible [33]. In another study, the authors observed the favorable adsorption of CO_2 on MCM-41 silica adsorbent from the binary blend of CO_2 with individual gases N_2, H_2 and CH_4 [34]. Tsai *et al.* [35] accomplished enhancement in the flux and selectivity of CO_2 through the use of dual-layer of microporous silica adsorbents [35]. A study by Himeno *et al.* [36] also observed beneficial adsorption selectivity of CO_2 from a binary mixture of CO_2 and CH_4 on a series of silica Deca-dodecasil 3R zeolites [36]. Bellussi *et al.* [37] studied the bulk adsorption behavior of CO_2 contained in sour natural gas on silica-alumina adsorbent. The reported adsorbent performed well in terms of both specific and working capacity and selectivity [37]. Xomeritakis *et al.* [38] developed nickel doped sol-gel silica adsorbent to enrich the adsorptivity and selectivity of CO_2 in the flue gas. Also, the nickel doping of silica adsorbent was useful to reduce the densification [38].

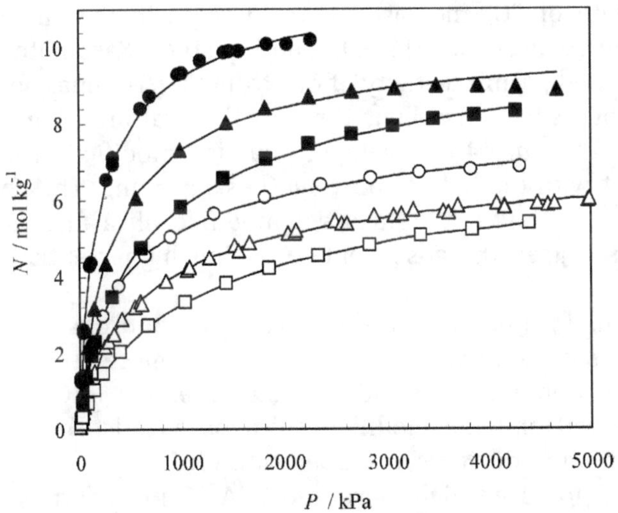

Figure 12.1 Adsorption isotherms of CH_4 and CO_2 on activated carbon. O: methane at 273 K; △: methane at 298 K; □: methane at 323 K; ●: carbon dioxide at 273 K; ▲: carbon dioxide at 298 K; ■: carbon dioxide at 323 K. Reproduced from Reference 20 with permission from American Chemical Society.

Leal *et al.* [39] reported chemisorption of CO_2 on amine-surface bonded silica gel at ambient temperature and observed a capability to hold 10 STP cm^3 of dry CO_2 per g. Also, desorption of the stored CO_2 was observed during heating at 100 °C [39]. Instead of post-synthetic grafting of amine moieties on mesoporous silica, a study by Kim *et al.* [40] reported the development of amine modified mesoporous silica adsorbent through anionic surfactant-mediated synthesis. Better adsorption of CO_2 as well as greater stability upon continues cycles of adsorption and desorption was achieved. Yue *et al.* [41] achieved improvement in CO_2 adsorption on amine functionalized silica by the introduction of hydroxyl group. The generation of comparatively thermally weak intermediate at the interface of CO_2 and amine led to simple desorption of CO_2 molecules [41].

Knowles *et al.* [42] reported the existence of notable reversible adsorptivity of CO_2 on aminopropyl-functionalized mesoporous silica even in the influence of water [42]. Hiyoshi *et al.* [43] also reported the application of aminosilane functionalized SBA-15 silica adsorbent for achieving the intensified adsorption capacity of CO_2. The strong adsorption of CO_2 occurred by the generation of alkylammonium carbamate even in the presence of water molecules. Also, the adsorptivity of CO_2 increased with increase in the amine surface density on the adsorbent [43]. In another study, Sakamoto *et al.* [44] accomplished permselectivity of CO_2 from CO_2/N_2 binary mixture on mesoporous silica adsorbent through the grafting of amino-silane [44]. Knofel *et al.* [45] developed diamine modified SBA-16 silica adsorbent with the help of post-synthesis grafting. The positive interactions of CO_2 with the surface of modified silica adsorbent at low pressures caused the adsorbent to exhibit higher enthalpy values [45].

Liu *et al.* [46] modified the surface of mesoporous silica, SBA-15 using triethanolamine with an aim to enrich the selectivity of CO_2 in a mixture containing CO_2 and CH_4. Improvement in CO_2 selectivity was achieved and the modified silica based adsorbent exhibited greater stability during cyclic operation [46]. In another study, Liu *et al.* [47] prepared highly ordered SBA-15 mesoporous silica adsorbent possessing versatile textural characteristics and observed beneficial CO_2 adsorption [47]. Zelenak *et al.* [48] evaluated the influence of amine functionalization of SBA-12 mesoporous silica towards the adsorption of CO_2. Amine functionalization using 3-aminopropyl and 3-(methylamino)propyl resulted in greater capacity of CO_2 adsorption while 3-(phenylamino)propyl presented rapid desorption characteristics [48]. Another study by Jang *et al.* [49] reported the functionalization of mesoporous silica using amine comprising polysilsesquioxane, which resulted in improved permeability and selectivity of CO_2 from the mixture of gases [49]. Hicks *et al.* [50] observed substantial CO_2 adsorptivity on hyperbranched aminosilica adsorbent along with consistency upon cyclic operations [50].

Song *et al.* [51] reported the use of chemically functionalized silica adsorbents for the purpose of CO_2 adsorption. Chemical modification of silica adsorbent was carried out by developing base moieties on the surface through either treating with calcium acetate or synthesis in an alcoholic solution comprising of TEOS and calcium acetate. Improved adsorptivity of CO_2 gas on the resultant adsorbent was observed through chemisorption. The adsorptivity of CO_2 was enhanced significantly by about 275% at 100 °C and 312% at 250 °C [51].

The impact of polyethyleneimine (PEI) impregnation on a series of silica adsorbents such as MCM-41, MCM-48, SBA-15, SBA-16 and KIT-6 was investigated by Son *et al.* [52]. The appearance of PEI in all silica adsorbents enriched the CO_2 adsorptivity along with stabilized adsorption-desorption reversibility with a significant recovery of more than 99%. Among all studied silica adsorbents, modified KIT-6 silica based adsorbent generated optimal CO_2 adsorption and stability during cyclic operation. Goeppert *et al.* [53] revealed the feasibility of PEI impregnated fumed silica based adsorbent for effective CO_2 capture from the atmospheric air [53]. In another study, Xu *et al.* [54] reported about 24 folds improvement in CO_2 adsorption capacity of MCM-41 silica adsorbent through PEI modification [54]. The authors also suggested the effect of temperature on the absorption process (Figure 12.2), which differed due to the availability of adsorption sites as a function of temperature. In order to further enrich the performance of PEI impregnated silica adsorbent, Rezaei *et al.* [55] introduced aminosilane functionalization, which was useful to improve adsorption of CO_2. Also, PEI impregnated silica adsorbents with and without aminosilane functionalization exhibited decreasing tendency of CO_2 adsorptivity with increase in temperature [55]. A study by Kim *et al.* [18] reported superior rate of CO_2 adsorption on monomeric aminopropyl functionalized MCM-48 silica adsorbent compared to polymeric aminopropyl functionalized one due to the easy approachability of amine modified adsorption spots [18].

Kusakabe *et al.* [56] observed greater permeance of CO_2 on silica/polyimide composite membrane comprising of 60 wt% of silica compared to pure polyimide membrane. The permselectivity of 30 at 30 °C and 13 at 100 °C was observed [56]. In another study, Zornoza *et al.* [57] developed mixed matrix adsorbent comprising of mesoporous silica spheres in respective polyimide and polysulfone matrices. The occupancy of silica on the polymer matrices led to the enrichment of the permeability and selectivity of CO_2 on the adsorbents, which was attributed to the increased interfacial bonding and

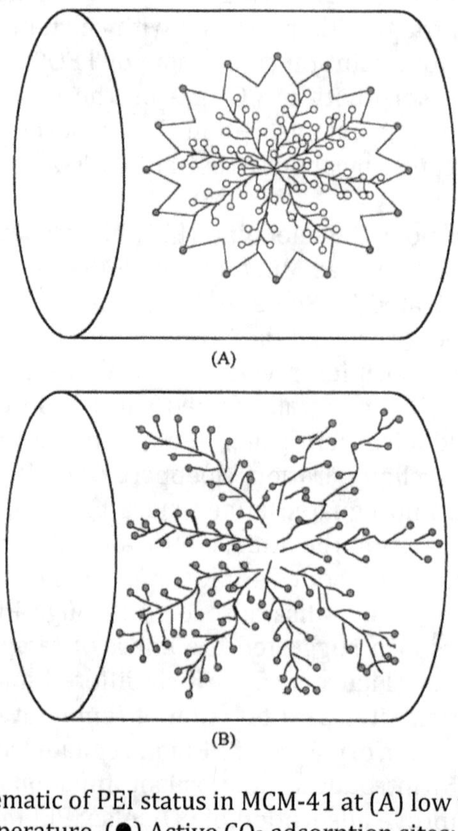

Figure 12.2 Schematic of PEI status in MCM-41 at (A) low temperature, and (B) high temperature. (●) Active CO_2 adsorption sites; (○) hidden CO_2 adsorption sites. Reproduced from Reference 54 with permission from American Chemical Society.

the mesoporous nature of silica [57]. Another study by Kim *et al.* [58] also reported the existence of advanced permeability and permselectivity of CO_2 over silica/poly(amide-6-b-ethylene oxide) (PEBAX) hybrid owing to the appreciable interaction of CO_2 molecules with SiO_2 groups along with the presence of extra active moieties and organic/inorganic interphase [58].

12.2.2 Permselectivity towards H_2

A study by Morooka *et al.* [59] obtained H_2 permeance of 10^{-8} mol.m^{-2}s^{-1}Pa^{-1} on silica membrane at 600 °C and which was developed through the thermal decomposition of tetraethylorthosilicate at 600-650 °C [59]. In another study, Sea *et al.* [60] reported a permeance of 3 x 10^{-8} mol.m^{-2}s^{-1}Pa^{-1} for H_2 on silica membranes generated by chemical vapor deposition. Also, the existence of H_2/N_2 selectivity of 7.6 was noted at a temperature of 400 °C [60]. A study by Vos *et al.* [61] also reported silica adsorbent possessing reduced amount of defects and enhanced characteristics using sol-gel method. The existence of much better H_2 permeance of 2 x 10^{-6} mol m^{-2} s^{-1} Pa^{-1} and permselectivity of more than 700 was noted [61]. Boffa *et al.* [62] reported the existence of H_2 permeance of 6 x 10^{-7} mol.m^{-2}s^{-1}Pa^{-1} at 473 K on templated mesoporous silica adsorbent [62]. Zornoza *et al.* [63] made use of mixed matrix adsorbents consisting of hollow silicalite spheres in respective polysulfone and polyimide matrices for H_2 adsorption. Adsorbents with 8 wt% of silicalite spheres exhibited maximal H_2 permeability of 38.4 Barrer and H_2/CH_4 selectivity of 180 [63].

12.2.3 Adsorption of H_2S

To remove H_2S from natural gas, Huang *et al.* [64] made use of amine functionalized silica xerogel and MCM-48 silica adsorbents. Both adsorbents were observed to be beneficial with respect to adsorption capacity as well as rate [64] (Figure 12.3). Belmabkhout *et al.* [65] also reported effective removal of H_2S from natural gas and biogas through the use of triamine modified mesoporous silica adsorbent. Better adsorption capacity and selectivity of H_2S on the resultant adsorbent was observed even under the influence of moisture (Figure 12.4). Also, the adsorbent exhibited greater regenerability [65].

Wang *et al.* [66] examined the behavior of ZnO functionalized aluminum containing mesoporous silica towards the adsorption of H_2S. The resultant adsorbent comprising of 2.1 wt% of ZnO exhibited optimal adsorptivity and therein micropores and mesopores served as energetic adsorption areas for H_2S. The presence of ZnO in the adsorbent composition encouraged the growth of pore structure along with the desulphurization features, which consequently increased the adsorptivity of H_2S [66]. In another study, Wang *et al.*

[67] utilized ZnO nanoparticles comprising SBA-15 mesoporous silica adsorbents for H_2S removal and the resultant system comprising of 3.04 wt% ZnO possessed high adsorptivity [67]. Adsorbent based on PEI impregnated SBA-15 mesoporous silica also exhibited higher H_2S capture capacity even under the influence of moisture along with the adequate stability and regenerability [68].

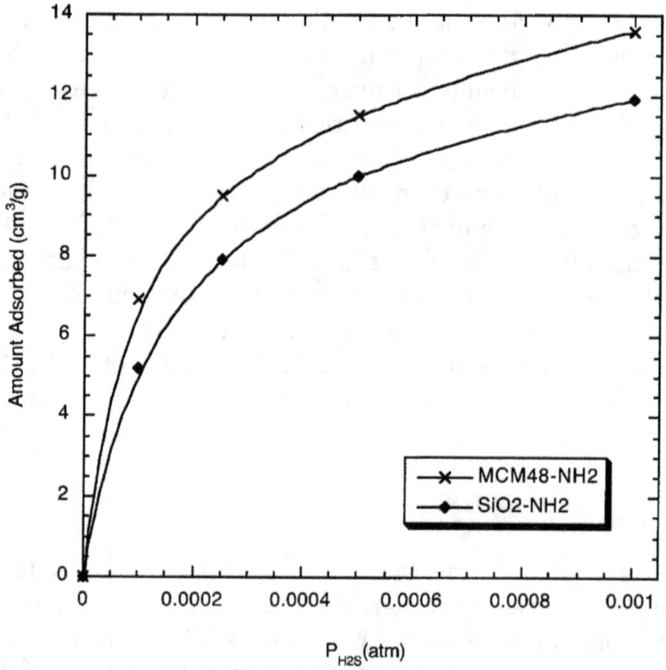

Figure 12.3 Adsorption isotherms of H_2S on amine-modified silica materials at room temperature. Reproduced from Reference 64 with permission from American Chemical Society.

12.2.4 Permselectivity of N_2

Vos *et al.* [69] prepared hydrophobic silica based adsorbent by the application of methyl-tri-ethoxy-silane onto silica sol. The resultant adsorbent presented N_2 permeance in the order of 4×10^{-7} mol.m^{-2}s^{-1}Pa^{-1}. The greater hydrophobicity of the modified silica adsorbent offered good stability even in the humid environments [69]. Sitter *et al.* [70] observed improvement in permeability and reduction in selectivity of N_2 with increase in concentration of silica in the silica/poly(1-trimethylsilyl-1-propyne) nanocomposite based adsorbent [70]. In another study, the authors achieved the stability in N_2 selectivity through the use of silica/poly(4-methyl-2-pentyne) nanocomposite adsorbent [71].

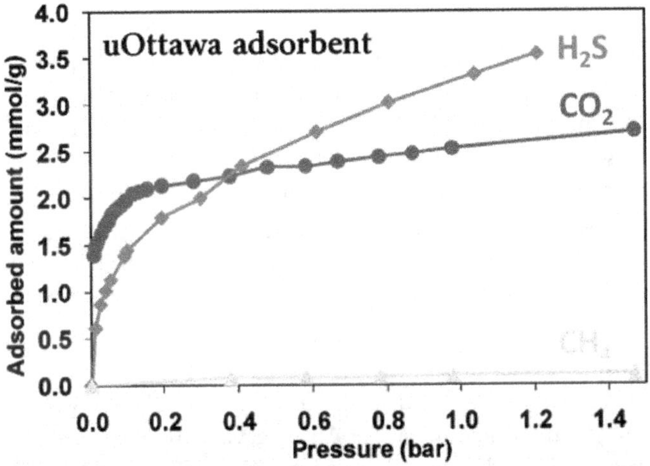

Figure 12.4 Adsorption isotherms of H_2S (and other gases) on the modified adsorbent. Reproduced from Reference 65 with permission from American Chemical Society.

12.2.5 Adsorption of Other Gases

Newalkar *et al.* [72] made use of adsorbent based on SBA-15 mesoporous silica to investigate the adsorptivity of methane, ethane, ethylene, acetylene, propane and propylene. Among the studied light hydrocarbons, ethylene and propylene exhibited greater adsorption capacity and acetylene possessed strong adsorption on adsorbent of SBA-15 mesoporous silica [72]. Montanari *et al.* [73] examined the adsorption behavior of silica adsorbents towards hexamethylcyclotrisiloxane (HMCTS), a usual impurity presented in the biogases. Formation of strong hydrogen bond of HMCTS with surface hydroxyl group of silica led to enhanced adsorptivity [73].

12.3 Clay Based Adsorbents

Clay based materials have been used to develop gas adsorbents possessing high adsorption capacities owing to their substantial specific surface area, porous volume, resistance to chemicals, mechanical stability, versatile surface as well as structural features, etc. [74].

12.3.1 Adsorption of H_2S

Stepova *et al.* [75] studied the adsorption of H_2S on adsorbent based on bentonite. The authors also made use of iron and copper chloride functionalization of bentonite to enhance the adsorptivity. Bentonite modified with iron presented pronounced increase in H_2S adsorptivity whereas copper chloride modified bentonite could hold H_2S for long term. Also, the generation of sulfides on the functionalized bentonite surface through the reaction of metal hydroxide with H_2S was observed. The adsorbent could be regenerated when exposed to open air [75]. To accomplish H_2S adsorption, Thanh *et al.* [76] carried out functionalization of montmorillonite (MMT) using iron using different routes such as interchanging of sodium in the sodium rich MMT with iron, incorporation of iron oxocations and iron doping of aluminum-pillared MMTs. Among various iron modifications, iron doping exhibited maximal adsorptivity of H_2S on MMT whereas iron oxocations resulted in lower adsorption capacity. The difference in adsorption behavior of various iron modifications of MMT resulted from the comprehensive effect of material features, dispersion and pore volume [76].

12.3.2 Adsorption of Light Hydrocarbon Gases

Ji *et al.* [77] studied methane adsorption on a series of clay adsorbents, specifically montmorillonite, kaolinite, illite, chlorite and interstratified illite/smectite. Successful methane adsorption on all clay adsorbents was reported *via* physisorption. Adsorbents based on MMT and interstratified illite/smectite clay possessed more volume of micro- and mesopores, which consequently exhibited greater methane adsorption capacity. The methane adsorptivity of various clay adsorbents was noted in the order of MMT > interstratified illite/smectite > kaolinite > chlorite > illite [77]. Pires *et al.* [78] reported the substantial selectivity of methane from CO_2 and ethane on porous smectite clay adsorbent fabricated through gallery templated approach with the help of quaternary ammonium cation and neutral amines [78].

The impact of vapor pressure on the adsorption behavior of C1–C6 hydrocarbon gases on MMT and kaolinite based adsorbents was examined by Cheng *et al.* [79]. An interesting increase was observed in the adsorbed gas concentration with increase in partial pressure of the hydrocarbon gases. Adsorption of methane was higher in kaolinite when compared to MMT, whereas the overall gas adsorption was lower [79]. Choudary *et al.* [80] made use of Ag^{+} modified clay based alkene-selective adsorbent to study the adsorption responses of light hydrocarbon gases like ethylene (C_2H_4), ethane (C_2H_6), propylene (C_3H_6) and propane (C_3H_8). Alkenes exhibited greater heats of adsorption whereas alkanes had superior adsorption selectivity on the adsorbent [80]. A study by Pires *et al.* [74] reported beneficial separation of C_2H_6/CH_4 binary mixture on adsorbent based on porous clay heterostructures (PCH) comprising of higher amount of phenyltriethoxysilane in smectite clay.

12.3.3 Adsorption of Volatile Organic Compounds (VOCs)

Zaitan *et al.* [81] revealed the suitability of bentonite based adsorbent to capture volatile organic compound (VOC), namely *o*-xylene. Bentonite exhibited an adsorptivity of 1042 μmol/g for xylene at 300 K and also presented an ability to regenerate about 90% of adsorbed xylene. The authors concluded the use of the adsorbent for effective as well as economic removal of VOCs [81]. In another study, Qu *et al.* [82] developed PCH *via* functionalization of bentonite using cetyltrimethylammonium bromide and dodecylamine to achieve substantial adsorption of VOCs such as acetone, toluene, ethylbenzene, *o*-xylene, *m*-xylene and *p*-xylene. Among the various VOCs, acetone (aliphatic hydrocarbon) presented higher adsorptivity on PCH than aromatic hydrocarbons due to their differences in HOMO energy values [82]. In another study, Zuo *et al.* [83] made comparison of VOCs adsorption behavior on Na-MMT, Al and AlCe pillared Na-MMT adsorbents. Al pillaring of Na-MMT adsorbent led to an increase in the surface area and pore volume, which correspondingly increased the adsorptivity of VOCs [83]. A study by Pires and co-workers [78] reported considerable adsorption of VOCs on adsorbent based on porous smectite clay prepared by gallery templated approach using quaternary ammonium cation and neutral amines . In another study, Nunes *et al.* [84] developed PCH *via* chemical modification of natural clay by bis(triethoxysilyl)benzene and tetraethyl orthosilicate to acquire VOCs adsorption. Resultant adsorbent exhibited effective adsorption of VOCs due to the hydrophobic nature [84].

12.3.4 Adsorption of CO_2

Pires *et al.* [74] examined the adsorption behavior of CO_2 from binary mixtures of CO_2/CH_4 and CO_2/C_2H_6 on PCH developed by impregnating tetraethoxysilane and phenyltriethoxysilane individually in smectite clay. High selectivity and separation of CO_2/CH_4 was observed for PCH adsorbents consisting of tetraethoxysilane and lower concentration of phenyltriethoxysilane [74]. In another study, Pires *et al.* [85] reported optimal adsorption of CO_2 on specific Al_2O_3 and ZrO_2 pillared MMT clay adsorbents compared to other gases such as C_2H_6, CH_4 and N_2. Clay with Al_2O_3 pillaring exhibited greater surface area whereas ZrO_2 pillaring exhibited better separation feasibility.

12.4 Activated Carbon Based Adsorbents

Activated carbons are one of the widely accepted adsorbents for large extent of gas adsorption purposes due to their interactions with the gas molecules resulting from the well-ordered pore structure, along with the tunable surface chemistry [86].

12.4.1 Adsorption of CO_2

A study by Presser *et al.* [87] observed the existence of ultra-high CO_2 adsorptivity of about 7.1mol/kg on carbide derived activated carbon adsorbent at 0 °C and ambient pressure. Also, the adsorbent with pore size smaller than 0.8 nm presented greater CO_2 adsorption feasibility than the larger pores [87]. Valer *et al.* [88] made use of anthracites derived activated carbon for the purpose of CO_2 capture and observed a storage capacity of 65.7 mg-CO_2/g-adsorbent for carbon activated at 800 °C for 2 h. Further improvement in the adsorption features of activated carbon was achieved through PEI infusion and high temperature ammonia treatment [88]. In another study, Goel *et al.* [89] reported optimal CO_2 adsorption capacity for activated mesoporous carbon derived at 700 °C from melamine formaldehyde resin [89]. Siriwardane *et al.* [90] also observed the generation of greater CO_2 adsorptivity on activated carbon compared to molecular sieves at pressure greater than 25 psi.

In another study, Drage *et al.* [91] reported the use of activated carbon based adsorbent in gasification to adsorb CO_2 and CO_2 adsorptivity of about 12 mmol/g under a pressure of 4 MPa was reported. Ganesan *et al.* [92] reported the derivation of activated carbon from graphene exhibiting enormous specific area of about 3240 m^2/g and substantial porous nature. The adsorbent presented high adsorptivity for CO_2 along with swift kinetics. A study by Sui *et al.* [93] observed greater adsorption of CO_2 over carbon based adsorbent derived from graphene aerogels *via* steam activation.

Zhang *et al.* [94] developed KOH activated carbon based adsorbent from polyaniline possessing comparatively smaller micropores favorable for superior CO_2 capture. The resultant adsorbent exhibited adsorption capacities of about 1.86 mmol/g at 75 °C and 1 bar and 1.39 mmol/g at 25 °C and 0.15 bar [94]. Adsorbent based on metal-organic coordination polymers derived activated carbon with a surface area of 2368 m^2/g also exhibited suitability for CO_2 capture with a storage capability of 2.9 mmol/g at ambient pressure and 300 K [95]. Hu *et al.* [96] prepared activated carbon with nanopores from petroleum coke and observed an exceptional adsorptivity of about 15.2 wt% at 1 bar for the material activated at 700 °C [96]. The feasibility of activated carbon sourced from pine nut shell for CO_2 capture was revealed by Deng *et al.* [97]. Alberto *et al.* observed exceptional feasibility of activated carbon molecular sieves to adsorb significant amount of CO_2 at ambient temperature [98]. In another study, Daud *et al.* [99] obtained selectivity of 16 for CO_2 from CH_4 by the amine functionalization of activated carbon molecular sieves [99].

Wickramaratne *et al.* [100] reported the application of activated carbon spheres derived from the carbonization of phenolic resin as adsorbent for CO_2 capture (Figure 12.5). The resultant adsorbent exhibited significantly superior adsorption capacities of 4.55 mmol/g at 25 °C and 8.05 mmol/g at 0 °C [100]. In another study, the authors concluded the CO_2 adsorption capacity of 8.9 mmol/g for phenolic-resin derived activated carbon based adsorbent with a substantial surface area of 2400 m^2/g [101]. Another study by Sevilla *et al.* [102] observed insignificant effect of N-doping of activated carbon microspheres towards the adsorption of CO_2 (Figure 12.6).

Loser *et al.* [103] reported the use of activated carbon monoliths based adsorbent for CO_2 capture. Volumetric adsorption capacity of 440 g/l was attained in the study. In another study, Ribeiro *et al.* [104] obtained the adsorption capacity of the activated carbon honeycomb monolith as $CO_2 > CH_4 > N_2$. In another study, Nandi *et al.* [105] accomplished CO_2 adsorption of 5.14 mmol/g at room temperature and pressure by the use of adsorbent based on nitrogen doped activated carbon monoliths derived from the carbonization of polyacrylonitrile. Zhong *et al.* [106] reported the development of polyacrylonitrile based block copolymer templated nitrogen supplemented activated carbon as a better adsorbent for selective CO_2 capture [106].

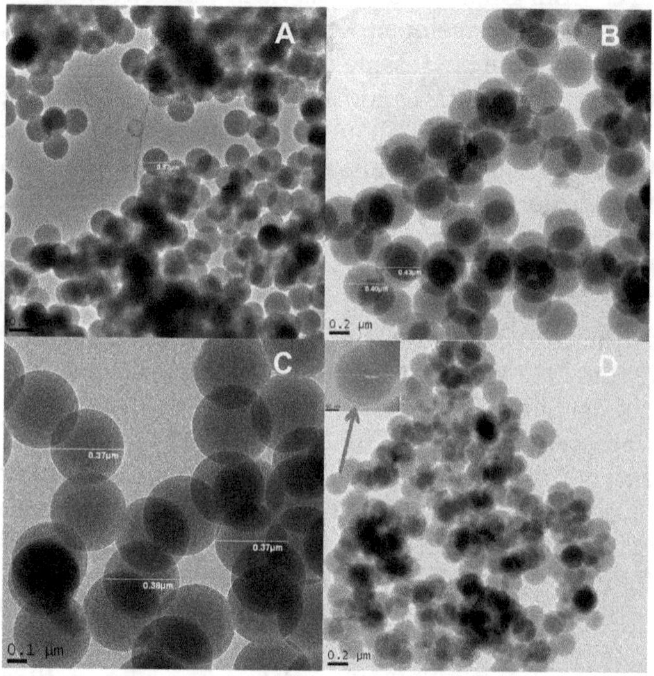

Figure 12.5 TEM images of the activated carbon spheres. Reproduced from Reference 100 with permission from American Chemical Society.

Xing *et al.* [107] obtained substantial CO_2 adsorption capacity of activated carbon through nitrogen doping. Nitrogen doping of activated carbon was useful to enhance the interactions with CO_2 *via* hydrogen bonding. In another study, Zhao *et al.* [108] made use of activated carbon with high amount (13 wt%) of nitrogen doping to enrich the performance. As a result, the adsorbent system presented an improved adsorption capacity of 10.53 mmol/g at 25 °C and 8 bar. Further, the large amount of nitrogen led to the enhancement of the adsorption selectivity of CO_2/N_2 to about 42 [108]. Saleh *et al.* [109] reported the generation of nitrogen impregnated activated microporous carbon based adsorbents from polyindole with outstanding CO_2 adsorption selectivity and stability [109]. A study by Sevilla *et al.* [110] reported eminently porous nitrogen doped activated carbon based adsorbent from hydrothermal carbons of algae and glucose. The surface area in the range of 1300–2400 m^2/g and pore volume of about 1.2 cm^3/g for resultant adsorbent was reported. The adsorbent exhibited an adsorption capacity of about 7.4 mmol/g at 0 °C and 1 bar. Zhao *et al.* [111] introduced extra-framework cations onto nitrogen enriched microporous activated carbon with an aim to widen adsorption characteristics. The modification resulted in the improvement in adsorptivity and selectivity of CO_2 on nitrogen doped activated carbon, as demonstrated in Figure 12.7.

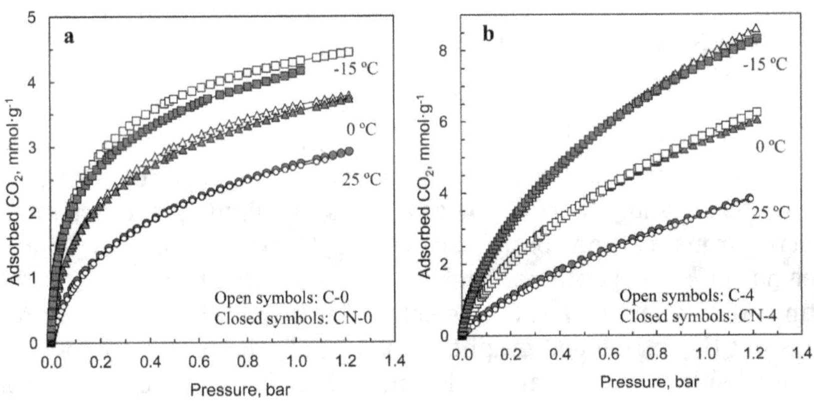

Figure 12.6 Comparison of CO_2 isotherms at several temperatures for (a) non-activated and (b) activated carbons (CN indicates carbon with N-doping). Reproduced from Reference 102 with permission from American Chemical Society.

The formation of nitrogen impregnated microporous activated carbon from chitosan with the help of K_2CO_3 was also reported by Fan *et al.* [112] to achieve the efficient adsorption of CO_2. An adequate CO_2 adsorption capacity of 3.86 mmol/g was observed at 1 atm and 25 °C for activated carbon prepared at 635 °C with K_2CO_3/chitosan ratio of 2. Also, the system offered significantly improved performance in terms of CO_2 selectivity over N_2, stability and recyclability [112]. Another study by Chen *et al.* [113] observed greater CO_2 adsorptivity, selectivity and recyclability on highly ordered nitrogen doped activated carbon with micro-mesoporous framework. Activated N-doped carbon derived from polypyrrole modified graphene exhibited significant CO_2 uptake of 4.3 mmol/g along with higher selectivity and regeneration [114].

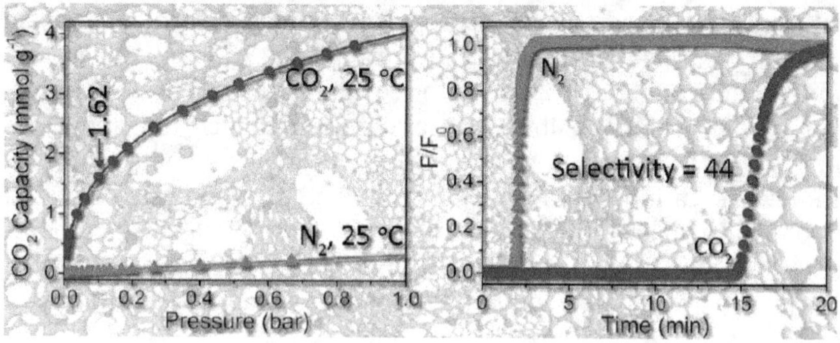

Figure 12.7 Improvement in adsorptivity and selectivity of CO_2 of the adsorbent. Reproduced from Reference 111 with permission from American Chemical Society.

Another study by Xia *et al.* [115] demonstrated the generation of noticeable adsorption energy of 59 KJ/mol and 22 KJ/mol at respective inferior and superior CO_2 coverages on sulfur doped activated microporous carbon based adsorbent [115]. In another study, Seema *et al.* [116] obtained high CO_2 adsorptivity of 4.5 mmol/g at 1 atm and 298 K as well as CO_2 selectivity atop other gases namely, N_2, CH_4 and H_2 on reduced graphene oxide/poly-thiophene composite-derived sulfur impregnated microporous activated carbon. The resultant adsorbent material presented a peak surface area of 1567 m^2/g and exhibited maximal adsorption at 0.6 nm pore gauge [116].

In another study, Builes *et al.* [117] observed superior adsorptivity of CO_2 on adsorbent based on zeolite templated activated carbon. Xia *et al.* [118] made use of zeolite templated nitrogen impregnated activated microporous carbon based adsorbent. CO_2 adsorption capacities of 6.9 mmol/g at 273 K and 4.4 mmol/g at room temperature and ambient pressure conditions were reported. Also, the resultant adsorbent exhibited satisfactory adsorption energy along with good selectivity and stability. In another study, Sevilla *et al.* [119] reported the synthesis of cubic and hexagonal mesostructured silica templated activated carbon for the purpose of enriched CO_2 adsorption performances [119].

Pevida *et al.* [120] attained slight enhancement in CO_2 capacity of activated carbon through the surface modification *via* ammonia treatment at 200-800 °C temperature range [120]. Enriched CO_2 adsorptivity of granular activated carbon was accomplished by Shafeeyan *et al.* [121] with the help of oxidation and subsequent ammonia treatment at a high temperature of 800 °C [121]. Zhang *et al.*

[122] accomplished an increase of about 28% in CO_2 capture *via* modifications using microwave irradiation under N_2 atmosphere. In another study, Plaza *et al.* [123] analyzed the influence of amine coating over activated carbon towards CO_2 adsorption and observed increment in microporous volume and adsorptivity of activated carbon due to the embellished nitrogen content and basicity. Table 12.1 summarizes the adsorption behavior of selected activated carbon based adsorbents used for CO_2 capture.

12.4.2 Adsorption of H_2S

To accomplish efficient adsorption of H_2S, Guo *et al.* [124] used chemically activated carbon sourced from oil-palm shell. Adsorbents based on chemically activated carbon exhibited greater adsorption degree than thermally activated carbon. Also, the existence of three adsorption mechanisms such as physisorption, chemisorption and H_2S oxidation on the activated carbon was observed on the basis of method and agent used for the activation [124]. Yan *et al.* [125] observed greater capacity of alkaline activated carbon to capture H_2S due to the existence of simultaneous chemisorption and physisorption mechanisms [125].

Seredych *et al.* [126] obtained considerable improvement in H_2S adsorption through nitrogen impregnation on activated carbon procured from titanium carbide. Optimal adsorption capability was observed for 2% nitrogen impregnated activated carbon derived at 800 °C. Also, the strong influence of pore size and volume of the adsorbent on the adsorptivity of H_2S was noted (it was higher for the material with sizeable pore volume). In another study, Bagreev *et al.* [127] reported about 10 folds improvement in H_2S adsorptivity on functionalized bituminous coal derived activated carbon through the combination of nitrogen impregnation and heat treatment processes at 850 °C [127].

Huang *et al.* [128] synthesized copper impregnated activated carbon with an aim to upgrade H_2S adsorption characteristics. About 10 folds improvement in H_2S adsorptivity was achieved through copper impregnation. Also, the detrimental effect of moisture on H_2S adsorptivity was noted. Thanh *et al.* [129] reported the development of metal constituting bentonite binders functionalized activated carbon adsorbent to study the adsorption responses of H_2S. Copper containing activated carbon adsorbent exhibited increase in H_2S adsorptivity and overall performance in contrast to iron and zinc [129]. In another study, Xiao *et al.* [130] observed about 3 folds enhanced adsorption capacity of H_2S by the use of sodium carbonate impregnated activated carbon adsorbent [130].

Table 12.1 Summary of the performance of the selected activated carbon based adsorbents used for CO_2 capture

Authors	**activated carbon used**	**adsorption/ uptake capacity**	**conditions**
Presser *et al.* [87]	Carbide derived activated carbon	7.1 mol/kg	0 °C, ambient pressure
Zhang *et al.* [94]	Polyaniline derived activated carbon	1.86 mmol/g	25 °C, 0.15 bar
Deng *et al.* [95]	Metal-organic coordination polymers derived activated carbon	2.9 mmol/g	300 K, ambient pressure
Hu *et al.* [96]	Petroleum coke derived activated nanopores carbon	15.2 wt%	Room temperature, 1 bar
Wickramaratne *et al.* [100]	Phenolic resin derived activated carbon spheres	4.55 m mol/g 8.05 m mol/g	25 °C, ambient pressure 0 °C, ambient pressure
Nandi *et al.* [105]	Polyacrylonitrile derived nitrogen doped activated carbon monoliths	5.14 mmol/g	Room temperature and pressure
Zhao *et al.* [108]	Activated carbon with extremely high nitrogen doping of 13 wt%	10.53 mmol/g	25 °C, 8 bar
Sevilla *et al.* [110]	Eminently porous nitrogen doped activated carbon	7.4 mmol/g	0 °C, 1 bar
Fan *et al.* [112]	Chitosan derived nitrogen impregnated microporous activated carbon	3.86 mmol/g	25 °C, 1 atm
Chen *et al.* [113]	Activated N-doped carbon adsorbent derived from	4.3 mmol/g	Room temperature and pressure

	polypyrrole modified graphene sheets		
Seema *et al.* [116]	Reduced graphene oxide/polythiophene composite-derived sulfur impregnated microporous activated carbon	4.5 mmol/g	298 K, 1 atm
Xia *et al.* [118]	Zeolite templated nitrogen impregnated activated microporous carbon based adsorbent	6.9 mmol/g 4.4 mmol/g	273 K, ambient pressure Room temperature and pressure

12.4.3 Adsorption of H_2

Jorda-Beneyto *et al.* [131] used chemically activated carbon to accomplish H_2 adsorption and revealed the dependency of pore volume and adsorption ambience on the H_2 adsorption rate. Peak adsorptivity of 38.8 g H_2/l was observed for the adsorbent based on chemically activated carbon possessing greater specific surface area at 77 K and 4 MPa [131]. Another study by Juan *et al.* [132] reported favorable capability of activated carbon based adsorbent to accumulate more H_2 in volumetric premise than metal organic framework [132]. In another study, Lillo *et al.* [133] obtained H_2 storage capacity of up to 1 wt% at 10 MPa on activated carbon [133]. Kim *et al.* [134] obtained activated carbon based adsorbent from molybdenum carbide and reported about 4.3 wt% H_2 storability. Jord-Beneyto *et al.* [135] also achieved H_2 adsorptivity of 29.7 g H_2/l under the conditions of 77 K and 4 MPa by the application of activated carbon monoliths [135].

Xia *et al.* [115] utilized sulfur impregnated microporous activated carbon possessing highly ordered structure to establish H_2 capture and storage. Sulfur doping of carbon was useful to attain pronounced adsorption heat of 9.2 KJ/mol and adsorption density of 14.3×10^{-3} mmol/m^2 at -196 °C and 20 bar [115]. Almasoudi *et al.* [136] used zeolite templated activated carbon based adsorbent possessing a surface area of 900–1100 m^2/g and a pore volume of about 0.7 cm^3 to accomplish effective H_2 adsorption. Significant H_2 storage capacity in the order of 13-15.5 μmol H_2/m^2 was obtained due to their exceptional microporosity [136].

12.4.4 Adsorption of Hydrocarbon Gases

Sun *et al.* [137] observed substantial methane adsorption on adsorbent based on granular activated carbon obtained from Illinois bituminous coal through KOH activation compared to the one based on steam-activated carbons. This was attributed to the increased pore volume of granular activated carbon. In another study, activated carbon acquired from carbide with a specific surface area of 3360 m^2/g exhibited higher methane storage capacity of 18.5 wt% at 60 bar and 25 °C [138]. Prauchner *et al.* [139] also reported about 95 V/v adsorptivity for methane for granular activated carbon based adsorbent. In another study, activated carbon sourced from graphene oxide exhibited greater feasibility for methane uptake owing to the existence of substantial specific surface area of about 1900 m^2/g [140]. Yeon *et al.* [141] discussed the use of carbide derived activated carbon monoliths for methane storage purpose and the resultant adsorbent presented a storage capacity of 219 V(STP)/v in volumetric terms under 60 bar pressure and 25 °C temperature [141]. In another study, Castello *et al.* [142] used adsorbents based on activated carbon monoliths to attain methane capture and storage. Comparing to pristine activated carbons, monoliths exhibited reduction in methane adsorptivity, but had no diffusional problems [142].

The application of benzene functionalized activated carbon with a surface area of 1065m^2/g for methane capture was reported by Adinata *et al.* [143]. The resultant adsorbent had enhanced methane adsorption capacity and selectivity when compared to CO_2, N_2 and O_2 [143]. Aroua *et al.* [144] generated PEI impregnated activated carbon with enhanced methane adsorption capacity, which was higher for the adsorbent containing 0.26 wt% of PEI.

12.4.5 Adsorption of Volatile Organic Compounds (VOCs)

Chiang *et al.* [145] observed adsorption of benzene on activated carbon compared to other VOCs like carbon tetrachloride, chloroform and methylene chloride. This was ascribed to the superior heats of adsorption and entropy change of benzene on activated carbon [145]. Das *et al.* [146] obtained considerable feasibility of activated carbon fiber based adsorbent towards adsorption and desorption of VOC under dynamic conditions of temperature, gas concentration, gas flow rate and adsorbent weight [146]. Kim *et al.* [147] reported advantageous utility of 1 wt% H_3PO_4 impregnated activated carbon based adsorbent for repeated VOCs capture operations [147].

12.5 Graphene/Graphite Oxide (GO) Based Adsorbents

Adsorbents based on graphene or GO have acquired great research interest in the last few years, as compared to other adsorbents because of their outstanding adsorption performances, which results from the distinctive features such as peculiar molecular structure, vast surface area, adjustable porosity, outstanding mechanical properties, high chemical permanence, good thermal conductivity, etc. [22,23,148].

12.5.1 Adsorption of CO_2 on Graphene

Mishra *et al.* [149] showcased the capability of graphene based adsorbent to capture CO_2 *via* physisorption. The existence of peak adsorption capacity of 21.6 mmol/g for the resultant graphene based adsorbent at 25 °C and 11 bar was observed. Meng *et al.* [150] reported the generation of low temperature (150-400 °C) exfoliated graphene nanoplates based adsorbent for CO_2 capture. The resultant graphene nanoplates exhibited predominant interlayer spacing and extensive pore volume, which consequently led to greater adsorption capacity of 248 wt% at 30 bar and 25 °C [150]. Hierarchically porous graphene derived through the CO_2 activation of GO at 850 °C exhibited an adsorptivity of 1.76 mmol/g for CO_2 at 1 bar and 273 K [151]. In another study, Ning *et al.* [152] obtained favorable selectivity of CO_2 on corrugated graphene porous structure. Adsorbent based on 3-dimensional graphene generated a consequential adsorption capacity of 2.98 wt% for CO_2 under the conditions of 273 K and 106.6 kPa [153].

To enrich the CO_2 adsorption features of porous graphene based adsorbent, Shan *et al.* [154] carried out chemical functionalization and obtained improvement in uptake capacity and selectivity compared to N_2 owing to better electrostatic interactions [154]. Carrillo *et al.* [155] observed adsorption of CO_2 on titanium enriched graphene structure and the corresponding dissociation into CO and O

[155]. Mishra *et al.* [156] reported the use of polyaniline impregnated graphene nanosheets as adsorbent for CO_2 capture and displayed an ultra-high adsorption capacity of up to 75 mmol/g at 25 °C and 11 bar [156]. In another study, Dasgupta *et al.* [157] reported the improvement in CO_2 adsorption and selectivity on graphene nanoribbons through COOH functionalization.

Kumar *et al.* [158] detailed the fabrication of porous graphene framework through iodobenzene treatment of reduced GO and subsequent C-C coupling. The resultant adsorbent exhibited up to 112 wt% of storage capacity for CO_2 under the conditions of 1 atm and 195 K due to the characteristic large surface area and porosity [158]. In another study, Ding *et al.* [159] synthesized graphene/manganese oxide composite *via* hydrothermal treatment of $GO/MnO(OH)_2$ blend with an aim to achieve significant CO_2 uptake. The resultant adsorbent had a specific surface area of up to 680 m^2/g and which consequently created an optimal storage capacity of 11 wt% [159].

Kemp *et al.* [160] utilized nitrogen doped graphene/polyaniline composite with notable surface area of up to 1336 m^2/g for effective capture of CO_2. At 1 atm pressure and 273 K temperature, resultant adsorbent exhibited an adsorption capacity of 2.7 mmol/g along with the adequate recyclability and high selectivity over H_2, N_2, Ar and CH_4 [160]. In another study, Saleh *et al.* [161] reported the application of nitrogen doped graphene/polyindole hybrid as a better CO_2 adsorbent and observed a high storage capacity of 3.0 mmol/g at 1 atm and 25 °C. Also, the system exhibited excellent stability up to 10 cycles and substantial selectivity compared to H_2, N_2, and CH_4 [161]. Table 12.2 summarizes the adsorption behavior of selected graphene based adsorbents used for CO_2 capture.

12.5.2 Adsorption of CO_2 on GO

Sui *et al.* [162] yielded hydrothermally reduced GO based adsorbent manifesting three dimensional porous network structure along with sizeable surface area and pore volume. The excellent properties of GO led to considerable adsorption of CO_2 with an adsorption capacity of 2.4 mmol/g at 273K and 1 bar [162]. Chen *et al.* [163] demonstrated enhanced CO_2 capture on GO through the impregnation of light metals like Li and Al [163]. Tsoufis *et al.* [164] claimed pronounced adsorptivity of CO_2 even under the wet conditions on GO by the intercalation of poly(propylene imine) dendrimer, which led to strong crosslinking between vicinal sheets of GO [164]. In another study, Sui *et al.* [165] used PEI impregnated GO based adsorbent, as represented schematically in Figure 12.8, for CO_2 uptake and exhibited a high storage capacity of 11.2 wt% at 1 bar pressure and 273 K temperature [165].

Table 12.2 Summary of selected graphene based adsorbents used for CO_2 capture

Authors	adsorbent used	Adsorption /uptake	conditions
Mishra *et al.* [149]	Graphene sheets	21.6 mmol/g	25 °C, 11 bar
Meng *et al.* [150]	Low temperature (150-400°C) exfoliated graphene nanoplates	248 wt%	25 °C, 30 bar
Xia *et al.* [151]	Hierarchically porous graphene	1.76 mmol/g	273 K, 1 bar
Wang *et al.* [153]	3-dimensional graphene	2.98 wt%	273 K, 106.6 kPa
Mishra *et al.* [156]	Polyaniline impregnated graphene nanosheets	75 mmol/g	25 °C; 11 bar
Kumar *et al.* [158]	Porous graphene framework	112 wt%	195 K; 1 atm
Kemp *et al.* [160]	Nitrogen doped graphene/polyaniline hybrid	2.7 mmol/g	273 K; 1 atm
Saleh *et al.* [161]	Nitrogen doped graphene/polyindole hybrid	3.0 mmol/g	25 °C; 1 atm

In order to upgrade the CO_2 adsorption features, Zhao *et al.* [166] carried out amine intercalation of GO using various amines such as ethylenediamine, diethylenetriamine and triethylenetetramine. Aminated GO presented better adsorption of CO_2, especially the GO with ethylenediamine intercalation exhibited an optimal adsorptivity of 53.62 mg CO_2/g-adsorbent [166]. In another study, Zhao *et al.* [167] achieved notable CO_2 adsorption capacity of 1.06 mmol/g on aminated GO/metal-organic framework composite based adsorbent, even under the influence of moisture. The observed behavior resulted from the combined effect of the features of the composites such as substantial surface area and pore volume, presence of additional active moieties after functionalization and compatibility of pore dimensions with adsorbate [167].

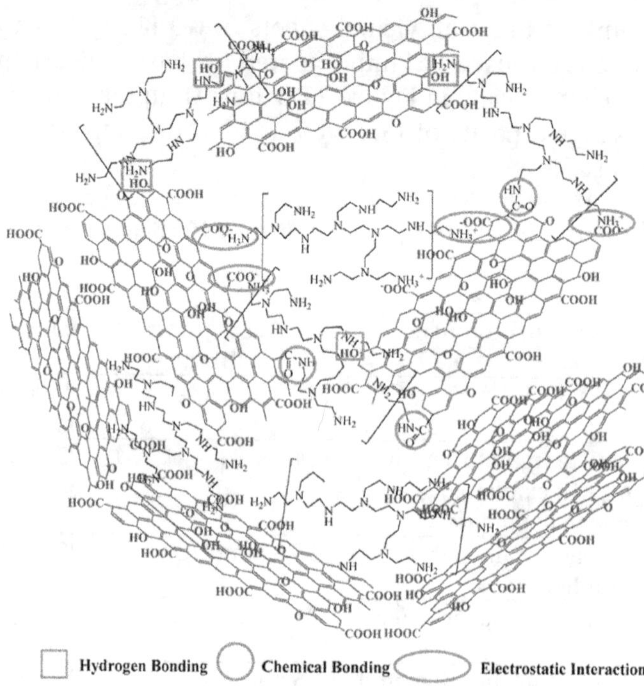

Figure 12.8 Schematic representation of GO-PEI hydrogel, exhibiting various interactions like hydrogen bonding, chemical bonding and electrostatic interactions. Reproduced from Reference 165 with permission from American Chemical Society.

Cao *et al.* [168] reported the fabrication of GO/zirconium metal organic framework composite with an ultimate objective to enhance the CO_2 adsorption capacity. Under the conditions of 298 K and 1 bar, GO based composite exhibited an optimal CO_2 adsorption capacity of 3.37 mmol/g, which was about 48% greater than zirconium metal organic framework [168]. In another study, Chowdhury *et al.* [169] fabricated GO/TiO_2 composite based adsorbent and obtained a satisfactory adsorptivity of 1.88 mmol/g for CO_2 under ambient temperature along with significantly high selectivity from exhaust gases, thus, confirming the functional nature of the adsorbent.

12.5.3 Adsorption of H_2S

With an aim to achieve adsorption of H_2S, Mabayoje *et al.* [170] developed GO/cobalt oxide composite based adsorbent. The existence of meaningful adsorption of H_2S through the generation of sulfites on the composite surface was noted even in the presence of moisture [170]. In another study, Seredych *et al.* [171] also obtained appreciable H_2S adsorption by the use of adsorbent based on GO/zirconium hydroxide composite owing to the substantial basicity and the newly generated active centers on the composite structure. The enhanced H_2S adsorption of composite resulted from the existing reactions of H_2S with epoxy and carboxylic functional groups of GO. Adsorbent based on GO/metal-organic framework composite also exhibited beneficial H_2S uptake capacity through the creation of miniature pores [172]. Seredych *et al.* [173] also used GO based ammonia adsorbent as H_2S adsorbent using high temperature treatment at 950 °C. The existence of nitrogen functionalities on the treated GO surface led to enhanced H_2S adsorptivity. These studies confirmed the high potential of graphene and modified GO based adsorbents for the selective adsorption of H_2S from the gas mixtures.

12.5.4 Adsorption of Ammonia

Seredych *et al.* [174] revealed the possibility of successful ammonia capture on graphene through the sequential intercalation and the reactive adsorption mechanisms [174]. In another study, the authors revealed greater ammonia capture for GO adsorbent derived from the amorphous carbon rich graphite [175]. Seredych *et al.* [176] examined the adsorption behavior of ammonia on exfoliated GO/bentonite hybrid based adsorbent under dynamic conditions. The layered composite structure offered the consumption of considerable concentration of ammonia *via* Brønsted and Lewis active centers on the composite framework as well as interlayered hybrid structures [176]. Composites of GO and manganese oxide were also developed as adsorbent for ammonia and appreciable adsorption capacity for composite composed of immensely oxidized GO was observed [177].

In another study, Seredych *et al.* [178] reported the functionalization of GO using aluminum and zirconium-aluminum polyoxycations separately and the subsequent calcination at 300 °C to enrich ammonia adsorption characteristics. Enhanced adsorption performance was reported for zirconium-aluminum polyoxycation functionalized GO owing to the superior acidity [178]. The intensity of ammonia adsorption degree on the zirconium-aluminum polyoxycation functionalized GO based adsorbent was improved by prehumidification [179]. The reduction in adsorptivity for ammonia on GO/aluminum polyoxycation composite under humid environment was also reported [180].

12.5.5 Adsorption of H_2

Srinivas *et al.* [181] obtained H_2 adsorptivies of 1.2 and 0.1 wt% at temperatures corresponding to 77 K and 298 K by the use of graphene sheets as an adsorbent with a surface area of 640 m^2/g. Favorable adsorption heats of up to 5.9 KJ/mol on graphene sheets owing to their strong interaction with H_2 were also reported [181]. Yuan *et al.* [182] reported the development of graphene sheets through the reduction of GO in the presence of glucose for the purpose of effective H_2 adsorption. The resultant graphene was characterized with a surface area of 1205.8 m^2/g, which consequently led to an effective H_2 storage capacity of 2.7 wt% at 25 bar and 298 K [182]. In another study, Xia *et al.* [151] also reported an adsorptivity of about 3.76 mmol/g for H_2 at 1 bar and 77 K on highly ordered micro-meso-macroporous graphene obtained from the CO_2 activation of GO at a high temperature of 850 °C. The adsorbent exhibited a high surface area of about 532 m^2/g along with a sizeable pore volume of around 1.67 cm^3/g [151].

Ghosh *et al.* [183] obtained H_2 capture of 3.1 wt% at 298 K and 100 atm on graphene based adsorbent derived from the exfoliation of GO. An interesting improvement in adsorption performance was reported by decreasing the mean graphene layers, a salso suggested in Figure 12.9. Ma *et al.* [184] reported poor adsorption of H_2 on single layer graphene sheets owing to their inadequate surface area and poor interaction with H_2 [184], which underlined the need of high surface area and interfacial interactions for better adsorbent performance. Wang *et al.* [153] revealed the use of 3-dimensional graphene with a specific surface area of 477 m^2/g along with the mesoporosity for effective H_2 capture and observed a storage capacity of 1.40 wt% at 106.6 kPa and 77 K [153]. In another study, Lyth *et al.* [185] achieved H_2 storage capacity of 2.1 wt% on graphene foam based adsorbent derived from sodium ethoxide with a specific surface area of 1200 m^2/g [185].

Studies of Burress *et al.* [23] and Srinivas *et al.* [186] reported the development of GO framework based adsorbent with the aid of diboronic acid for enhanced H_2 capture. Adsorption performance such as adsorptivity and heat of adsorption of the adsorbents were about two folds greater when compared to carbon based adsorbents and competitive when compared to metal organic frameworks. Also, GO frameworks exhibited higher thermal stability than GO [186]. Petit *et al.* [187] observed an improved H_2 capture on GO/copper metal organic framework composite due to the presence of miniature pores [187].

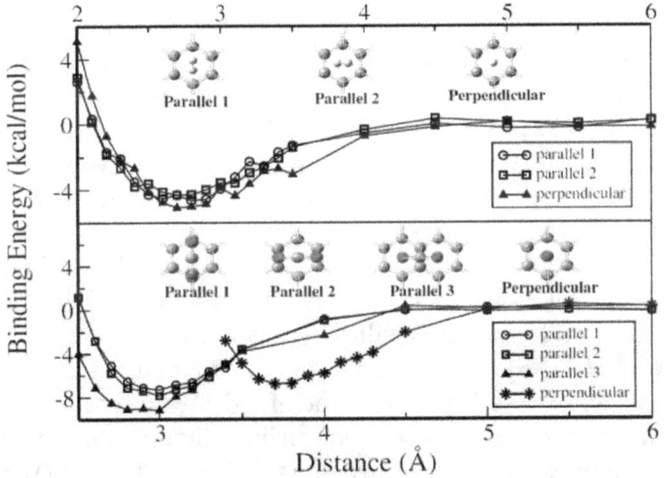

Figure 12.9 Representation of the binding energies of a single molecule on a graphene supercell as a function of distance with different adsorption orientations (shown in insets) for hydrogen (upper panel) and carbon dioxide (lower panel). Reproduced from Reference 183 with permission from American Chemical Society.

12.6 Nanotubes Based Adsorbents

Nanotubes are one-dimensional porous nanomaterials possessing tubular honeycomb like network structure. Better adsorption performance of nanotubes is derived from their porous structure and these materials provide five possible active adsorption sites such as cavities, interstitial channels, two ends and exterior surfaces for successful gas capture [188-190].

12.6.1 Adsorption of CO_2

Babu *et al.* [191] used the ordered as well as vertically positioned double-walled carbon nanotubes (double-walled CNTs) based adsorbent exhibiting substantial surface area and purity for CO_2 uptake. The adsorbent exhibited competitive adsorption properties when compared to other conventional adsorbents such as mesoporous molecular sieves and metal organic frameworks under elevated pressure. Also, the impact of oxygen plasma induced chemical modifications and diameter reduced structural modifications on the CO_2 adsorption features of these materials were investigated. The former modification caused significant performance at nether pressure through C-O moieties generation, while the latter modification resulted in greater capture capacity under elevated pressure [191]. Cinke *et al.* [192] observed CO_2 capture on single-walled CNTs (SWCNTs) *via* physisorption route and noted up to two folds greater storage capacity than adsorbent based on activated carbon [192]. Hu *et al.* [193] reported the CO_2 adsorption on SWCNTs through interior as well as exterior surfaces. In another study, Golkhar *et al.* [194] also achieved high rate of CO_2 uptake on adsorbent based on nanofluids of CNTs than nanosilica.

Casco *et al.* [195] made comparison of adsorption characteristics of multi-walled CNTs (MWCNTs), acidified MWCNTs and unzipped MWCNTs of carbon and N-doped carbon. Significant influence of surface roughness and curved regions of MWCNTs on CO_2 adsorption was revealed, with the materials rich in these parameters exhibiting superior adsorption capacities. In the study, adsorbent based on unzipped MWCNTs of N-doped carbon exhibited optimal adsorption features for CO_2 [195]. Adsorbent based on boron CNTs exhibited appreciable adsorption for CO_2 over N_2, H_2 and CH_4, which was attributed to the creation of strong interactions. For successful CO_2 adsorption, Upender *et al.* [197] used sodium and potassium titanate nanotubes as adsorbent. Comparing with pristine titanate, these materials exhibited higher extent of adsorption degree and recyclability. Also, an interesting improvement in CO_2 adsorptivity was observed in the presence of moisture [197].

Hsu *et al.* [198] revealed the usefulness of 3-aminopropyltriethoxysilane (APTS) functionalized CNTs for continuous operations without affecting the performance. The decreasing tendency of adsorption intensity was observed with increasing temperature. In another study, Su *et al.* [199] reported the generation of better CO_2 adsorption features of APTS functionalized MWCNTs over conventional silica or activated carbon based adsorbents at 20 °C. Also, the workable regeneration of captured CO_2 under low temperature conditions was obtained. Lu *et al.* [200] also reported greater CO_2 adsorption capacity of APTS modified CNTs over granular activated carbon and zeolite based adsorbents. The authors also revealed accomplishable regeneration of adsorbed CO_2 under low temperature conditions for APTS modified CNTs. Figure 12.10 shows morphological changes of CNTs before and after APTS modification [200].

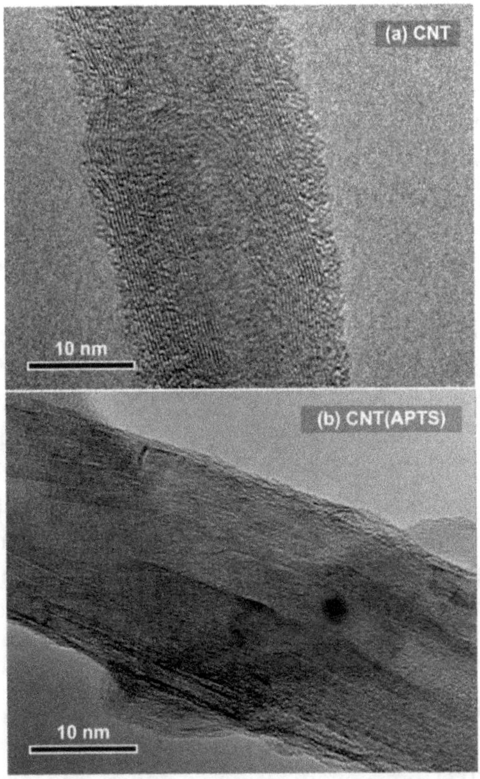

Figure 12.10 TEM images of CNTs (up) before and (down) after APTS modification. Reproduced from Reference 200 with permission from American Chemical Society.

To achieve efficient adsorption of CO_2 from atmosphere, Jana *et al.* [201] evaluated the feasibility of halloysite and APTS-functionalized halloysite nanotubes. Amine modification of halloysite nanotubes was useful to upgrade the CO_2 uptake and successful regeneration under ambient air conditions. APTS- functionalized halloysite nanotubes also exhibited superior recyclability along with outstanding stability, even under the oxidative surroundings [201]. Effectiveness of amine impregnated TiO_2 nanotubes towards CO_2 capture even under the wet conditions was demonstrated by Song *et al.* [202]. At 60 °C, TiO_2 nanotubes based adsorbent containing tetraethylenepentamine exhibited maximal storage capacity of 4.37 mmol/g for CO_2, which was improved to 5.24 mmol/g in the presence of moisture. Also, the adsorbent exhibited greater recycling stability under the dry conditions. In another study, Ko *et al.* [203] obtained significant CO_2 capture on adsorbent based on amine functionalized double walled silica nanotubes. A reduction in efficiency of the resultant adsorbent was also observed with the corresponding increase in temperature.

Liu *et al.* [204] reported the use of PEI impregnated protonated titanate nanotubes as a beneficial adsorbent for CO_2 capture. Optimal adsorptivity of 130.8 mg/g for modified protonated titanate nanotubes with 50 wt% of PEI at 100 °C was observed. Also, adsorbent based on the modified nanotubes exhibited greater recycling stability. In another study, Dillon *et al.* [205] developed PEI impregnated SWCNTs for CO_2 capture. A peak adsorption capacity of 9.2 wt% for CO_2 on modified SWCNTs at 27 °C was observed. Moreover, the adsorbed CO_2 was easily desorbed from the modified SWCNTs by heating at 75 °C [205]. Cong *et al.* [206] reported the application of composite composed of CNTs and brominated poly(2,6-diphenyl-1,4-phenylene oxide) as CO_2 adsorbent, which presented a greater adsorption degree than the pure polymer. The adsorption degree was further enhanced through the use of carboxylic acid treated SWCNTs [206].

12.6.2 Adsorption of H_2

Zuttel *et al.* [207] employed SWCNTs for H_2 adsorption and observed good adsorption and desorption characteristics. Successful adsorption of H_2 on SWCNTs through either cavities or monolayered structure was observed. The intensity of H_2 adsorption on SWCNTs enhanced on increasing the CNT diameter, with the CNTs with 2.2 nm diameter reported to have the potential to hold about 5 mass% of H_2 [207]. In another study, Ma *et al.* [208] examined the feasibility of multiwalled boron nitride nanotubes for H_2 capture. Adsorption capacity of up to 2.6 wt% for H_2 under ambient temperature and 10 MPa was observed. Nour *et al.* [209] attained adsorption selectivity of H_2 from CH_4 on MWCNT/polydimethylsiloxane composite based adsorbent and an improvement in effectiveness was observed with increase in concentration of MWCNTs in the composite.

12.6.3 Adsorption of Sulfur Gases

Wang *et al.* [210] investigated the adsorption behavior of sulfur gases such as H_2S and SO_2 from gas mixtures on SWCNTs arrays based adsorbent. SWCNTs with larger diameter exhibited optimal adsorption capacities of 16.31 mmol/g for H_2S and 16.03 mmol/g for SO_2 at a temperature of 303 K. At the same time, substantial selectivity of sulfur gases from gas mixtures was reported for SWCNTs possessing smaller diameter. Further, the functionalization of SWCNTs surfaces with carbonyl moieties enhanced the adsorption selectivity of sulfur gases from the gas mixture containing CH_4 to about two folds [210]. Mohamadalizadeh *et al.* [211] carried out a comparative study concerning the adsorption behavior of H_2S on various MWCNTs based adsorbents such as pristine MWCNTs, tungsten impregnated MWCNTs and amino/amido functionalized MWCNTs at the conditions of 20 °C and 1-10 bar. The adsorptivity of MWCNTs was enhanced through amine functionalization, whereas the impregnation of tungsten nanoparticles reduced it.

12.6.4 Adsorption of VOCs

Hussain *et al.* [212] reported strong adsorption and desorption characteristics of VOCs on both single and multiwalled carbon nanotubes based adsorbent owing to high aspect ratio. Also, the materials offered mass transfer resistance free conditions in consequence to their non-porous nature. In another study, Li *et al.* [213] further confirmed the potential of MWCNTs for effective VOCs capture and subsequent successful recovery. For obtaining improved VOCs adsorption characteristics of MWCNTs, Wang *et al.* [214] reported the use of MWCNTs/porous silica gel composites. In another study, Zheng *et al.* [215] reported the application of SWCNTs in the paper form for effective adsorption and subsequent desorption of VOCs like methyl ethyl ketone, toluene and dimethyl methylphosphonate.

12.6.5 Adsorption of Other Gases

Delavar *et al.* [216] used MWCNTs with a specific surface area of 294 m^2/g and a mean pore volume of 0.62 cm^3/g for methane capture and attained adsorption capacity of about 34 wt% at 283 K and 50 bar. In another study, Cao *et al.* [217] theoretically evaluated the adsorption characteristics of methane on triangular arrays of

SWCNTs and obtained a volumetric adsorptivity of 216 V/V and a gravimetric adsorptivity of 215 g CH_4/kg at ambient temperature and 4.1 MPa.

Kim *et al.* [218] reported the development of CNTs/poly(imide siloxane) composite based adsorbent for capture of O_2, N_2 and CH_4. The occupancy of CNTs on copolymer matrix caused to enrich the adsorption performance, which was higher for the material containing more amount of CNTs. Wang *et al.* [219] fabricated titania nanotubes *via* hydrothermal route and applied successively for mercury capture from exhaust gases. In another study, Ding *et al.* [220] also detailed the feasibility of CNTs for advantageous adsorption of SF_4 and SOF_2.

Foroutan *et al.* [221] achieved higher capability of CNT bundles to capture considerable amount of noble gas mixture comprising of argon, krypton and xenon. Jalili *et al.* [222] also observed substantial adsorption of xenon and krypton through internal and external surfaces of SWCNTs. In another study, the adsorptivity of neon on open ended SWCNTs exhibited decreasing tendency with increase in temperature [223]. Foroutan *et al.* [224] also reported the adsorption degree of heavier noble gas on SWCNTs based adsorbent to be greater as compared to the lighter one.

12.7 Zeolite Based Adsorbents

Zeolites are crystalline microporous materials with three dimensional structure comprising of hydrated aluminosilicate minerals. These materials offer wide range of gas adsorption and storage applications owing to their structural features [225,226].

12.7.1 Adsorption of CO_2

Montanari *et al.* [227] investigated the adsorption of CO_2 from synthetic biogas on adsorbents based on 4A and 13X zeolites. Significant adsorptivity of CO_2 on 13X zeolites was reported and the adsorption intensity was additionally improved under wet conditions. Adsorbent based on 4A zeolites exhibited poor adsorptivity of CO_2 even under wet conditions due to the methane co-adsorption. 4A zeolites presented swift regeneration compared to 13X zeolites under nitrogen environment at ambient temperature. In another study, adsorbent based on Y-type zeolite was observed to exhibit greater permeance for CO_2 than N_2 owing to the substantial adsorption selectivity [228]. The permeance of CO_2 on Y-type zeolite adsorbent was more than 10^{-7} mol.m^{-2}s^{-1}Pa^{-1}. Also, a good permselectivity of 20-100 for CO_2 to N_2 was reported. Saha *et al.* [229] obtained sizeable adsorption of CO_2 and N_2O from atmospheric air as well as significant CO_2 uptake from methane on zeolite 5A based adsorbent compared to metal organic frameworks (Figure 12.11).

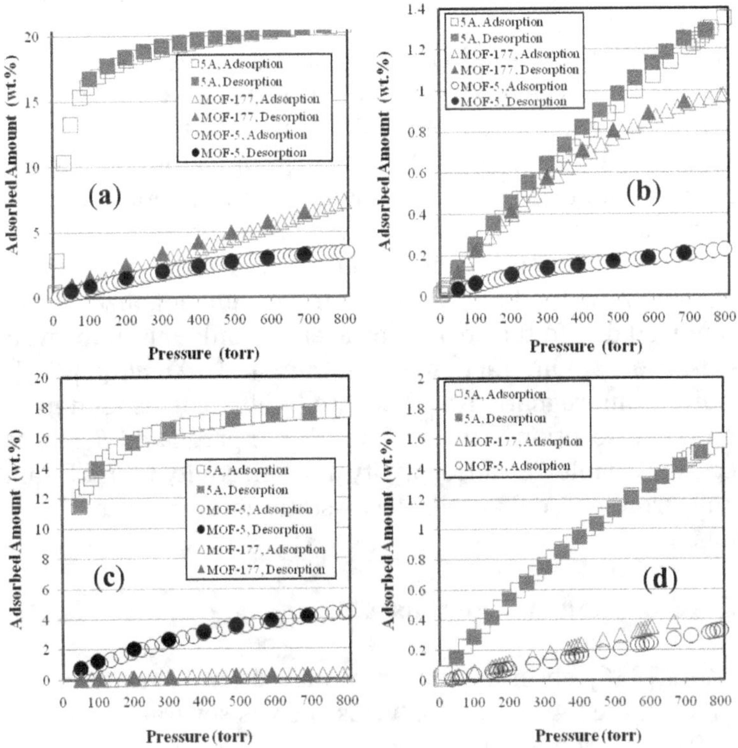

Figure 12.11 Adsorption isotherms of (a) CO_2, (b) CH_4, (c) N_2O, and (d) N_2 on MOF-5, MOF-177, and zeolite 5A at 298 K and pressures up to 800 Torr. Reproduced form reference 229 with permission from American Chemical Society.

Chatti *et al.* [230] applied 13X zeolite and its amine functionalized versions for successful CO_2 capture. Adsorbent based on 13X zeolite had an adsorption capacity of 16.01mg/g, which enhanced to 22.78 mg/g and 19.98 mg/g after amine functionalization using isopropanol amine and monoethanol amine, respectively [230]. In another study, Zhao *et al.* [226] achieved improved performance of 13X zeolite towards CO_2 capture through alkali (NaOH) modification. The creation of enlarged active adsorption surfaces and the reduced diffusion barrier after alkali treatment of 13X zeolite resulted in intensifying CO_2 adsorption rate as well as capacity. Tagliabue *et al.* [231] used sodium impregnated FAU zeolite based adsorbent for the mass adsorption of CO_2 from natural gas. Mixed matrix membrane based adsorbent consisting of 50 wt% of zeolite 4A and poly(vinyl acetate) was also developed by Adams *et al.* [232] for efficient removal of CO_2 from CH_4.

Li *et al.* [233] reported considerable adsorption of CO_2 over β-zeolite based adsorbents compared to CH_4, N_2 and O_2 [233]. Xu *et al.* [234] also executed monoethanol amine functionalization of β-zeolite in order to enhance the adsorption performance. Amine functionalization of β-zeolite improved the adsorption selectivity of CO_2, with the material having 40 wt% of monoethanol amine exhibiting the maximum enhancement. Maximal selectivity of 25.67 for CO_2/N_2 and 7.70 for CO_2/CH_4 over amine functionalized β-zeolite was reported due to the strong steric effect and, thus, improved adsorbent-adsorbate interaction. In another study, Xu *et al.* [235] carried out Na impregnation of β-zeolite, which correspondingly enhanced the adsorption features. Thus, Na impregnated β-zeolite exhibited appreciable CO_2 adsorptivity and selectivity than CH_4 and H_2 resulting from the enhanced electrostatic interaction of CO_2 with Na^+ ions.

12.7.2 Adsorption of Other Gases

Diaz *et al.* [236] observed superior adsorption capacity of Zeolite 13X over Zeolite 5A for various VOCs such as several alkanes, cyclic hydrocarbons, aromatic hydrocarbons and chlorinated compounds owing to the extensive pore dimensions. Sen *et al.* [237] fabricated zeolite 4A/polycarbonate mixed matrix adsorbent for achieving efficient capture of gas molecules like N_2, H_2, O_2, CO_2 and CH_4. Increase in zeolite 4A concentration in the composition increased the selectivity of all gas molecules at the cost of adsorption permeabilities. For adsorption of gases like O_2, CO_2, N_2 and He as well as gas mixture of CO_2/CH_4, Husain *et al.* [238] reported the use of functionalized HSSZ-13 zeolite/PEI hybrid based adsorbents. Functionalization of zeolite was carried out through two different routes, namely PEI sizing and Grignard treatment. PEI sizing had no significant effect on the adsorption selectivity of gases whereas Grignard treatment was advantageous for improved adsorption selectivity. In another study,

Sen *et al.* [239] also achieved improvement in adsorption selectivities of H_2 and CO_2 on polycarbonate/*p*-nitroaniline blend through the incorporation of zeolite 4A.

12.8 Polymer Based Adsorbents

In this section, only polymer based adsorbents not mentioned in the earlier sections have been reviewed. Many polymeric materials offer effective gas adsorption and storage feasibility based on their structural characteristics. Rolker *et al.* [240] revealed the usefulness of branched polymers of polyethers, polyesters and polyamines as adsorbent for CO_2 uptake. Branched polymers exhibited large CO_2 adsorption capacity along with CO_2/N_2 selectivity. Bali *et al.* [241] used polyamines such as PEI and poly(allylamine) along with alumina as adsorbent for CO_2 capture. The oxidation stability of adsorbents is a critical parameter which decides their reusability. In this study, the adsorbent based on poly(allylamine) exhibited greater oxidation stability than PEI due to the absence of secondary amines and the presence of primary amines.

An *et al.* [242] generated an ionic polymer namely, poly(1,1,3,3-tetramethylguanidine acrylate) based adsorbent to achieve effective SO_2 adsorption and successive desorption. The ionic polymer exhibited high adsorption capacity, adsorption rate and selectivity for SO_2 compared to monomer based adsorbents. Successful desorption of SO_2 with the aid of elevated temperature or vacuum was reported, which correspondingly offered great potential of copolymer based adsorbent for reuse. Kikkinides *et al.* [243] reported adsorption of SO_2 molecules over nitrogen monoxide (NO) modified polystyrene based adsorbent. The hydrophobic NO-modified polystyrene exhibited excellent SO_2 storage capacity along with high diffusion rate and SO_2/CO_2 selectivity. Up to 90% removal of SO_2 from the flue gases was accomplished using the adsorbent. In another study, Deshmukh *et al.* [244] generated Hofmann-type porous coordination polymer $\{Fe(Pz)[Pt(CN)_4]\}_n$ (PCP) with three dimensional structure (Figure 12.12) to examine the adsorption behavior of CO_2 and CS_2. Compared to CO_2, gas molecules of CS_2 exhibited efficient adsorption on PCP owing to superior dispersion interactions generated through Pt and Pz moieties of PCP with CS_2. Noro *et al.* [245] also reported an outstanding adsorption density of 0.21 gm/L for methane on the porous coordination polymer $\{CuSiF_6(4,4'\text{-bipyridine})_2\}_n$ based adsorbent.

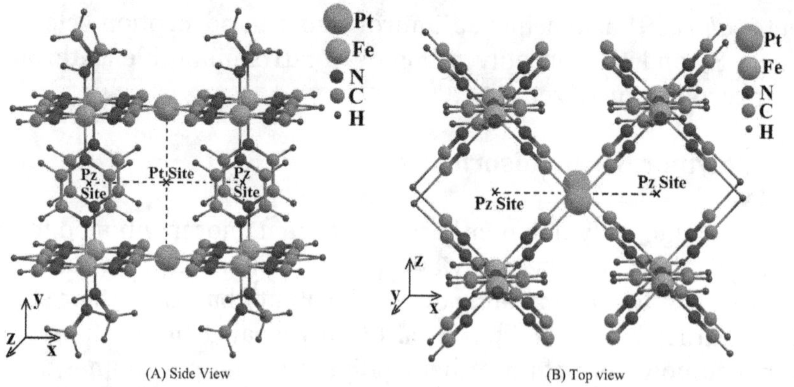

Figure 12.12 Large realistic model (LRM) of the Hofmann-type PCP. Reproduced from Reference 244 with permission from American Chemical Society.

12.9 Conclusions

Adsorption of diverse pollutant gases with the help of miscellaneous adsorbents is immensely important research field in terms of environmental and energy aspects. The selection of suitable adsorbent is the critical parameter which decides the performance during the adsorption of gas molecules. In this study, the adsorption performance of various adsorbents based on silica, clay, activated carbon, graphene, graphite oxide, nanotubes, zeolites and polymers have been thoroughly compared. The adsorption potential of each adsorbent for different gas molecules including CO_2, N_2, H_2, sulfur gases, hydrocarbon gases, ammonia, volatile organic compounds and other gases has been evaluated with respect to adsorption capacity, selectivity, stability, recyclability/reusability, heat of adsorption, etc. The modifications of adsorbent materials using chemical treatment, physical methods, elemental doping, polymer impregnation, metal/metal oxide infusion, *etc.* are useful to upgrade the ultimate performance of the adsorbents, which has also been thoroughly reviewed.

References

1. Yu, K. M. K., Curcic, I., Gabriel, J., and Tsang, S. C. E. (2008) Recent advances in CO_2 capture and utilization. *ChemSusChem*, **1**, 893-899.

2. Basic Research Needs for Carbon Capture: Beyond 2020 (2010) *Report of the Basic Energy Sciences Workshop for Carbon Capture: Beyond 2020,* USA. Online: https://science.energy.gov/~/media/bes/pdf/reports/files/Basic_Research_Needs_for_Carbon_Capture_rpt.pdf [assessed 21st April 2017].
3. Griggs, D. J., and Noguer, M. (2002) Climate change 2001: the scientific basis. Contribution of working group I to the third assessment report of the intergovernmental panel on climate change. *Weather,* **57**, 267-269.
4. Lee, S.-Y. and Park, S.-J. (2015) A review on solid adsorbents for carbon dioxide capture. *Journal of Industrial and Engineering Chemistry,* **23**, 1-11.
5. Wang, Q., Luo, J., Zhong, Z., and Borgna, A. (2011) CO_2 capture by solid adsorbents and their applications: current status and new trends. *Energy & Environmental Science,* **4**, 42-55.
6. Olivares-Marin, M., and Maroto-Valer, M. M. (2012) Development of adsorbents for CO_2 capture from waste materials: a review. *Greenhouse Gases: Science and Technology,* **2**, 20-35.
7. D'Alessandro, D. M., Smit, B., and Long, J. R. (2010) Carbon dioxide capture: prospects for new materials. *Angewandte Chemie International Edition,* **49**, 6058-6082.
8. Srivastava, N., and Eames, I. (1998) A review of adsorbents and adsorbates in solid-vapour adsorption heat pump systems. *Applied Thermal Engineering,* **18**, 707-714.
9. Suzuk, M. (1990) Adsorption engineering. *Kodansha,* **551**, 128.
10. Yang, R. T. (2013) *Gas Separation by Adsorption Processes,* Butterworth-Heinemann, USA.
11. Bhown, A. S., and Freeman, B. C. (2011) Analysis and status of post-combustion carbon dioxide capture technologies. *Environmental Science & Technology,* **45**, 8624-8632.
12. Ben, T., Li, Y., Zhu, L., Zhang, D., Cao, D., Xiang, Z., Yao, X., and Qiu, S. (2012) Selective adsorption of carbon dioxide by carbonized porous aromatic framework (PAF). *Energy & Environmental Science,* **5**, 8370-8376.
13. Sayari, A., Belmabkhout, Y., and Serna-Guerrero, R. (2011) Flue gas treatment via CO_2 adsorption. *Chemical Engineering Journal,* **171**, 760-774.
14. Choi, S., Drese, J. H., and Jones, C. W. (2009) Adsorbent materials for carbon dioxide capture from large anthropogenic point sources. *ChemSusChem,* **2**, 796-854.
15. Florin, N. H., Blamey, J., and Fennell, P. S. (2010) Synthetic CaO-based sorbent for CO_2 capture from large-point sources. *Energy & Fuels,* **24**, 4598-4604.
16. Drage, T., Arenillas, A., Smith, K. M., and Snape, C. E. (2008) Therm-

al stability of polyethylenimine based carbon dioxide adsorbents and its influence on selection of regeneration strategies. *Microporous and Mesoporous Materials,* **116**, 504-512.

17. Bae, Y. S., and Snurr, R. Q. (2011) Development and evaluation of porous materials for carbon dioxide separation and capture. *Angewandte Chemie International Edition,* **50**, 11586-11596.
18. Kim, S., Ida, J., Guliants, V. V., and Lin, Y. (2005) Tailoring pore properties of MCM-48 silica for selective adsorption of CO_2. *The Journal of Physical Chemistry B,* **109**, 6287-6293.
19. Pereira, P., Pires, J., and Brotas de Carvalho, M. (1998) Zirconium pillared clays for carbon dioxide/methane separation. 1. Preparation of adsorbent materials and pure gas adsorption. *Langmuir,* **14**, 4584-4588.
20. Himeno, S., Komatsu, T., and Fujita, S. (2005) High-pressure adsorption equilibria of methane and carbon dioxide on several activated carbons. *Journal of Chemical & Engineering Data,* **50**, 369.
21. Thiruvenkatachari, R., Su, S., An, H., and Yu, X. X. (2009) Post combustion CO_2 capture by carbon fibre monolithic adsorbents. *Progress in Energy and Combustion Science,* **35**, 438-455.
22. Balasubramanian, R., and Chowdhury, S. (2015) Recent advances and progress in the development of graphene-based adsorbents for CO_2 capture. *Journal of Materials Chemistry A,* **3**, 21968-21989.
23. Burress, J. W., Gadipelli, S., Ford, J., Simmons, J. M., Zhou, W., and Yildirim, T. (2010) Graphene oxide framework materials: theoretical predictions and experimental results. *Angewandte Chemie International Edition,* **49**, 8902-8904.
24. Zhao, J., Buldum, A., Han, J., and Lu, J. P. (2002) Gas molecule adsorption in carbon nanotubes and nanotube bundles. *Nanotechnology,* **13**, 195.
25. Ren, X., Chen, C., Nagatsu, M., and Wang, X. (2011) Carbon nanotubes as adsorbents in environmental pollution management: A review. *Chemical Engineering Journal,* **170**, 395-410.
26. Rouquerol, J., Rouquerol, F., Llewellyn, P., Maurin, G., and Sing, K. S. (2013) *Adsorption by Powders and Porous Solids: Principles, Methodology and Applications.* 2nd edition, Academic Press, USA.
27. LeVan, M. D., Carta, G., and Yon, C. M. (1997) Adsorption and ion exchange. *Energy,* **16**, 17.
28. Farha, O. K., Spokoyny, A. M., Hauser, B. G., Bae, Y.-S., Brown, S. E., Snurr, R. Q., Mirkin, C. A., and Hupp, J. T. (2009) Synthesis, properties, and gas separation studies of a robust diimide-based microporous organic polymer. *Chemistry of Materials,* **21**, 3033-3035.
29. Yong, Z., Mata, V., and Rodrigues, A. E. (2000) Adsorption of carbon dioxide on basic alumina at high temperatures. *Journal of Chemical & Engineering Data,* **45**, 1093-1095.
30. Bhatta, L. K. G., Subramanyam, S., Chengala, M. D., Olivera, S., and

Venkatesh, K. (2015) Progress in hydrotalcite like compounds and metal-based oxides for CO_2 capture: A review. *Journal of Cleaner Production,* **103**, 171-196.

31. Liu, J., Thallapally, P. K., McGrail, B. P., Brown, D. R., and Liu, J. (2012) Progress in adsorption-based CO_2 capture by metal-organic frameworks. *Chemical Society Reviews,* **41**, 2308-2322.
32. Hoffmann, F., Cornelius, M., Morell, J., and Froba, M. (2006) Silica-based mesoporous organic-inorganic hybrid materials. *Angewandte Chemie International Edition,* **45**, 3216-3251.
33. Belmabkhout, Y., Serna-Guerrero, R., and Sayari, A. (2009) Adsorption of CO_2 from dry gases on MCM-41 silica at ambient temperature and high pressure. 1: Pure CO_2 adsorption. *Chemical Engineering Science,* **64**, 3721-3728.
34. Belmabkhout, Y., and Sayari, A. (2009) Adsorption of CO_2 from dry gases on MCM-41 silica at ambient temperature and high pressure. 2: Adsorption of CO_2/N_2, CO_2/CH_4 and CO_2/H_2 binary mixtures. *Chemical Engineering Science,* **64**, 3729-3735.
35. Tsai, C.-Y., Tam, S.-Y., Lu, Y., and Brinker, C. J. (2000) Dual-layer asymmetric microporous silica membranes. *Journal of Membrane Science,* **169**, 255-268.
36. Himeno, S., Tomita, T., Suzuki, K., and Yoshida, S. (2007) Characterization and selectivity for methane and carbon dioxide adsorption on the all-silica DD3R zeolite. *Microporous and Mesoporous Materials,* **98**, 62-69.
37. Bellussi, G., Broccia, P., Carati, A., Millini, R., Pollesel, P., Rizzo, C., and Tagliabue, M. (2011) Silica-aluminas for carbon dioxide bulk removal from sour natural gas. *Microporous and Mesoporous Materials,* **146**, 134-140.
38. Xomeritakis, G., Tsai, C., Jiang, Y., and Brinker, C. (2009) Tubular ceramic-supported sol–gel silica-based membranes for flue gas carbon dioxide capture and sequestration. *Journal of Membrane Science,* **341**, 30-36.
39. Leal, O., Bolivar, C., Ovalles, C., Garcia, J. J., and Espidel, Y. (1995) Reversible adsorption of carbon dioxide on amine surface-bonded silica gel. *Inorganica Chimica Acta,* **240**, 183-189.
40. Kim, S.-N., Son, W.-J., Choi, J.-S., and Ahn, W.-S. (2008) CO_2 adsorption using amine-functionalized mesoporous silica prepared via anionic surfactant-mediated synthesis. *Microporous and Mesoporous Materials,* **115**, 497-503.
41. Yue, M. B., Sun, L. B., Cao, Y., Wang, Z. J., Wang, Y., Yu, Q., and Zhu, J. H. (2008) Promoting the CO_2 adsorption in the amine-containing SBA-15 by hydroxyl group. *Microporous and Mesoporous Materials,* **114**, 74-81.
42. Knowles, G. P., Graham, J. V., Delaney, S. W., and Chaffee, A. L. (2005) Aminopropyl-functionalized mesoporous silicas as CO_2 ad-

sorbents. *Fuel Processing Technology,* **86**, 1435-1448.

43. Hiyoshi, N., Yogo, K., and Yashima, T. (2005) Adsorption characteristics of carbon dioxide on organically functionalized SBA-15. *Microporous and Mesoporous Materials,* **84**, 357-365.
44. Sakamoto, Y., Nagata, K., Yogo, K., and Yamada, K. (2007) Preparation and CO_2 separation properties of amine-modified mesoporous silica membranes. *Microporous and Mesoporous Materials,* **101**, 303-311.
45. Knofel, C., Descarpentries, J., Benzaouia, A., Zelenak, V., Mornet, S., Llewellyn, P., and Hornebecq, V. (2007) Functionalised micro-/mesoporous silica for the adsorption of carbon dioxide. *Microporous and Mesoporous Materials,* **99**, 79-85.
46. Liu, X., Zhou, L., Fu, X., Sun, Y., Su, W., and Zhou, Y. (2007) Adsorption and regeneration study of the mesoporous adsorbent SBA-15 adapted to the capture/separation of CO_2 and CH_4. *Chemical Engineering Science,* **62**, 1101-1110.
47. Liu, X., Li, J., Zhou, L., Huang, D., and Zhou, Y. (2005) Adsorption of CO_2, CH_4 and N_2 on ordered mesoporous silica molecular sieve. *Chemical Physics Letters,* **415**, 198-201.
48. Zelenak, V., Halamova, D., Gaberova, L., Bloch, E., and Llewellyn, P. (2008) Amine-modified SBA-12 mesoporous silica for carbon dioxide capture: Effect of amine basicity on sorption properties. *Microporous and Mesoporous Materials,* **116**, 358-364.
49. Jang, K.-S., Kim, H.-J., Johnson, J., Kim, W.-g., Koros, W. J., Jones, C. W., and Nair, S. (2011) Modified mesoporous silica gas separation membranes on polymeric hollow fibers. *Chemistry of Materials,* **23**, 3025-3028.
50. Hicks, J. C., Drese, J. H., Fauth, D. J., Gray, M. L., Qi, G., and Jones, C. W. (2008) Designing adsorbents for CO_2 capture from flue gas-hyperbranched aminosilicas capable of capturing CO_2 reversibly. *Journal of the American Chemical Society,* **130**, 2902-2903.
51. Song, H.-K., Cho, K. W., and Lee, K.-H. (1998) Adsorption of carbon dioxide on the chemically modified silica adsorbents. *Journal of Non-crystalline Solids,* **242**, 69-80.
52. Son, W.-J., Choi, J.-S., and Ahn, W.-S. (2008) Adsorptive removal of carbon dioxide using polyethyleneimine-loaded mesoporous silica materials. *Microporous and Mesoporous Materials,* **113**, 31-40.
53. Goeppert, A., Czaun, M., May, R. B., Prakash, G. S., Olah, G. A., and Narayanan, S. (2011) Carbon dioxide capture from the air using a polyamine based regenerable solid adsorbent. *Journal of the American Chemical Society,* **133**, 20164-20167.
54. Xu, X., Song, C., Andresen, J. M., Miller, B. G., and Scaroni, A. W. (2002) Novel polyethylenimine-modified mesoporous molecular sieve of MCM-41 type as high-capacity adsorbent for CO_2 capture. *Energy & Fuels,* **16**, 1463-1469.

55. Rezaei, F., Lively, R. P., Labreche, Y., Chen, G., Fan, Y., Koros, W. J., and Jones, C. W. (2013) Aminosilane-grafted polymer/silica hollow fiber adsorbents for CO_2 capture from flue gas. *ACS Applied Materials & Interfaces,* **5**, 3921-3931.
56. Kusakabe, K., Ichiki, K., Hayashi, J.-i., Maeda, H., and Morooka, S. (1996) Preparation and characterization of silica-polyimide composite membranes coated on porous tubes for CO_2 separation. *Journal of Membrane Science,* **115**, 65-75.
57. Zornoza, B., Tellez, C., and Coronas, J. (2011) Mixed matrix membranes comprising glassy polymers and dispersed mesoporous silica spheres for gas separation. *Journal of Membrane Science,* **368**, 100-109.
58. Kim, J. H., and Lee, Y. M. (2001) Gas permeation properties of poly (amide-6-b-ethylene oxide)-silica hybrid membranes. *Journal of Membrane Science,* **193**, 209-225.
59. Morooka, S., Yan, S., Kusakabe, K., and Akiyama, Y. (1995) Formation of hydrogen-permselective SiO_2 membrane in macropores of α-alumina support tube by thermal decomposition of TEOS. *Journal of Membrane Science,* **101**, 89-98.
60. Sea, B.-K., Watanabe, M., Kusakabe, K., Morooka, S., and Kim, S.-S. (1996) Formation of hydrogen permselective silica membrane for elevated temperature hydrogen recovery from a mixture containing steam. *Gas Separation & Purification,* **10**, 187-195.
61. De Vos, R. M., and Verweij, H. (1998) Improved performance of silica membranes for gas separation. *Journal of Membrane Science,* **143**, 37-51.
62. Boffa, V., Ten Elshof, J., and Blank, D. (2007) Preparation of templated mesoporous silica membranes on macroporous α-alumina supports via direct coating of thixotropic polymeric sols. *Microporous and Mesoporous Materials,* **100**, 173-182.
63. Zornoza, B., Esekhile, O., J. Koros, W., Tellez, C., and Coronas, J. (2011) Hollow silicalite-1 sphere-polymer mixed matrix membranes for gas separation. *Separation and Purification Technology,* **77**, 137-145.
64. Huang, H. Y., Yang, R. T., Chinn, D., and Munson, C. L. (2003) Amine-grafted MCM-48 and silica xerogel as superior sorbents for acidic gas removal from natural gas. *Industrial & Engineering Chemistry Research,* **42**, 2427-2433.
65. Belmabkhout, Y., De Weireld, G., and Sayari, A. (2009) Amine-bearing mesoporous silica for CO_2 and H_2S removal from natural gas and biogas. *Langmuir,* **25**, 13275-13278.
66. Wang, X., Jia, J., Zhao, L., and Sun, T. (2008) Chemisorption of hydrogen sulphide on zinc oxide modified aluminum-substituted SBA-15. *Applied Surface Science,* **254**, 5445-5451.
67. Wang, X., Sun, T., Yang, J., Zhao, L., and Jia, J. (2008) Low-temperat-

ure H_2S removal from gas streams with SBA-15 supported ZnO nanoparticles. *Chemical Engineering Journal,* **142**, 48-55.

68. Wang, X., Ma, X., Sun, L., and Song, C. (2007) A nanoporous polymeric sorbent for deep removal of H_2S from gas mixtures for hydrogen purification. *Green Chemistry,* **9**, 695-702.
69. De Vos, R. M., Maier, W. F., and Verweij, H. (1999) Hydrophobic silica membranes for gas separation. *Journal of Membrane Science,* **158**, 277-288.
70. De Sitter, K., Winberg, P., D'Haen, J., Dotremont, C., Leysen, R. Martens, J. A., Mullens, S., Maurer, F. H., and Vankelecom, I. F. (2006) Silica filled poly (1-trimethylsilyl-1-propyne) nanocomposite membranes: relation between the transport of gases and structural characteristics. *Journal of Membrane Science,* **278**, 83-91.
71. De Sitter, K., Andersson, A., D'Haen, J., Leysen, R., Mullens, S., Maurer, F. H., and Vankelecom, I. F. (2008) Silica filled poly (4-methyl-2-pentyne) nanocomposite membranes: Similarities and differences with poly (1-trimethylsilyl-1-propyne)-silica systems. *Journal of Membrane Science,* **321**, 284-292.
72. Newalkar, B. L., Choudary, N. V., Kumar, P., Komarneni, S., and Bhat, T. S. (2002) Exploring the potential of mesoporous silica, SBA-15, as an adsorbent for light hydrocarbon separation. *Chemistry of Materials,* **14**, 304-309.
73. Montanari, T., Finocchio, E., Bozzano, I., Garuti, G., Giordano, A., Pistarino, C., and Busca, G. (2010) Purification of landfill biogases from siloxanes by adsorption: A study of silica and 13X zeolite adsorbents on hexamethylcyclotrisiloxane separation. *Chemical Engineering Journal,* **165**, 859-863.
74. Pires, J., Bestilleiro, M., Pinto, M., and Gil, A. (2008) Selective adsorption of carbon dioxide, methane and ethane by porous clays heterostructures. *Separation and Purification Technology,* **61**, 161-167.
75. Stepova, K. V., Maquarrie, D. J., and Krip, I. M. (2009) Modified bentonites as adsorbents of hydrogen sulfide gases. *Applied Clay Science,* **42**, 625-628.
76. Nguyen-Thanh, D., Block, K., and Bandosz, T. J. (2005) Adsorption of hydrogen sulfide on montmorillonites modified with iron. *Chemosphere,* **59**, 343-353.
77. Ji, L., Zhang, T., Milliken, K. L., Qu, J., and Zhang, X. (2012) Experimental investigation of main controls to methane adsorption in clay-rich rocks. *Applied Geochemistry,* **27**, 2533-2545.
78. Pires, J., Araujo, A., Carvalho, A., Pinto, M., Gonzalez-Calbet, J., and Ramırez-Castellanos, J. (2004) Porous materials from clays by the gallery template approach: synthesis, characterization and adsorption properties. *Microporous and Mesoporous Materials,* **73**, 175-180.

79. Cheng, A.-L., and Huang, W.-L. (2004) Selective adsorption of hydrocarbon gases on clays and organic matter. *Organic Geochemistry,* **35**, 413-423.
80. Choudary, N. V., Kumar, P., Bhat, T. S., Cho, S. H., Han, S. S., and Kim, J. N. (2002) Adsorption of light hydrocarbon gases on alkene-selective adsorbent. *Industrial & Engineering Chemistry Research,* **41**, 2728-2734.
81. Zaitan, H., Bianchi, D., Achak, O., and Chafik, T. (2008) A comparative study of the adsorption and desorption of o-xylene onto bentonite clay and alumina. *Journal of Hazardous Materials,* **153**, 852-859.
82. Qu, F., Zhu, L., and Yang, K. (2009) Adsorption behaviors of volatile organic compounds (VOCs) on porous clay heterostructures (PCH). *Journal of Hazardous Materials,* **170**, 7-12.
83. Zuo, S., Liu, F., Zhou, R., and Qi, C. (2012) Adsorption/desorption and catalytic oxidation of VOCs on montmorillonite and pillared clays. *Catalysis Communications,* **22**, 1-5.
84. Nunes, C. D., Pires, J., Carvalho, A. P., Calhorda, M. J., and Ferreira, P. (2008) Synthesis and characterisation of organo-silica hydrophobic clay heterostructures for volatile organic compounds removal. *Microporous and Mesoporous Materials,* **111**, 612-619.
85. Pires, J., Saini, V. K., and Pinto, M. s. L. (2008) Studies on selective adsorption of biogas components on pillared clays: approach for biogas improvement. *Environmental Science & Technology,* **42**, 8727-8732.
86. Sircar, S., Golden, T., and Rao, M. (1996) Activated carbon for gas separation and storage. *Carbon,* **34**, 1-12.
87. Presser, V., McDonough, J., Yeon, S.-H., and Gogotsi, Y. (2011) Effect of pore size on carbon dioxide sorption by carbide derived carbon. *Energy & Environmental Science,* **4**, 3059-3066.
88. Maroto-Valer, M. M., Tang, Z., and Zhang, Y. (2005) CO_2 capture by activated and impregnated anthracites. *Fuel Processing Technology,* **86**, 1487-1502.
89. Goel, C., Bhunia, H., and Bajpai, P. K. (2015) Mesoporous carbon adsorbents from melamine-formaldehyde resin using nanocasting technique for CO_2 adsorption. *Journal of Environmental Sciences,* **32**, 238-248.
90. Siriwardane, R. V., Shen, M.-S., Fisher, E. P., and Poston J. A. (2001) Adsorption of CO_2 on molecular sieves and activated carbon. *Energy & Fuels,* **15**, 279-284.
91. Drage, T. C., Blackman, J. M., Pevida, C., and Snape, C. E. (2009) Evaluation of activated carbon adsorbents for CO_2 capture in gasification. *Energy & Fuels,* **23**, 2790-2796.
92. Ganesan, A., and Shaijumon, M. M. (2016) Activated graphene-derived porous carbon with exceptional gas adsorption properties.

Microporous and Mesoporous Materials, **220**, 21-27.

93. Sui, Z.-Y., Meng, Q.-H., Li, J.-T., Zhu, J.-H., Cui, Y., and Han, B.-H. (2014) High surface area porous carbons produced by steam activation of graphene aerogels. *Journal of Materials Chemistry A,* **2**, 9891-9898.
94. Zhang, Z., Zhou, J., Xing, W., Xue, Q., Yan, Z., Zhuo, S., and Qiao, S. Z. (2013) Critical role of small micropores in high CO_2 uptake. *Physical Chemistry Chemical Physics,* **15**, 2523-2529.
95. Deng, H.-g., Jin, S.-l., Liang, Z., Wang, Y.-l., Lu, B.-h., Qiao, W.-m., and Ling, L.-c. (2012) Synthesis of porous carbons derived from metal-organic coordination polymers and their adsorption performance for carbon dioxide. *New Carbon Materials,* **27**, 194-199.
96. Hu, X., Radosz, M., Cychosz, K. A., and Thommes, M. (2011) CO_2-filling capacity and selectivity of carbon nanopores: synthesis, texture, and pore-size distribution from quenched-solid density functional theory (QSDFT). *Environmental Science & Technology,* **45**, 7068-7074.
97. Deng, S., Wei, H., Chen, T., Wang, B., Huang, J., and Yu, G. (2014) Superior CO_2 adsorption on pine nut shell-derived activated carbons and the effective micropores at different temperatures. *Chemical Engineering Journal,* **253**, 46-54.
98. Silvestre-Albero, J., Wahby, A., Sepulveda-Escribano, A., Martinez-Escandell, M., Kaneko, K., and Rodriguez-Reinoso, F. (2011) Ultra-high CO_2 adsorption capacity on carbon molecular sieves at room temperature. *Chemical Communications,* **47**, 6840-6842.
99. Daud, W. W., Ahmad, M., and Aroua, M. (2007) Carbon molecular sieves from palm shell: Effect of the benzene deposition times on gas separation properties. *Separation and Purification Technology,* **57**, 289-293.
100. Wickramaratne, N. P., and Jaroniec, M. (2013) Activated carbon spheres for CO_2 adsorption. *ACS Applied Materials & Interfaces,* **5**, 1849-1855.
101. Wickramaratne, N. P., and Jaroniec, M. (2013) Importance of small micropores in CO_2 capture by phenolic resin-based activated carbon spheres. *Journal of Materials Chemistry A,* **1**, 112-116.
102. Sevilla, M., Parra, J. B., and Fuertes, A. B. (2013) Assessment of the role of micropore size and N-doping in CO_2 capture by porous carbons. *ACS Applied Materials & Interfaces,* **5**, 6360-6368.
103. Marco-Lozar, J., Kunowsky, M., Suarez-Garcia, F., Carruthers, J., and Linares-Solano, A. (2012) Activated carbon monoliths for gas storage at room temperature. *Energy & Environmental Science,* **5**, 9833-9842.
104. Ribeiro, R. P., Sauer, T. P., Lopes, F. V., Moreira, R. F., Grande, C. A., and Rodrigues, A. r. E. (2008) Adsorption of CO_2, CH_4, and N_2 in activated carbon honeycomb monolith. *Journal of Chemical & Engine-*

ering Data, **53**, 2311-2317.
105. Nandi, M., Okada, K., Dutta, A., Bhaumik, A., Maruyama, J., Derks, D., and Uyama, H. (2012) Unprecedented CO_2 uptake over highly porous N-doped activated carbon monoliths prepared by physical activation. *Chemical Communications,* **48**, 10283-10285.
106. Zhong, M., Natesakhawat, S., Baltrus, J. P., Luebke, D., Nulwala, H., Matyjaszewski, K., and Kowalewski, T. (2012) Copolymer-templated nitrogen-enriched porous nanocarbons for CO_2 capture. *Chemical Communications,* **48**, 11516-11518.
107. Xing, W., Liu, C., Zhou, Z., Zhang, L., Zhou, J., Zhuo, S., Yan, Z., Gao, H., Wang, G., and Qiao, S. Z. (2012) Superior CO_2 uptake of N-doped activated carbon through hydrogen-bonding interaction. *Energy & Environmental Science,* **5**, 7323-7327.
108. Zhao, Y., Zhao, L., Yao, K. X., Yang, Y., Zhang, Q., and Han, Y. (2012) Novel porous carbon materials with ultrahigh nitrogen contents for selective CO_2 capture. *Journal of Materials Chemistry,* **22**, 19726-19731.
109. Saleh, M., Tiwari, J. N., Kemp, K. C., Yousuf, M., and Kim, K. S. (2013) Highly selective and stable carbon dioxide uptake in polyindole-derived microporous carbon materials. *Environmental Science & Technology,* **47**, 5467-5473.
110. Sevilla, M., Falco, C., Titirici, M.-M., and Fuertes, A. B. (2012) High-performance CO_2 sorbents from algae. *RSC Advances,* **2**, 12792-12797.
111. Zhao, Y., Liu, X., Yao, K. X., Zhao, L., and Han, Y. (2012) Superior capture of CO_2 achieved by introducing extra-framework cations into N-doped microporous carbon. *Chemistry of Materials,* **24**, 4725-4734.
112. Fan, X., Zhang, L., Zhang, G., Shu, Z., and Shi, J. (2013) Chitosan derived nitrogen-doped microporous carbons for high performance CO_2 capture. *Carbon,* **61**, 423-430.
113. Chen, C., Kim, J., and Ahn, W.-S. (2012) Efficient carbon dioxide capture over a nitrogen-rich carbon having a hierarchical micro-mesopore structure. *Fuel,* **95**, 360-364.
114. Chandra, V., Yu, S. U., Kim, S. H., Yoon, Y. S., Kim, D. Y., Kwon, A. H., Meyyappan, M., and Kim, K. S. (2012) Highly selective CO_2 capture on N-doped carbon produced by chemical activation of polypyrrole functionalized graphene sheets. *Chemical Communications,* **48**, 735-737.
115. Xia, Y., Zhu, Y., and Tang, Y. (2012) Preparation of sulfur-doped microporous carbons for the storage of hydrogen and carbon dioxide. *Carbon,* **50**, 5543-5553.
116. Seema, H., Kemp, K. C., Le, N. H., Park, S.-W., Chandra, V., Lee, J. W., and Kim, K. S. (2014) Highly selective CO_2 capture by S-doped microporous carbon materials. *Carbon,* **66**, 320-326.

117. Builes, S., Roussel, T., Ghimbeu, C. M., Parmentier, J., Gadiou, R., Vix-Guterl, C., and Vega, L. F. (2011) Microporous carbon adsorbents with high CO_2 capacities for industrial applications. *Physical Chemistry Chemical Physics,* **13**, 16063-16070.
118. Xia, Y., Mokaya, R., Walker, G. S., and Zhu, Y. (2011) Superior CO_2 adsorption capacity on N-doped, high-surface-area, microporous carbons templated from zeolite. *Advanced Energy Materials,* **1**, 678-683.
119. Sevilla, M., and Fuertes, A. B. (2012) CO_2 adsorption by activated templated carbons. *Journal of Colloid and Interface Science,* **366**, 147-154.
120. Pevida, C., Plaza, M., Arias, B., Fermoso, J., Rubiera, F., and Pis, J. (2008) Surface modification of activated carbons for CO_2 capture. *Applied Surface Science,* **254**, 7165-7172.
121. Shafeeyan, M. S., Daud, W. M. A. W., Houshmand, A., and Arami-Niya, A. (2011) Ammonia modification of activated carbon to enhance carbon dioxide adsorption: Effect of pre-oxidation. *Applied Surface Science,* **257**, 3936-3942.
122. Zhang, Z., Xu, M., Wang, H., and Li, Z. (2010) Enhancement of CO_2 adsorption on high surface area activated carbon modified by N_2, H_2 and ammonia. *Chemical Engineering Journal,* **160**, 571-577.
123. Plaza, M., Pevida, C., Arenillas, A., Rubiera, F., and Pis, J. (2007) CO_2 capture by adsorption with nitrogen enriched carbons. *Fuel,* **86**, 2204-2212.
124. Guo, J., Luo, Y., Lua, A. C., Chi, R.-a., Chen, Y.-l., Bao, X.-t., and Xiang, S.-x. (2007) Adsorption of hydrogen sulphide (H_2S) by activated carbons derived from oil-palm shell. *Carbon,* **45**, 330-336.
125. Yan, R., Liang, D. T., Tsen, L., and Tay, J. H. (2002) Kinetics and mechanisms of H_2S adsorption by alkaline activated carbon. *Environmental Science & Technology,* **36**, 4460-4466.
126. Seredych, M., Portet, C., Gogotsi, Y., and Bandosz, T. J. (2009) Nitrogen modified carbide-derived carbons as adsorbents of hydrogen sulfide. *Journal of Colloid and interface Science,* **330**, 60-66.
127. Bagreev, A., Menendez, J. A., Dukhno, I., Tarasenko, Y., and Bandosz, T. J. (2004) Bituminous coal-based activated carbons modified with nitrogen as adsorbents of hydrogen sulfide. *Carbon,* **42**, 469-476.
128. Huang, C.-C., Chen, C.-H., and Chu, S.-M. (2006) Effect of moisture on H_2S adsorption by copper impregnated activated carbon. *Journal of Hazardous Materials,* **136**, 866-873.
129. Nguyen-Thanh, D., and Bandosz, T. J. (2005) Activated carbons with metal containing bentonite binders as adsorbents of hydrogen sulfide. *Carbon,* **43**, 359-367.
130. Xiao, Y., Wang, S., Wu, D., and Yuan, Q. (2008) Experimental and simulation study of hydrogen sulfide adsorption on impregnated

activated carbon under anaerobic conditions. *Journal of Hazardous Materials,* **153**, 1193-1200.

131. Jorda-Beneyto, M., Suarez-Garcia, F., Lozano-Castello, D., Cazorla-Amoros, D., and Linares-Solano, A. (2007) Hydrogen storage on chemically activated carbons and carbon nanomaterials at high pressures. *Carbon,* **45**, 293-303.
132. Juan-Juan, J., Marco-Lozar, J., Suarez-Garcia, F., Cazorla-Amoros, D., and Linares-Solano, A. (2010) A comparison of hydrogen storage in activated carbons and a metal-organic framework (MOF-5). *Carbon,* **48**, 2906-2909.
133. De la Casa-Lillo, M., Lamari-Darkrim, F., Cazorla-Amoros, D., and Linares-Solano, A. (2002) Hydrogen storage in activated carbons and activated carbon fibers. *The Journal of Physical Chemistry B,* **106**, 10930-10934.
134. Kim, H. S., Singer, J. P., Gogotsi, Y., and Fischer, J. E. (2009) Molybdenum carbide-derived carbon for hydrogen storage. *Microporous and Mesoporous Materials,* **120**, 267-271.
135. Jorda-Beneyto, M., Lozano-Castello, D., Suarez-Garcia, F., Cazorla-Amoros, D., and Linares-Solano, A. (2008) Advanced activated carbon monoliths and activated carbons for hydrogen storage. *Microporous and Mesoporous Materials,* **112**, 235-242.
136. Almasoudi, A., and Mokaya, R. (2012) Preparation and hydrogen storage capacity of templated and activated carbons nanocast from commercially available zeolitic imidazolate framework. *Journal of Materials Chemistry,* **22**, 146-152.
137. Sun, J., Rood, M. J., Rostam-Abadi, M., and Lizzio, A. A. (1996) Natural gas storage with activated carbon from a bituminous coal. *Gas Separation & Purification,* **10**, 91-96.
138. Yeon, S.-H., Osswald, S., Gogotsi, Y., Singer, J. P., Simmons, J. M., Fischer, J. E., Lillo-Rodenas, M. A., and Linares-Solano, A. (2009) Enhanced methane storage of chemically and physically activated carbide-derived carbon. *Journal of Power Sources,* **191**, 560-567.
139. Prauchner, M. J., and Rodriguez-Reinoso, F. (2008) Preparation of granular activated carbons for adsorption of natural gas. *Microporous and Mesoporous Materials,* **109**, 581-584.
140. Srinivas, G., Burress, J., and Yildirim, T. (2012) Graphene oxide derived carbons (GODCs): Synthesis and gas adsorption properties. *Energy & Environmental Science,* **5**, 6453-6459.
141. Yeon, S.-H., Knoke, I., Gogotsi, Y., and Fischer, J. E. (2010) Enhanced volumetric hydrogen and methane storage capacity of monolithic carbide-derived carbon. *Microporous and Mesoporous Materials,* **131**, 423-428.
142. Lozano-Castello, D., Cazorla-Amoros, D., Linares-Solano, A., and Quinn, D. (2002) Activated carbon monoliths for methane storage: Influence of binder. *Carbon,* **40**, 2817-2825.

143. Adinata, D., Daud, W. M. A. W., and Aroua, M. K. (2007) Production of carbon molecular sieves from palm shell based activated carbon by pore sizes modification with benzene for methane selective separation. *Fuel Processing Technology,* **88**, 599-605.
144. Aroua, M. K., Daud, W. M. A. W., Yin, C. Y., and Adinata, D. (2008) Adsorption capacities of carbon dioxide, oxygen, nitrogen and methane on carbon molecular basket derived from polyethyleneimine impregnation on microporous palm shell activated carbon. *Separation and Purification Technology,* **62**, 609-613.
145. Chiang, Y.-C., Chiang, P.-C., and Huang, C.-P. (2001) Effects of pore structure and temperature on VOC adsorption on activated carbon. *Carbon,* **39**, 523-534.
146. Das, D., Gaur, V., and Verma, N. (2004) Removal of volatile organic compound by activated carbon fiber. *Carbon,* **42**, 2949-2962.
147. Kim, K.-J., Kang, C.-S., You, Y.-J., Chung, M.-C., Woo, M.-W., Jeong, W.-J., Park, N.-C., and Ahn, H.-G. (2006) Adsorption-desorption characteristics of VOCs over impregnated activated carbons. *Catalysis Today,* **111**, 223-228.
148. Kemp, K. C., Seema, H., Saleh, M., Le, N. H., Mahesh, K., Chandra, V., and Kim, K. S. (2013) Environmental applications using graphene composites: water remediation and gas adsorption. *Nanoscale,* **5**, 3149-3171.
149. Mishra, A. K., and Ramaprabhu, S. (2011) Carbon dioxide adsorption in graphene sheets. *AIP Advances,* **1**, 032152.
150. Meng, L.-Y., and Park, S.-J. (2012) Effect of exfoliation temperature on carbon dioxide capture of graphene nanoplates. *Journal of Colloid and interface Science,* **386**, 285-290.
151. Xia, K., Tian, X., Fei, S., and You, K. (2014) Hierarchical porous graphene-based carbons prepared by carbon dioxide activation and their gas adsorption properties. *International Journal of Hydrogen Energy,* **39**, 11047-11054.
152. Ning, G., Xu, C., Mu, L., Chen, G., Wang, G., Gao, J., Fan, Z., Qian, W., and Wei, F. (2012) High capacity gas storage in corrugated porous graphene with a specific surface area-lossless tightly stacking manner. *Chemical Communications,* **48**, 6815-6817.
153. Wang, Y., Guan, C., Wang, K., Guo, C. X., and Li, C. M. (2011) Nitrogen, hydrogen, carbon dioxide, and water vapor sorption properties of three-dimensional graphene. *Journal of Chemical & Engineering Data,* **56**, 642-645.
154. Shan, M., Xue, Q., Jing, N., Ling, C., Zhang, T., Yan, Z., and Zheng, J. (2012) Influence of chemical functionalization on the CO_2/N_2 separation performance of porous graphene membranes. *Nanoscale,* **4**, 5477-5482.
155. Carrillo, I., Rangel, E., and Magana, L. (2009) Adsorption of carbon dioxide and methane on graphene with a high titanium coverage.

Carbon, **47**, 2758-2760.

156. Mishra, A. K., and Ramaprabhu, S. (2012) Nanostructured polyaniline decorated graphene sheets for reversible CO_2 capture. *Journal of Materials Chemistry,* **22**, 3708-3712.
157. Dasgupta, T., Punnathanam, S. N., and Ayappa, K. (2015) Effect of functional groups on separating carbon dioxide from CO_2/N_2 gas mixtures using edge functionalized graphene nanoribbons. *Chemical Engineering Science,* **121**, 279-291.
158. Kumar, R., Suresh, V. M., Maji, T. K., and Rao, C. (2014) Porous graphene frameworks pillared by organic linkers with tunable surface area and gas storage properties. *Chemical Communications,* **50**, 2015-2017.
159. Zhou, D., Liu, Q., Cheng, Q., Zhao, Y., Cui, Y., Wang, T., and Han, B. (2012) Graphene-manganese oxide hybrid porous material and its application in carbon dioxide adsorption. *Chinese Science Bulletin,* **57**, 3059-3064.
160. Kemp, K. C., Chandra, V., Saleh, M., and Kim, K. S. (2013) Reversible CO_2 adsorption by an activated nitrogen doped graphene/polyaniline material. *Nanotechnology, 24,* 235703.
161. Saleh, M., Chandra, V., Kemp, K. C., and Kim, K. S. (2013) Synthesis of N-doped microporous carbon via chemical activation of polyindole-modified graphene oxide sheets for selective carbon dioxide adsorption. *Nanotechnology,* **24**, 255702.
162. Sui, Z.-Y., and Han, B.-H. (2015) Effect of surface chemistry and textural properties on carbon dioxide uptake in hydrothermally reduced graphene oxide. *Carbon,* **82**, 590-598.
163. Chen, C., Xu, K., Ji, X., Miao, L., and Jiang, J. (2014) Enhanced adsorption of acidic gases (CO_2, NO_2 and SO_2) on light metal decorated graphene oxide. *Physical Chemistry Chemical Physics,* **16**, 11031-11036.
164. Tsoufis, T., Katsaros, F., Sideratou, Z., Romanos, G., Ivashenko, O., Rudolf, P., Kooi, B., Papageorgiou, S., and Karakassides, M. (2014) Tailor-made graphite oxide-DAB poly (propylene imine) dendrimer intercalated hybrids and their potential for efficient CO_2 adsorption. *Chemical Communications,* **50**, 10967-10970.
165. Sui, Z.-Y., Cui, Y., Zhu, J.-H., and Han, B.-H. (2013) Preparation of three-dimensional graphene oxide-polyethylenimine porous materials as dye and gas adsorbents. *ACS Applied Materials & Interfaces,* **5**, 9172-9179.
166. Zhao, Y., Ding, H., and Zhong, Q. (2012) Preparation and characterization of aminated graphite oxide for CO_2 capture. *Applied Surface Science,* **258**, 4301-4307.
167. Zhao, Y., Ding, H., and Zhong, Q. (2013) Synthesis and characterization of MOF-aminated graphite oxide composites for CO_2 capture. *Applied Surface Science,* **284**, 138-144.

168. Cao, Y., Zhao, Y., Lv, Z., Song, F., and Zhong, Q. (2015) Preparation and enhanced CO_2 adsorption capacity of UiO-66/graphene oxide composites. *Journal of Industrial and Engineering Chemistry,* **27**, 102-107.
169. Chowdhury, S., Parshetti, G. K., and Balasubramanian, R. (2015) Post-combustion CO_2 capture using mesoporous TiO_2/graphene oxide nanocomposites. *Chemical Engineering Journal,* **263**, 374-384.
170. Mabayoje, O., Seredych, M., and Bandosz, T. J. (2012) Cobalt (hydr) oxide/graphite oxide composites: Importance of surface chemical heterogeneity for reactive adsorption of hydrogen sulfide. *Journal of Colloid and Interface Science,* **378**, 1-9.
171. Seredych, M. and Bandosz, T. J. (2011) Reactive adsorption of hydrogen sulfide on graphite oxide/$Zr(OH)_4$ composites. *Chemical Engineering Journal,* **166**, 1032-1038.
172. Petit, C., Levasseur, B., Mendoza, B., and Bandosz, T. J. (2012) Reactive adsorption of acidic gases on MOF/graphite oxide composites. *Microporous and Mesoporous Materials,* **154**, 107-112.
173. Seredych, M., and Bandosz, T. J. (2009) Adsorption of hydrogen sulfide on graphite derived materials modified by incorporation of nitrogen. *Materials Chemistry and Physics,* **113**, 946-952.
174. Seredych, M., and Bandosz, T. J. (2007) Removal of ammonia by graphite oxide via its intercalation and reactive adsorption. *Carbon,* **45**, 2130-2132.
175. Seredych, M., Petit, C., Tamashausky, A. V., and Bandosz, T. J. (2009) Role of graphite precursor in the performance of graphite oxides as ammonia adsorbents. *Carbon,* **47**, 445-456.
176. Seredych, M., Tamashausky, A. V., and Bandosz, T. J. (2008) Surface features of exfoliated graphite/bentonite composites and their importance for ammonia adsorption. *Carbon,* **46**, 1241-1252.
177. Seredych, M. and Bandosz, T. J. (2012) Manganese oxide and graphite oxide/MnO_2 composites as reactive adsorbents of ammonia at ambient conditions. *Microporous and Mesoporous Materials,* **150**, 55-63.
178. Seredych. M. and Bandosz, T. J. (2008) Adsorption of ammonia on graphite oxide/aluminium polycation and graphite oxide/zirconium-aluminium polyoxycation composites. *Journal of Colloid and Interface Science,* **324**, 25-35.
179. Seredych, M. and Bandosz, T. J. (2009) Graphite oxide/AlZr polycation composites: Surface characterization and performance as adsorbents of ammonia. *Materials Chemistry and Physics,* **117**, 99-106.
180. Seredych, M., and Bandosz, T. J. (2010) Adsorption of ammonia on graphite oxide/Al 13 composites. *Colloids and Surfaces A: Physicochemical and Engineering Aspects,* **353**, 30-36.

181. Srinivas, G., Zhu, Y., Piner, R., Skipper, N., Ellerby, M., and Ruoff, R. (2010) Synthesis of graphene-like nanosheets and their hydrogen adsorption capacity. *Carbon,* **48**, 630-635.
182. Yuan, W., Li, B., and Li, L. (2011) A green synthetic approach to graphene nanosheets for hydrogen adsorption. *Applied Surface Science,* **257**, 10183-10187.
183. Ghosh, A., Subrahmanyam, K., Krishna, K. S., Datta, S., Govindaraj, A., Pati, S. K., and Rao, C. (2008) Uptake of H_2 and CO_2 by graphene. *The Journal of Physical Chemistry C,* **112**, 15704-15707.
184. Ma, L.-P., Wu, Z.-S., Li, J., Wu, E.-D., Ren, W.-C., and Cheng, H.-M. (2009) Hydrogen adsorption behavior of graphene above critical temperature. *International Journal of Hydrogen Energy,* **34**, 2329-2332.
185. Lyth, S. M., Shao, H., Liu, J., Sasaki, K., and Akiba, E. (2014) Hydrogen adsorption on graphene foam synthesized by combustion of sodium ethoxide. *International Journal of Hydrogen Energy,* **39**, 376-380.
186. Srinivas, G., Burress, J. W., Ford, J., and Yildirim, T. (2011) Porous graphene oxide frameworks: synthesis and gas sorption properties. *Journal of Materials Chemistry,* **21**, 11323-11329.
187. Petit, C., Burress, J., and Bandosz, T. J. (2011) The synthesis and characterization of copper-based metal–organic framework/graphite oxide composites. *Carbon,* **49**, 563-572.
188. Krungleviciute, V., Heroux, L., Migone, A. D., Kingston, C. T., and Simard, B. (2005) Isosteric heat of argon adsorbed on single-walled carbon nanotubes prepared by laser ablation. *The Journal of Physical Chemistry B,* **109**, 9317-9320.
189. Valcarcel, M., Cardenas, S., Simonet, B. M., Moliner-Martinez, Y., and Lucena, R. (2008) Carbon nanostructures as sorbent materials in analytical processes. *TrAC Trends in Analytical Chemistry,* **27**, 34-43.
190. Fujiwara, A., Ishii, K., Suematsu, H., Kataura, H., Maniwa, Y., Suzuki, S., and Achiba, Y. (2001) Gas adsorption in the inside and outside of single-walled carbon nanotubes. *Chemical Physics Letters,* **336**, 205-211.
191. Babu, D. J., Lange, M., Cherkashinin, G., Issanin, A., Staudt, R., and Schneider, J. J. (2013) Gas adsorption studies of CO_2 and N_2 in spatially aligned double-walled carbon nanotube arrays. *Carbon,* **61**, 616-623.
192. Cinke, M., Li, J., Bauschlicher, C. W., Ricca, A., and Meyyappan, M. (2003) CO_2 adsorption in single-walled carbon nanotubes. *Chemical Physics Letters,* **376**, 761-766.
193. Hu, Y. H., and Ruckenstein, E. (2006) Applicability of Dubinin–Astakhov equation to CO_2 adsorption on single-walled carbon nanotubes. *Chemical Physics Letters,* **425**, 306-310.

194. Golkha A., Keshavar P., and Mowla, D. (2013) Investigation of CO_2 removal by silica and CNT nanofluids in microporous hollow fiber membrane contactors. *Journal of Membrane Science,* **433**, 17-24.
195. Casco, M. E., Morelos-Gomez, A., Vega-Diaz, S. M., Cruz-Silva, R., Tristan-Lopez. F., Muramatsu, H., Hayashi, T., Martinez-Escandell, M., Terrones, M., and Endo, M. (2014) CO_2 adsorption on crystalline graphitic nanostructures. *Journal of CO_2 Utilization,* **5**, 60-65.
196. Sun, Q., Wang, M., Li, Z., Ma, Y., and Du, A. (2013) CO_2 capture and gas separation on boron carbon nanotubes. *Chemical Physics Letters,* **575**, 59-66.
197. Upendar, K., Kumar, A. S. H., Lingaiah, N., Rao, K. R., and Prasad, P. S. (2012) Low-temperature CO_2 adsorption on alkali metal titanate nanotubes. *International Journal of Greenhouse Gas Control,* **10**, 191-198.
198. Hsu, S.-C., Lu, C., Su, F., Zeng, W., and Chen, W. (2010) Thermodynamics and regeneration studies of CO_2 adsorption on multiwalled carbon nanotubes. *Chemical Engineering Science,* **65**, 1354-1361.
199. Su, F., Lu, C., Cnen, W., Bai, H., and Hwang, J. F. (2009) Capture of CO_2 from flue gas via multiwalled carbon nanotubes. *Science of the Total Environment,* **407**, 3017-3023.
200. Lu, C., Bai, H., Wu, B., Su, F., and Hwang, J. F. (2008) Comparative study of CO_2 capture by carbon nanotubes, activated carbons, and zeolites. *Energy & Fuels,* **22**, 3050-3056.
201. Jana, S., Das, S., Ghosh, C., Maity, A., and Pradhan, M. (2015) Halloysite nanotubes capturing isotope selective atmospheric CO_2. *Scientific Reports,* **5**, 2015.
202. Song, F., Zhao, Y., Cao, Y., Ding, J., Bu, Y., and Zhong, Q. (2013) Capture of carbon dioxide from flue gases by amine-functionalized TiO_2 nanotubes. *Applied Surface Science,* **268**, 124-128.
203. Ko, Y. G., Lee, H. J., Oh, H. C., and Choi, U. S. (2013) Amines immobilized double-walled silica nanotubes for CO_2 capture. *Journal of Hazardous Materials,* **250**, 53-60.
204. Liu, J., Liu, Y., Wu, Z., Chen, X., Wang, H., and Weng, X. (2012) Polyethyleneimine functionalized protonated titanate nanotubes as superior carbon dioxide adsorbents. *Journal of Colloid and Interface Science,* **386**, 392-397.
205. Dillon, E. P., Crouse, C. A., and Barron, A. R. (2008) Synthesis, characterization, and carbon dioxide adsorption of covalently attached polyethyleneimine-functionalized single-wall carbon nanotubes. *ACS Nano,* **2**, 156-164.
206. Cong, H., Zhang, J., Radosz, M., and Shen, Y. (2007) Carbon nanotube composite membranes of brominated poly (2, 6-diphenyl-1, 4-phenylene oxide) for gas separation. *Journal of Membrane Science,* **294**, 178-185.
207. Zuttel, A., Sudan, P., Mauron, P., Kiyobayashi, T., Emmenegger, C.,

and Schlapbach, L. (2002) Hydrogen storage in carbon nanostructures. *International Journal of Hydrogen Energy,* **27**, 203-212.

208. Ma, R., Bando, Y., Zhu, H., Sato, T., Xu, C., and Wu, D. (2002) Hydrogen uptake in boron nitride nanotubes at room temperature. *Journal of the American Chemical Society,* **124**, 7672-7673.
209. Nour, M., Berean, K., Balendhran, S., Ou, J. Z., Du Plessis, J., McSweeney, C., Bhaskaran, M., Sriram, S., and Kalantar-zadeh, K. (2013) CNT/PDMS composite membranes for H_2 and CH_4 gas separation. *International Journal of Hydrogen Energy,* **38**, 10494-10501.
210. Wang, W., Peng, X., and Cao, D. (2011) Capture of trace sulfur gases from binary mixtures by single-walled carbon nanotube arrays: a molecular simulation study. *Environmental Science & Technology,* **45**, 4832-4838.
211. Mohamadalizadeh, A., Towfighi, J., Rashidi, A., Mohajeri, A., and Golkar, M. (2011) Modification of carbon nanotubes for H_2S sorption. *Industrial & Engineering Chemistry Research,* **50**, 8050-8057.
212. Hussain, C. M., Saridara, C., and Mitra, S. (2008) Microtrapping characteristics of single and multi-walled carbon nanotubes. *Journal of Chromatography A,* **1185**, 161-166.
213. Li, Q.-L., Yuan, D.-X., and Lin, Q.-M. (2004) Evaluation of multi-walled carbon nanotubes as an adsorbent for trapping volatile organic compounds from environmental samples. *Journal of Chromatography A,* **1026**, 283-288.
214. Wang, L., Liu, J., Zhao, P., Ning, Z., and Fan, H. (2010) Novel adsorbent based on multi-walled carbon nanotubes bonding on the external surface of porous silica gel particulates for trapping volatile organic compounds. *Journal of Chromatography A,* **1217**, 5741-5745.
215. Zheng, F., Baldwin, D. L., Fifield, L. S., Anheier, N. C., Aardahl, C. L., and Grate, J. W. (2006) Single-walled carbon nanotube paper as a sorbent for organic vapor preconcentration. *Analytical Chemistry,* **78**, 2442-2446.
216. Delavar, M., Ghoreyshi, A. A., Jahanshahi, M., Khalili, S., and Nabian, N. (2012) Equilibria and kinetics of natural gas adsorption on multi-walled carbon nanotube material. *RSC Advances,* **2**, 4490-4497.
217. Cao, D., Zhang, X., Chen, J., Wang, W., and Yun, J. (2003) Optimization of single-walled carbon nanotube arrays for methane storage at room temperature. *The Journal of Physical Chemistry B,* **107**, 13286-13292.
218. Kim, S., Pechar, T. W., and Marand, E. (2006) Poly (imide siloxane) and carbon nanotube mixed matrix membranes for gas separation. *Desalination,* **192**, 330-339.
219. Wang, H., Zhou, S., Xiao, L., Wang, Y., Liu, Y., and Wu, Z. (2011) Titania nanotubes - A unique photocatalyst and adsorbent for eleme-

ntal mercury removal. *Catalysis Today,* **175**, 202-208.

220. Ding, W., Hayashi, R., Ochi, K., Suehiro, J., Imasaka, K., Hara, M., Sano, N., Nagao, E., and Minagawa, T. (2006) Analysis of PD-generated SF 6 decomposition gases adsorbed on carbon nanotubes. *IEEE Transactions on Dielectrics and Electrical Insulation,* **13**, 1200-1207.
221. Foroutan, M., and Nasrabadi, A. T. (2010) Adsorption behavior of ternary mixtures of noble gases inside single-walled carbon nanotube bundles. *Chemical Physics Letters,* **497**, 213-217.
222. Jalili, S., and Majidi, R. (2007) Study of Xe and Kr adsorption on open single-walled carbon nanotubes using molecular dynamics simulations. *Physica E: Low-dimensional Systems and Nanostructures,* **39**, 166-170.
223. Foroutan, M., and Nasrabadi, A. T. (2010) Molecular dynamics simulation study of neon adsorption on single-walled carbon nanotubes. *Physica E: Low-dimensional Systems and Nanostructures,* **43**, 261-265.
224. Foroutan, M., and Nasrabadi, A. T. (2011) Adsorption and separation of binary mixtures of noble gases on single-walled carbon nanotube bundles. *Physica E: Low-dimensional Systems and Nanostructures,* **43**, 851-856.
225. Kiselev, A. V., and Yashin, Y. I. (2013) *Gas-adsorption Chromatography,* Springer, USA.
226. Zhao, Z., Cui, X., Ma, J., and Li, R. (2007) Adsorption of carbon dioxide on alkali-modified zeolite 13X adsorbents. *International Journal of Greenhouse Gas Control,* **1**, 355-359.
227. Montanari, T., Finocchio, E., Salvatore, E., Garuti, G., Giordano, A., Pistarino, C., and Busca, G. (2011) CO_2 separation and landfill biogas upgrading: a comparison of 4A and 13X zeolite adsorbents. *Energy,* **36**, 314-319.
228. Kusakabe, K., Kuroda, T., Murata, A., and Morooka, S. (1997) Formation of a Y-type zeolite membrane on a porous α-alumina tube for gas separation. *Industrial & Engineering Chemistry Research,* **36**, 649-655.
229. Saha, D., Bao, Z., Jia, F., and Deng, S. (2010) Adsorption of CO_2, CH_4, N_2O, and N_2 on MOF-5, MOF-177, and zeolite 5A. *Environmental Science & Technology,* **44**, 1820-1826.
230. Chatti, R., Bansiwal, A. K., Thote, J. A., Kumar, V., Jadhav, P., Lokhande, S. K., Biniwale, R. B., Labhsetwar, N. K., and Rayalu, S. S. (2009) Amine loaded zeolites for carbon dioxide capture: Amine loading and adsorption studies. *Microporous and Mesoporous Materials,* **121**, 84-89.
231. Tagliabue, M., Rizzo, C., Onorati, N. B., Gambarotta, E. F., Carati, A., and Bazzano, F. (2012) Regenerability of zeolites as adsorbents for natural gas sweetening: A case-study. *Fuel,* **93**, 238-244.

232. Adams, R. T., Lee, J. S., Bae, T.-H., Ward, J. K., Johnson, J., Jones, C. W., Nair, S., and Koros, W. J. (2011) CO_2-CH_4 permeation in high zeolite 4A loading mixed matrix membranes. *Journal of Membrane Science,* **367**, 197-203.
233. Li, P., and Tezel, F. H. (2007) Adsorption separation of N_2, O_2, CO_2 and CH_4 gases by β-zeolite. *Microporous and Mesoporous Materials,* **98**, 94-101.
234. Xu, X., Zhao, X., Sun, L., and Liu, X. (2009) Adsorption separation of carbon dioxide, methane and nitrogen on monoethanol amine modified β-zeolite. *Journal of Natural Gas Chemistry,* **18**, 167-172.
235. Xu, X., Zhao, X., Sun, L., and Liu, X. (2008) Adsorption separation of carbon dioxide, methane, and nitrogen on Hβ and Na-exchanged β-zeolite. *Journal of Natural Gas Chemistry,* **17**, 391-396.
236. Diaz, E., Ordonez, S., Vega, A., and Coca, J. (2004) Adsorption characterisation of different volatile organic compounds over alumina, zeolites and activated carbon using inverse gas chromatography. *Journal of Chromatography A,* **1049**, 139-146.
237. Sen, D., Kalipcilar, H., and Yilmaz, L. (2006) Development of zeolite filled polycarbonate mixed matrix gas separation membranes. *Desalination,* **200**, 222-224.
238. Husain, S., and Koros, W. J. (2007) Mixed matrix hollow fiber membranes made with modified HSSZ-13 zeolite in polyetherimide polymer matrix for gas separation. *Journal of Membrane Science,* **288**, 195-207.
239. Sen, D., Kalipcilar, H., and Yilmaz, L. (2007) Development of polycarbonate based zeolite 4A filled mixed matrix gas separation membranes. *Journal of Membrane Science,* **303**, 194-203.
240. Rolker, J., Seiler, M., Mokrushina, L., and Arlt, W. (2007) Potential of branched polymers in the field of gas absorption: Experimental gas solubilities and modeling. *Industrial & Engineering Chemistry Research,* **46**, 6572-6583.
241. Bali, S., Chen, T. T., Chaikittisilp, W., and Jones, C. W. (2013) Oxidative stability of amino polymer-alumina hybrid adsorbents for carbon dioxide capture. *Energy & Fuels,* **27**, 1547-1554.
242. An, D., Wu, L., Li, B.-G., and Zhu, S. (2007) Synthesis and SO_2 absorption/desorption properties of poly (1,1,3,3-tetramethylguanidine acrylate). *Macromolecules,* **40**, 3388-3393.
243. Kikkinides, E., and Yang, R. (1993) Gas separation and purification by polymeric adsorbents: Flue gas desulfurization and sulfur dioxide recovery with styrenic polymers. *Industrial & Engineering Chemistry Research,* **32**, 2365-2372.
244. Deshmukh, M. M., Ohba, M., Kitagawa, S., and Sakaki, S. (2013) Absorption of CO_2 and CS_2 into the Hofmann-type porous coordination polymer: Electrostatic versus dispersion interactions. *Journal of the American Chemical Society,* **135**, 4840-4849.

245. Noro, S. i., Kitagawa, S., Kondo, M., and Seki, K. (2000) A new, methane adsorbent, porous coordination polymer [{CuSiF6 (4,4′-bipyridine)2}n]. *Angewandte Chemie International Edition*, **39**, 2081-2084.

13

Cobalt Ferrite Doped Graphene/Polyacrylamide Cryogel Nanocomposites: Synthesis and Characterization

13.1 Introduction

Cryogels have large and well-defined interconnected porous structure. These materials are heterogeneous and non-superimposable about few micrometer to few hundred micrometers [1,2]. Free-radical cryo-polymerization technique has been utilized for preparing cryogels from polymeric gel precursors below the solvent freezing point, where the function of the freeze-drying process is to maintain the interconnected structure of the resultant cryogel [3]. The properties of cryogels such as pore dimensions, pore wall thickness, mechanical toughness and elasticity depend on the conditions of preparation, amount of solvent and precursors, temperature gradients, cooling rate and other conditions [4,5]. High porosity is exhibited by the cryogels, thus, these can be utilized as an efficient adsorbent at the macroscopic level for the wastewater treatment or oil spill clean-up.

In this study, graphene-based carbonaceous monolithic material containing cobalt ferrite nanoparticles has been synthesized for adsorption applications. During the formation of polyacrylamide based cryogels, divalent metal ions (such as Co^{2+} and Fe^{2+}) perform as both crosslinker and catalyst due to their binding ability towards the oxygen groups [6] which leads to a crosslinking interaction between graphene oxide and polyacrylamide skeletons, thereby, enhancing the strength of the whole cryogel, resulting in an enhanced adsorption performance.

13.2 Materials and Methods

13.2.1 Materials

Natural graphite powder (10 mesh), H_2SO_4 (95-97%, Merck), H_3PO_4 (85%, Merck), $KMnO_4$ (Eurolab), HCl (37%, Panreac), hydrogen per-

Naman Arora and Vikas Mittal, Khalifa University of Science and Technology, Abu Dhabi, UAE oxide (35%, Merck), $CoCl_2.6H_2O$ (Sigma-Aldrich), iron(III) nitrate nanohydrate (≥98%, Sigma-Aldrich), acrylamide, polyacrylamide (PAAm), tetramethylethylenediamine (TEMED), N,N'-methylenebis(acrylamide) (MBA) and potassium persulphate (KPS) were used in the study.

13.2.2 Preparation of Graphene Oxide (GO)

Graphene oxide (GO) was prepared by using the improved Tour's method [7]. Graphite (10 g) was added to a 9:1 molar mixture of H_2SO_4 (540 mL) and H_3PO_4 (60 mL). The mixture was stirred for 30 min, followed by the gradual addition of $KMnO_4$ (56 g). The stirring was continued for further 72 h. The reaction was continued by slowly pouring H_2O_2 and subsequently allowing it to stir for 8 h until the mixture color turned yellow. Produced graphene oxide was washed with 1 M HCl solution until no sulfate ions were detected, followed by washing with excess distilled water to remove chloride ions. The GO solution was then poured into Teflon petri dishes for freeze drying.

13.2.3 Synthesis of Graphene based Polyacrylamide Cryogel containing Cobalt Ferrite Nanoparticles

GO exfoliation was carried out under ultra-sonication (2 h) by adding GO (0.1 g) in deionized water (50 mL). Later, $CoCl_2.6H_2O$ (0.6 g) and $Fe(NO_3)_3.9H_2O$ (0.5 g) were added as the Co^{2+} and Fe^{2+} sources, followed by stirring for 2 h. Polyacrylamide (PAAm) (10 g) and N,N'-methylenebis(acrylamide) (2 g) were subsequently added to the solution, followed by continuous stirring for 4 h to form a homogeneous solution. Finally, 100 μL TEMED and 0.2 g KPS were added as initiators. Hydrogel was obtained after 7 min, followed by washing with excess distilled water to remove the residual ions. Subsequently, the hydrogel was subjected to lyophilization process for 5 d at a temperature of -89 °C and vacuum of 10 Pa to obtain a GO based polyacrylamide cryogel. The cryogel was reduced by dipping into hydrazine hydrate solution for 2 h at 75 °C with slow agitation. Finally, graphene based polyacrylamide cryogel containing cobalt ferrite nanoparticles ($CoFe_2O_4$/graphene/PAAm) was obtained.

13.2.4 Characterization of Cryogel

Wide-angle X-ray diffraction (WAXRD) pattern of the sample was collected by employing analytical powder (X' Pert PRO) diffractometer, using CuKα (1.5406 Å wavelength) radiation in reflection mode. The sample was step-scanned between 5 and 60° at a step size of 2θ = 0.02° s^{-1}.

Infrared spectra of the sample was recorded by attenuated total reflectance Fourier transform infrared spectroscopy (FTIR-ATR), using a Bruker Vertex 70 spectrometer.

Scanning electron microscopy (SEM) and energy dispersive X-ray (EDX) analysis were carried out to ascertain the morphology of the sample, using FEI Quanta FEG250 equipment at accelerating voltages of 10-20 kV.

Transmission electron microscopy (TEM) analysis was also carried out to obtain the morphology of $CoFe_2O_4$/graphene/PAAm nanocomposite. The sample for TEM was prepared by dispersing 0.5 mg of adsorbent in 20 mL of dimethylformamide under sonication at room temperature for 10 min. Two drops of suspension were poured onto a 400 mesh copper electron microscopy grid covered with thin holey carbon film, followed by drying in air.

13.3 Results and Discussion

Figure 13.1a shows the digital image of the graphene oxide-polyacrylamide based cryogel containing divalent metal salts of Co^{2+} and Fe^{2+} (GO-PAAm). After carbonization process, the GO-PAAM based cryogel turned into black carbonaceous nanocomposite cryogel ($CoFe_2O_4$/graphene/PAAm), as shown in Figure 13.1b. To mention again, these cryogels have been synthesized by obtaining the initial suspensions as hydrogels and further lyophilization of the corresponding hydrogels in order to eliminate the water present in the structure, which results in dried monoliths/cryogels with desirable shapes, depending on the glass molds. It can be noticed that GO-PAAm was a homogeneous monolith with yellow-brownish color. After the reduction of GO-PAAM, divalent metal ions (Co^{2+} and Fe^{2+}) and GO are converted into cobalt ferrite nanoparticles and graphene, respectively. As a result, $CoFe_2O_4$/graphene/PAAm carbonaceous monolith was obtained.

WAXRD pattern of the resultant cryogel is represented in Figure 13.2. The observed diffraction peaks corresponded to the cubic $CoFe_2O_4$ spinel phase without any detectable impurities [8]. It indicated the effective synthesis of the desired cryogel structure. The SEM micrograph of $CoFe_2O_4$/graphene/PAAm is also represented in

Figure 13.1 (a) Graphene oxide based polyacrylamide (GO-PAAm) cryogel and (b) reduction of GO-PAAm cryogel to form $CoFe_2O_4$/graphene/PAAm cryogel.

Figure 13.3. Cobalt ferrite nanoparticles were observed to be present in a well-defined interconnected porous structure. Fourier transform infrared FTIR spectrum of the cryogel is also demonstrated in Figure 13.4. The amide groups in the cryogel were observed to exhibit two

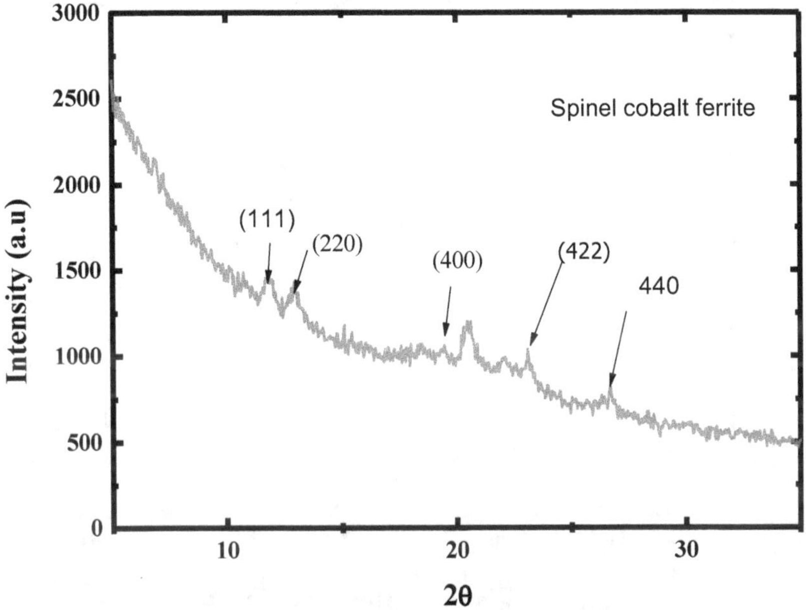

Figure 13.2 XRD spectrum of $CoFe_2O_4$/graphene/PAAm cryogel.

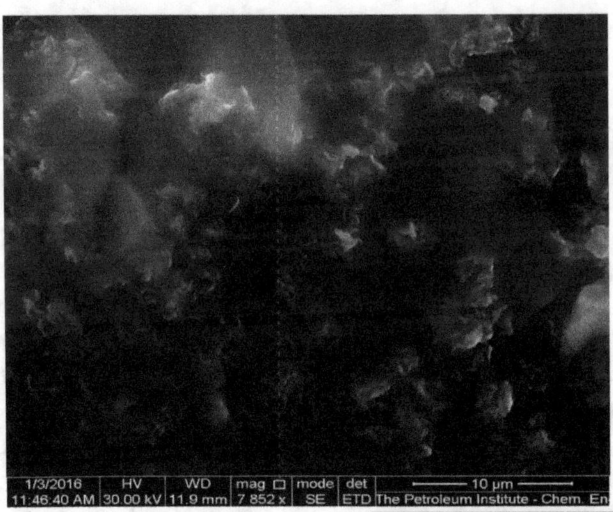

Element	*Wt %*	*At %*
C K	68.64	74.98
O K	30.18	24.75
FeK	00.36	00.08
CoK	00.82	00.18

Figure 13.3 SEM image and EDX analysis for the cryogel.

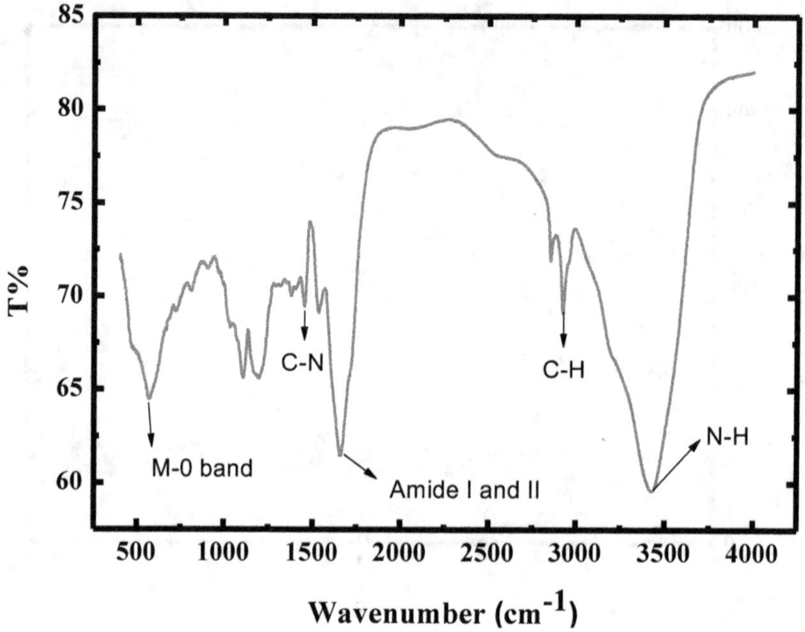

Figure 13.4 FTIR spectra of $CoFe_2O_4$/graphene/PAAm cryogel.

characteristic peaks, the first peak at 1672 cm^{-1} (amide I and II) representing the –C=O stretching (amide I) and $–NH_2$ bending vibrations of the amide groups (amide II), whereas the second peak at 1425 cm^{-1} was attributed to the –CN stretching vibration of the amide groups [9]. The peaks at 601 cm^{-1} and 1142 cm^{-1} were attributed to the v-ferrite structure (M-O) and Fe-Co alloys system, respectively [10].

TEM images of the resultant adsorbent are presented in Figure 13.5a. It can be observed that the carbonaceous matrix contained graphene sheets, thus, attributing enhanced surface area to the adsorbent. Additionally, cobalt ferrite nanoparticles were uniformly distributed onto graphene sheets, which corresponded to additional active adsorption sites and subsequently increased interactions with negatively charged contaminated ions. Uniform porosity could also be noticed in Figure 13.5b, thus, confirming that resultant adsorbent was a porous material.

13.4 Conclusion and Future Work

In this communication, $CoFe_2O_4$/graphene/PAAm nanocomposite cryogel has been synthesized by employing *in-situ* polymerization.

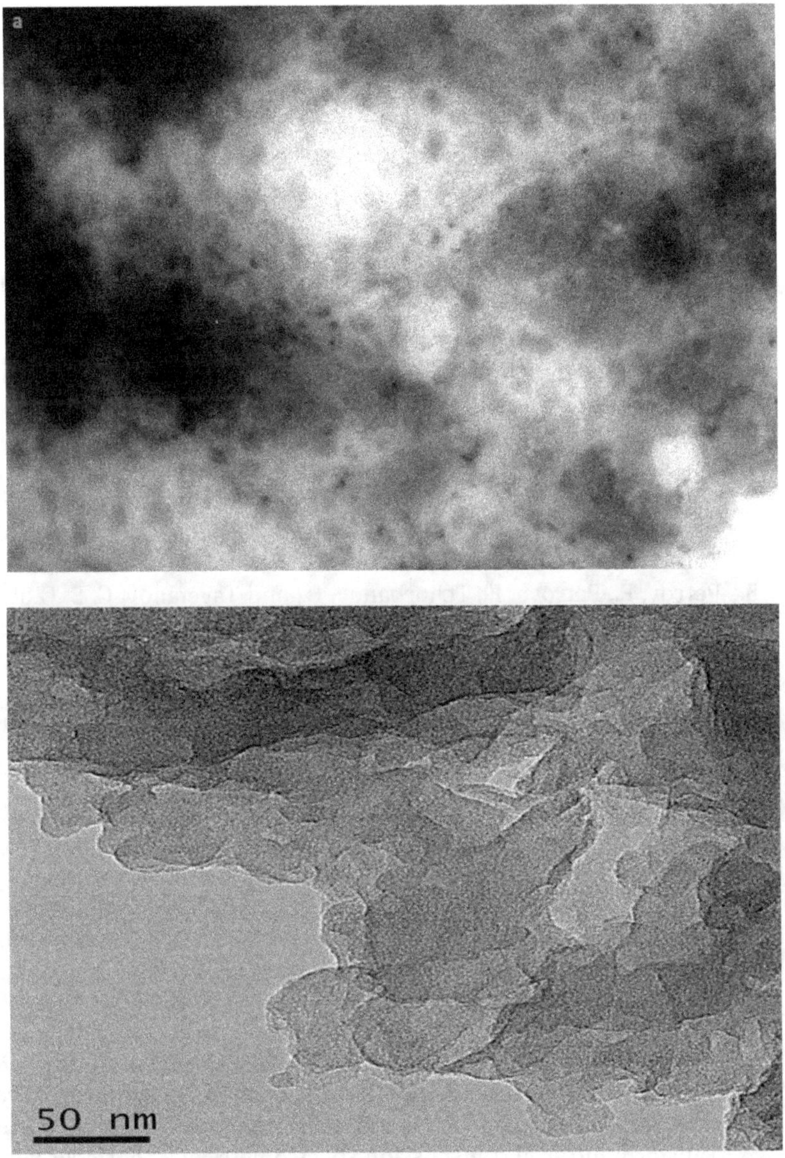

Figure 13.5 TEM images of $CoFe_2O_4$/graphene/PAAm cryogel.

The cryogel exhibited interconnected porous structure which enables the use of the developed material as promising adsorbent for the removal of contaminants from aqueous solutions, especially industrial water effluents. The monolithic structure also presents advantages in terms of efficiency and handling.

References

1. Bibi, N. S., Singh, N. K., Dsouza, R. N., Aasim, M., and Fernández-Lahore, M. (2013) Synthesis and performance of megaporous immobilized metal-ion affinity cryogels for recombinant protein capture and purification. *Journal of Chromatography A*, **1272**, 145-149.
2. Zheng, S., Wang, T., Liu, D., Liu, X., Wang, C., and Tong, Z. (2013) Fast deswelling and highly extensible poly (N-isopropylacrylamide)-hectorite clay nanocomposite cryogels prepared by freezing polymerization. *Polymer*, **54**, 1846-1852.
3. Sahiner, N., Seven, F., and Al-lohedan, H. (2015) Superporous cryogel-M (Cu, Ni, and Co) composites in catalytic reduction of toxic phenolic compounds and dyes from wastewaters. *Water, Air and Soil Pollution*, **226**, 1-12.
4. Reichelt, S., Becher, J., Weisser, J., Prager, A., Decker, U., Moeller, S., Berg, A., and Schnabelrauch, M. (2014) Biocompatible polysaccharide-based cryogels. *Materials Science and Engineering C*, **35**, 164-170.
5. Petrov, P., Petrova, E., Tchorbanov, B., and Tsvetanov, C. B. (2007) Synthesis of biodegradable hydroxyethylcellulose cryogels by UV irradiation. *Polymer*, **48**, 4943-4949.
6. Park, S., Lee, K.-S., Bozoklu, G., Cai, W., Nguyen, S. T., and Ruoff, R. S. (2008) Graphene oxide papers modified by divalent ions—enhancing mechanical properties via chemical cross-linking. *ACS Nano*, **2**, 572-578.
7. Marcano, D. C., Kosynkin, D. V., Berlin, J. M., Sinitskii, A., Sun, Z., Slesarev, A., Alemany, L. B., Lu, W., and Tour, J. M. (2010) Improved synthesis of graphene oxide. *ACS Nano*, **4**, 4806-4814.
8. Wang, W., Yang, H., Xian, T., and Jiang, J. (2012) XPS and magnetic properties of CoFe2O4 nanoparticles synthesized by a polyacrylamide gel route. *Materials Transactions*, **53**, 1586-1589.
9. Chiem, L. T., Huynh, L., Ralston, J., and Beattie, D. A. (2006) An in situ ATR–FTIR study of polyacrylamide adsorption at the talc surface. *Journal of Colloid and Interface Science*, **297**, 54-61.
10. Rana, S., Philip, J., and Raj, B. (2010) Micelle based synthesis of cobalt ferrite nanoparticles and its characterization using Fourier transform infrared transmission spectrometry and thermogravimetry. *Materials Chemistry and Physics*, **124**, 264-269.

Index

A

B

C

D

Index

E

F

G

H

I

L

M

N

O

P

R

S

T

U

V

W

X

Y

Z